Progress in Molecular and Subcellular Biology

Series Editors Werner E.G. Müller
Philippe Jeanteur, Robert E. Rhoads, Đurđica Ugarković,
Márcio Reis Custódio

54

Volumes Published in the Series

Progress in Molecular
and Subcellular Biology

Subseries:
Marine Molecular Biotechnology

Volume 36
Viruses and Apoptosis
C. Alonso (Ed.)

Volume 38
Epigenetics and Chromatin
Ph. Jeanteur (Ed.)

Volume 40
Developmental Biology of Neoplastic Growth
A. Macieira-Coelho (Ed.)

Volume 41
Molecular Basis of Symbiosis
J. Overmann (Ed.)

Volume 44
Alternative Splicing and Disease
Ph. Jeanlevr (Ed.)

Volume 45
Asymmetric Cell Division
A. Macieira Coelho (Ed.)

Volume 48
Centromere
Đurđica Ugarković (Ed.)

Volume 49
Viruses and Apoptosis
C.A. Navas and J.E. Carvalho (Eds.)

Volume 50
miRNA Regulation of the Translational Machinery
R.E. Rhoads (Ed.)

Volume 51
Long Non-Coding RNAs
Đurđica Ugarkovic (Ed.)

Volume 37
Sponges (Porifera)
W.E.G. Müller (Ed.)

Volume 39
Echinodermata
V. Matranga (Ed.)

Volume 42
Antifouling Compounds
N. Fusetani and A.S. Clare (Eds.)

Volume 43
Molluscs
G. Cimino and M. Gavagnin (Eds.)

Volume 46
Marine Toxins as Research Tools
N. Fusetani and W. Kem (Eds.)

Volume 47
Biosilica in Evolution, Morphogenesis, and Nanobiotechnology
W.E.G. Müller and M.A. Grachev (Eds.)

Volume 52
Molecular Biomineralization
W.E.G. Müller (Eds.)

Volume 53
Biology of Marine Fungi
C. Raghukumar (Ed.)

Volume 54
Biomedical Inorganic Polymers
W.E.G. Müller, Xiaohong Wang and H.C. Schröder (Eds.)

Werner E.G. Müller • Xiaohong Wang •
Heinz C. Schröder
Editors

Biomedical Inorganic Polymers

Bioactivity and Applications of Natural and Synthetic Polymeric Inorganic Molecules

Springer

Editors
Werner E.G. Müller
Xiaohong Wang
Heinz C. Schröder
ERC Advanced Investigator Group
Institute for Physiological Chemistry
University Medical Center of the
Johannes Gutenberg University
Mainz
Germany

ISSN 0079-6484
ISBN 978-3-642-41003-1 ISBN 978-3-642-41004-8 (eBook)
DOI 10.1007/978-3-642-41004-8
Springer Heidelberg New York Dordrecht London

Library of Congress Control Number: 2013956611

Printed on acid-free paper

Springer is part of Springer Science+Business Media (www.springer.com)

Preface

Inorganic polymers are a relatively small group of molecules as compared to the large group of organic polymers. They comprise both short-chain polymers and long-chain polymers with one type of atom or two or more types of atoms in the polymer backbone. Examples are homochain polymers, e.g., polysilanes, polygermanes, polystannanes, and polysulfides, and heterochain polymers, e.g., polysilazanes, polyphosphazenes, polyborazylenes, and polythiazyls. Many of these polymers are not soluble or not stable in water, but some of them can be formed in aqueous solution even by enzymatic reactions, for example, inorganic polyphosphates and polysilicates. Previous studies on the biological effects of polymeric compounds mainly focused on organic polymers. Recent results revealed that also inorganic polymers may possess biological activity. Much effort on the study of inorganic polymers focuses on the application of such polymers in nanotechnology or, in the field of biomedicine, on their application in drug delivery, e.g., of silica. Inorganic polymers which are biologically active can be formed by living organisms, e.g., arsenicin A, a polyarsenic compound isolated from a marine sponge, or polymeric silicate ("biosilica") formed by diatoms and siliceous sponges, or polyphosphates that have been identified in numerous organisms, from bacteria and yeast to plants and humans. These polymers often have multiple functions, for example, inorganic polyphosphates can be used as antimicrobial compounds, as a source of energy-rich phosphate, as a modulator of gene expression, as a chelator for metal cations, or in mineralization of bone tissue and in blood coagulation and fibrinolysis, and silica as skeletal element with unique property combinations (mechanical stability and light transmission). The research on these compounds is currently in a rapid development. Several European consortia are concerned with the investigation and development of products made of such polymers, in particular polyphosphates and polymeric silica. In this volume of the series "Progress in Molecular and Subcellular Biology" recent developments in the state of knowledge on selected inorganic polymers are summarized. These polymers include poly(arsenic) compounds, inorganic polyphosphates, polyoxometalates, polyvanadates, and polysilicates (biosilica). The biocompatibility, bioactivity, and stability of the latter polymers even allow a possible application in rapid prototyping procedures for the production of customized implants in surgery and dentistry.

Mainz, Germany

Werner E.G. Müller
Xiaohong Wang
Heinz C. Schröder

Contents

Contributors

Carlo Cerrano Dipartimento di Scienze dell'Ambiente e della Vita, Universitá Politecnica delle Marche, Ancona, Italy

Carolina M. Cronemberger Section Computational Science, Faculty of Science, University of Amsterdam, Amsterdam, The Netherlands

Andrea Defant Laboratorio di Chimica Bioorganica, Dipartimento di Fisica, Universitá di Trento, Povo-Trento, Italy

Qingling Feng Department of Materials Science and Engineering, Tsinghua University, Beijing, People's Republic of China

Marco Giovine Dipartimento di Scienze della Terra, dell'Ambiente e della Vita, University of Genova, Genoa, Italy

Narjes Javaheri Section Computational Science, Faculty of Science, University of Amsterdam, Amsterdam, The Netherlands

Jaap A. Kaandorp Section Computational Science, Faculty of Science, University of Amsterdam, Amsterdam, The Netherlands

Igor Kulaev Skryabin Institute of Biochemistry and Physiology of Microorganisms, Russian Academy of Sciences, Pushchino, Moscow Region, Russia

Tatyana Kulakovskaya Skryabin Institute of Biochemistry and Physiology of Microorganisms, Russian Academy of Sciences, Pushchino, Moscow Region, Russia

Werner E. G. Müller ERC Advanced Investigator Group, Institute for Physiological Chemistry, University Medical Center of the Johannes Gutenberg University, Mainz, Germany; NanotecMARIN GmbH, Mainz, Germany

Ines Mancini Laboratorio di Chimica Bioorganica, Dipartimento di Fisica, Universitá di Trento, Povo-Trento, Italy

Irene M. Mavridis National Center for Scientific Research Demokritos, Athens, Greece

Antonella Penna Dipartimento di Scienze Biomolecolari, Universitá di Urbino "Carlo Bo", Pesaro, Italy

Marina Pozzolini Dipartimento di Scienze della Terra, dell'Ambiente e della Vita, University of Genova, Genoa, Italy

Sonia Scarfì Dipartimento di Scienze della Terra, dell'Ambiente e della Vita, University of Genova, Genoa, Italy

Ute Schloßmacher ERC Advanced Investigator Group, Institute for Physiological Chemistry, University Medical Center of the Johannes Gutenberg University, Mainz, Gcrmany

Heinz C. Schröder ERC Advanced Investigator Group, Institute for Physiological Chemistry, University Medical Center of the Johannes Gutenberg University, Mainz, Germany;NanotecMARIN GmbH, Mainz, Germany

Zhijian Shen Department of Materials Science and Engineering, Tsinghua University, Beijing, People's Republic of China

Kui Wang State Key Laboratory of Natural and Biomimetic Drugs, School of Pharmaceutical Sciences, Peking University, Beijing, People's Republic of China; Department of Chemical Biology, School of Pharmaceutical Sciences, Peking University, Beijing, People's Republic of China

Xiaohong Wang ERC Advanced Investigator Group, Institute for Physiological Chemistry, University Medical Center of the Johannes Gutenberg University, Mainz, Germany

Matthias Wiens ERC Advanced Investigator Group, Institute for Physiological Chemistry, University Medical Center of the Johannes Gutenberg University, Mainz, Germany

Toshihiro Yamase MO device, Kanazawa, Japan;Chemical Resources Laboratory, Tokyo Institute of Technology, Midori-ku, Yokohama, Japan

Xiao-Gai Yang State Key Laboratory of Natural and Biomimetic Drugs, School of Pharmaceutical Sciences, Peking University, Beijing, People's Republic of China; Department of Chemical Biology, School of Pharmaceutical Sciences, Peking University, Beijing, People's Republic of China

Chapter 1
Chemical, Biochemical, and Biological Behaviors of Vanadate and Its Oligomers

Xiao-Gai Yang and Kui Wang

Contents

Abstract Vanadate is widely used as an inhibitor of protein tyrosine phosphatases (PTPase) and is routinely applied in cell lysis buffers or immunoprecipitations of phosphotyrosyl proteins. Additionally, vanadate has been extensively studied for its antidiabetic and anticancer effects. In most studies, orthovanadate or metavanadate was used as the starting compound, whereas these "vanadate" solutions may contain more or less oligomerized species. Whether and how different species of vanadium compounds formed in the biological media exert specific biological effect is still a

X.-G. Yang • K. Wang (✉)
State Key Laboratory of Natural and Biomimetic Drugs, School of Pharmaceutical Sciences, Peking University, Beijing 100191, People's Republic of China

Department of Chemical Biology, School of Pharmaceutical Sciences, Peking University, Beijing 100191, People's Republic of China
e-mail: wangkuipku@bjmu.edu.cn

W.E.G. Müller et al. (eds.), *Biomedical Inorganic Polymers*, Progress in Molecular and Subcellular Biology 54, DOI 10.1007/978-3-642-41004-8_1,

mystery. In the present commentary, we focus on the chemical, biochemical, and biological behaviors of vanadate. On the basis of species formation of vanadate in chemical and biological systems, we compared the biological effects and working mechanism of monovanadate with that of its oligomers, especially the decamer. We propose that different oligomers may exert a specific biological effect, which depends on their structures and the context of the cell types, by different modes of action.

Since the 1980s, vanadium compounds have been considered as a new class of metal-based drugs. Their diverse biological effects have attracted considerable interest. Studies have ranged from elucidation of the underlying mechanisms to the synthesis and screening of new vanadium compounds, from speciation analysis in solution to their fundamental biochemical reactions, from in vitro study to animal model and preclinical study. Multiple reviews have covered these topics in recent years (Chen and Owens 2008; Crans et al. 2004; Evangelou 2002; Faneca et al. 2009; Goc 2006; Kiss et al. 2008; Scior et al. 2005; Thompson and Orvig 2006). However, whether and how different species of vanadium compounds formed in the biological media exert specific biological effect is still a mystery. In the present commentary, we focus on the biological effects of vanadate, which is the classical and the most widely studied vanadium compound. In view of its species formation in chemical, biochemical, and biological systems, we will extend the discussions to the effects of various species and provide clues to the question stated as above.

1.1 The Speciation of Vanadate in the Chemical, Biochemical, and Biological System

There are three kinds of reactions to dominate vanadate aqueous chemistry: self-condensation reactions, coordination reactions, and electron transfer reactions. These reactions are influenced by pH and concentration of vanadate.

1.1.1 Self-Condensation Reactions of Vanadate

The formation of isopolyoxometalate ions, $[M_xO_y]^n$, via oxygen-bridged M–(O)–M is the general behavior of early transition metals:

$$xMO_4{}^{a-} + 2yH^+ \rightleftharpoons \left[M_xO_{4x-y}\right]^{n-} + yH_2O$$

The oligomerization from monomer to decamer comprises a cascade of protonation coupled with condensation (dehydration). The reverse reactions, hydrolysis of the oligomers, will lead to deoligomerization (decomposition). During this process, a series of intermediate oligomers, i.e., V_2, V_3, V_4, V_5, and several others may coexist in the solution. The distribution of these oligomers depends on pH level and the initial concentration of vanadate:

$$2VO_4{}^{3-} + 2H^+ \rightleftharpoons 2[HVO_4]^{2-} \rightleftharpoons [V_2O_7]^{4-} + H_2O$$
$$2VO_4{}^{3-} + 4H^+ \rightleftharpoons 2[H_2VO_4]^{2-} \rightleftharpoons [HV_2O_7]^{3-} + H^+ + H_2O$$
$$3[H_2VO_4]^{2-} \rightleftharpoons [V_3O_9]^{3-} + 3H_2O$$
$$4[H_2VO_4]^{2-} \rightleftharpoons [V_4O_{12}]^{3-} + 4H_2O$$
$$10[H_2VO_4]^{2-} + 4H^+ \rightleftharpoons [V_{10}O_{28}]^{6-} + 12H_2O$$
$$[V_{10}O_{28}]^{6-} + H^+ \rightleftharpoons [HV_{10}O_{28}]^{5-}$$
$$[HV_{10}O_{28}]^{5-} + H^+ \rightleftharpoons [H_2V_{10}O_{28}]^{4-}$$

For example, in the range of pH 7–9, the main species in dilute vanadate solution are V_1, V_2, V_4, and V_5 (Faneca et al. 2009; Heath and Howarth 1981; Larson 1995; Tracey et al. 1995), and the presence of V_3 and V_4 was confirmed later by ^{51}V and ^{7}O NMR and potentiometry (Andersson et al. 1996), while at pH 2–6 and in concentrated solutions, the decamer (V_{10}) and its protonated species become predominant (Aureliano and Gândara 2005; Crans and Tracey 1998; Faneca et al. 2009). By comparing the ^{51}V NMR of sodium metavanadate, Iannuzzi et al. noted that at pH 8.71, in 100 mM solution of metavanadate, the main species are the V_1, V_2, V_4, and V_5, while in 1–5 mM solution,V_1 is predominant (Iannuzzi et al. 2006).

The foregoing discussions are based on equilibrium state. In fact, the solutions used are in metastable state and the species may be transformed during cell incubation. The formation of the lower oligomers V_2, V_4, and V_5 from monomer is rapid; hence the solution of metavanadate or orthovanadate contains small amounts of V_2, V_4, and V_5 inevitably (Crans et al. 1990). However, formation and decomposition of V_{10} are slow in neutral solution under ambient temperature (Aureliano and Gândara 2005; Druskovich and Kepert 1975). The metavanadate or orthovanadate solution turns to yellow color slowly due to the formation of V_{10}. Conversely, the solution prepared from solid decavanadate decomposes gradually on standing (Rubinson 1981).

Since oligomerization is initiated by protonation of vanadate anions, acidification favors oligomer formation. Kalyani et al. followed the oligomerization of 30 mM ammonium metavanadate solutions with varied pH by ^{51}V NMR spectra (Kalyani and Ramasarma 1992). By the variation of NMR peaks, they found at pH 7.0, V_4 is the major species accompanied by traces of V_1, V_2, and V_5, but none of V_{10}. After incubation at pH 5.0 for 4 weeks at room temperature, all vanadium was in the form of V_{10}. On this ground, the decavanadate solution was suggested to be prepared by acidifying and then re-neutralizing the metavanadate solution subsequently before adding to the buffered culture media. However, there are still V_2, V_4, and V_5 present in this process (Iannuzzi et al. 2006; Turner et al. 2011). A kinetic study by means of UV–visible spectroscopy showed that the decomposition of V_{10} at 25 °C in the saline solution fits to a first-order kinetic process, with half-life of 16 h (Soares et al. 2007b). Since the deoligomerization is due to hydrolysis of the oligomers, dilution and high pH favor the decomposition.

As stated above, in biological studies, the solutions used are in metastable state and the species may undergo transformation during cell incubation. For this sake, studies at cellular level are influenced by how to prepare the solutions and how to run the experiments. In nonequilibrium state, the speciation at a particular time point depends on the rate of oligomerization and deoligomerization, which is determined by pH level and the initial concentration of vanadate.

In addition to the chemical consideration, the compartmentalization and dynamic state of the living cells may create dynamic intracellular microenvironments and thus alter the speciation of vanadate.

As is well known, the establishment and maintenance of an appropriate pH inside individual cellular compartment is critical for cellular functions. In eukaryotes, cytosolic pH was maintained roughly at pH ~ 7.4, but actually it varies in the range of pH 5–8 in different organelles as shown in Fig. 1.1. In the course of endocytic pathway, the pH value decreases successively in early endosomes, late endosomes, and lysosomes (Casey et al. 2010; Weisz 2003). The pH value of the intermembrane space is also generally regarded as significantly lower than that of the cytosol (Porcelli et al. 2005). Furthermore, several pathological conditions cause variation of intracellular pH, such as metabolic acidosis (Adrogue 2006; Kraut and Madias 2010). The altered pH was reported in transformed cells as well. It is observed that tumor cells typically have elevated organelle pH (Weisz 2003). Thus, it is conceivable that pH variations in a certain cellular compartment would cause speciation changes. Decavanadate formation would be possible in acidic vesicles, in muscle cells upon acidification, and even in intermembrane space of mitochondria (Aureliano and Crans 2009; Aureliano et al. 2002). In the local acidic environment, a series of oligomers formation of vanadate would also be more favored (Iannuzzi et al. 2006).

1.1.2 Coordination Reactions of Vanadate

Vanadate can react with a variety of mono- and polydentate ligands. Several reviews and monographs have covered this topic (Crans et al. 2004; Jakusch et al. 2011; Rehder 2008; Tracey et al. 2007). Given that vanadate has been dogmatically recognized as a structural and electronic analogue of phosphate, the two classes of vanadium compounds, vanadate esters and vanadate anhydrides, have been given more attention. These types of compounds are commonly used to explain the inhibitory and stimulatory effects of vanadate on ribonucleases, phosphatases, ATPase, some phosphorylases, and glucose dehydrogenase (Crans et al. 2004). Consequently, on the basis of well-characterized chemical model systems in test tubes, numerous extrapolations have been drawn to elucidate the mechanism concerning the in vivo effects of vanadate. However, currently it is

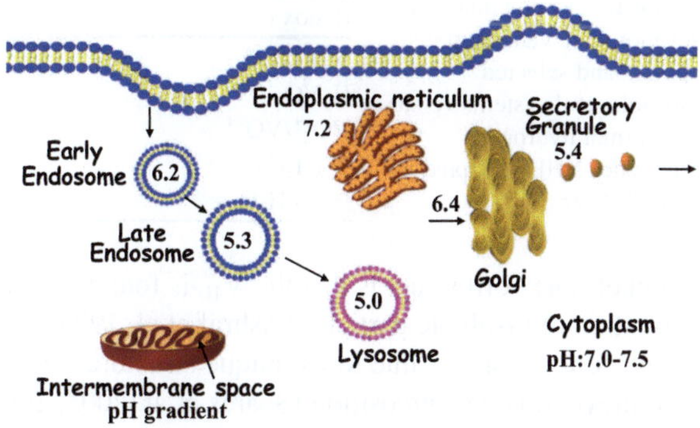

Fig. 1.1 A *cartoon diagram* showing some cellular organelles with typical pH values. Data adapted from Grabe and Oster (2001)

difficult to verify these reactions observed in test tubes occurred in vivo. Therefore, in this section, we focus on the coordination chemistry of vanadate in the scenario of biological systems.

It is found that vanadate, after oral uptake, in addition to the portions excreted with the feces, is mainly bound to serum proteins such as albumins and transferrin with a small amount of amino acids and other ligands (Jakusch et al. 2011; Zorzano et al. 2009). It also subjects to redox reactions (discussed in Sect. 1.1.3) and is partially reduced to vanadyl (VO^{2+}) (Macara et al. 1980).

Most portions of vanadate and vanadyl ions are bound to transferrin (Chasteen et al. 1986; Kiss et al. 2008; Sanna et al. 2009). Like Fe^{3+}, two VO^{2+} ions bind to the two iron-binding sites by the aid of HCO_3^- to stabilize the binding (Cannon and Chasteen 1975). V(V) species, as suggested by Kiss et al. (2012), bind to transferrin in form of VO_2^+ without requiring bicarbonate (Harris and Carrano 1984; Heinemann et al. 2002; Jakusch et al. 2009; Saponja and Vogel 1996).

VO^{2+} and VO_3^- are shown to bind to serum albumin as well (Ahmed-Ouameur et al. 2006; Ferrer et al. 2008). Human serum albumin binds VO^{2+} stronger than VO_3^-, with two binding sites, one strong and one weak, whereas VO_3^- binds by only one weak site (Purcell et al. 2001).

Computer modeling calculations were also used to understand the speciation of vanadate in serum. Based on the formation constants of twenty vanadate complexes with glycine, lactate, phosphate, citrate, histidine, albumin, transferrin, and also the ligands maltol and picolinate and the individual concentration, Pettersson's group calculated the distribution of these species in blood. The results illustrated that if the proteins were excluded, most of vanadate is present as $H_2VO_4^-$ and HVO_4^{2-}, ~12 % bound to glycine, and ~1–2 % to other constituents (mostly phosphate). If the proteins were included, more than 98% of the total vanadium is bound to transferrin (Gorzsás et al. 2006). This is confirmed by later experimental results (Jakusch et al. 2009, 2011), but some other species were recognized at low levels (Gorzsás et al. 2006).

In addition to the binding properties of monovanadate as stated above, more and more studies demonstrate that their oligomers should not be disregarded. In fact, the presence of a small portion of V_2, V_4, and V_5 may contribute to the protein binding.

Table 1.1 Redox potentials for inorganic vanadium species and selected physiological systems versus the normal hydrogen electrode (NHE), adapted from Rehder (2008)

Redox couples	E^0	$E^{pH=7}$
VO^{2+}/V^{3+}	+0.359	−0.462
$H_2VO_4^-/VO^{2+}$	+1.31	−0.34
VO_2^+/VO^{2+}	+1.016	+0.19
O_2/H_2O_2	+0.695	+0.295
½ O_2/H_2O	+1.23	+0.815

Another noteworthy fact is that the V_{10} is found to bind to albumin, leading to the formation of insoluble particles (Ashraf et al. 1995). The V_{10} binding to proteins is expected to be strong due to its unique structure (*vide infra*), and the binding may also prevent its decomposition (Soares et al. 2003, 2006).

Moreover, ligands that stabilize a metal complex in a lower oxidation state can promote the reduction of oxidized form. Therefore, the coordination reactions of vanadate can also affect the reducing tendency of vanadate in a biological system.

1.1.3 Electron Transfer Reactions of Vanadate

Vanadium shows a range of oxidation states (−1, 0, +1, +2, +3, +4, and +5). However, only three oxidation states, V(III), V(IV), and V(V), are relevant for biological systems (Rehder 2008).

Based on the standard reduction potentials (E^0) for inorganic vanadium species, there would be a lot of potent biogenic reducing agents capable of converting V (V) to V(IV) in acid solution (Baran 2008). However, the redox potentials of all the vanadium redox couples decrease dramatically with increasing pH. As seen in Table 1.1, under physiological system, in neutral solution (pH = 7) the oxidative ability of V(V) declined rapidly, so $H_2VO_4^-$ is less easily reduced.

However, in biological systems, as stated in Sect. 1.2, the reducing tendency of a mild reductant may be elevated by stabilizing V(IV) through complexation with biogenic ligands such as GSH and ATP, which are present in millimolar concentration in the cytosol.

As shown in Fig. 1.2, when vanadate encounters the cells, the membrane molecules stand in the frontier against the attacking vanadates. It is shown that vanadate is reduced to V(IV) by membrane-bound protein thiols during transport across the membrane (Yang et al. 2003; Zhang et al. 1997a). Recently, a two-electron transfer process in the hydrophobic environment in the membrane is proposed, i.e., the two-electron V(V)/V(III) redox chemistry may occur in the membrane (Crans et al. 2011).

In addition, vanadate may also stimulate nonenzymatic vanadate-dependent oxidation of NAD(P)H by using xanthine-oxidase reaction as a source of superoxide (Kalyani et al. 1992; Khandke et al. 1986; Liochev and Fridovich 1986), which will possibly lead to the formation of ROS or RONS. But it is still uncertain whether nonenzymatic vanadate-dependent NADH oxidation is actually coupled with NADPH oxidase-mediated superoxide production.

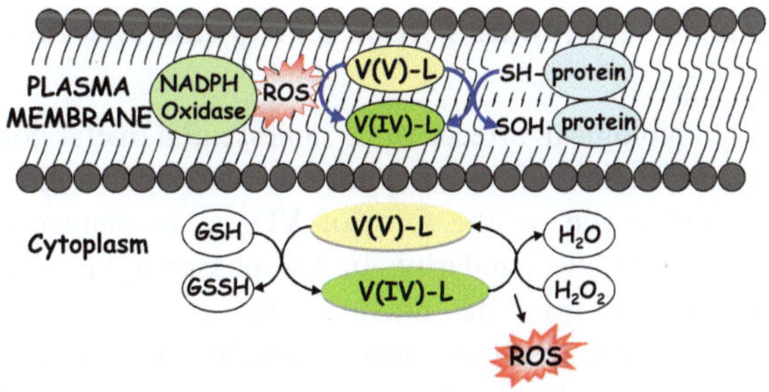

Fig. 1.2 Possible events happened when vanadate across membrane

In the cytosol, V(V) was found to be reduced to V(IV) by cytoplasmic glutathione(GSH) in a nonenzymatic way (Macara et al. 1980). The V(V) and V(IV) in the cytoplasm may interact with biomolecules in the cytoplasm and thus may result in the variation of redox potentials of V(V)/V(IV). Crans et al. also addressed that in V (V)-thiol redox reactions, the electron transfer reaction is promoted by the complexation of vanadium(IV)/(V) to the thiols (Crans et al. 2010).

In summary, the speciation of vanadate is determined by pH, vanadium concentration, and various biomolecules as ligands or reductants present in the intra- and extracellular fluids. Especially, the compartmentalization of organelles makes the vanadate speciation in biological system extremely complicated.

Vanadate is widely used as an inhibitor of protein tyrosine phosphatases (PTPase) (Gordon 1991) and is routinely applied in cell lysis buffers or immunoprecipitations of phosphotyrosyl protein. Additionally, vanadate has been extensively studied for its antidiabetic and anticancer effects. In most studies, orthovanadate or metavanadate was used as the starting compound, whereas these "vanadate" solutions may contain more or less oligomerized species, but the effects were usually attributed to monovanadate. This is reasonable in many cases, because pH was kept neutral and the concentration of vanadate was very low. However, as indicated above, in the local acidic environment, the formation of a series of oligomers would be favored. Additionally, by recent studies, the vanadate oligomers may be more reactive and may cause effects different from monomers. Thus in the following paragraphs, a brief account will be given to generalize the results obtained from the studies both with vanadate monomers and its oligomers.

1.2 Chemical and Biochemical Basis Underlying the Biological Effects of *Vanadate Monomers*

The mechanisms underlying the biological effects of vanadate are generally attributed to its two chemical properties: one is based on the structural analogy of vanadate to phosphate, and one is on the basis of the electron transfer reaction between V(IV) and V(V) in biological media.

1.2.1 Vanadate Ion as Phosphate Analogue

The biological effects of a nonessential metal ion can be considered as the manifestation of its similarity to corresponding essential ions (Wang 1997). Vanadate ion, in the form of VO_4^{3-}, is analogous to PO_4^{3-} by its structural and electronic similarity. In the phosphorylation reaction, a phosphate ion is transferred from the donor to the acceptor via a pentacoordinated trigonal bipyramidal transition state. Vanadate acts as a competitive inhibitor of the phosphatase by forming an analogue to mimic the substrate or mimic the intermediate complex between the donor and acceptor (Brandão et al. 2010; Davies and Hol 2004). Thus, it affects a great variety of phosphorylation-related physiological processes. Among them, the inhibition of protein tyrosine phosphatase (PTP), especially PTP1B, has been well studied and the results indicated that VO_4^{3-} inhibits PTP1B competitively (Huyer et al. 1997). X-ray crystal structures of vanadate complexes with PTP (Zhang et al. 1997b), acid phosphatases (Felts et al. 2006), and alkaline phosphatases (Holtz et al. 1999) suggested the mechanism with VO_4^{3-} binding in the transition state in place of PO_4^{3-}. The V–S linkage with a cysteine residue in the active site is similar to the thiol-phosphate linkage formed in the normal phosphatase catalysis (Zhang et al. 1997b). The inhibition of the PTP by vanadate enhances protein tyrosine phosphorylation and thus leads to intervention in the insulin-signaling pathway.

Another manifestation of vanadate as a phosphate analogue lies in the binding of vanadate monomer with ADP, which produces an ATP analogue, ADPV, and thus inhibits the activities of proteins (ATPase) (Zhang et al. 1996). Additionally, vanadate is considered as a substitute of phosphate ion in the crystal lattice of hydroxyapatite:

$$Ca_{10}(PO_4)_6(OH)_2 + nVO_4^{3-} \rightleftharpoons Ca_{10}(PO_4)_{6-n}(VO_4)_n(OH)_2 + nPO_4^{3-}$$

However, the incorporation may only occur when the hydroxyapatite is in an amorphous state and a small amount of vanadate ion does not produce any lattice distortion and has little effect on the strength of the P–O and O–H bonds (Etcheverry et al. 1984).

1.2.2 ROS Formation via Electron Transfer Reactions Between V(IV) and V(V)

The one-electron redox of V(V)/V(IV) generates superoxide radical $^{\bullet}O_2^-$ and then by dismutation $^{\bullet}O_2^-$ is converted to H_2O_2. Subsequently, a Fenton-like reaction between H_2O_2 and VO^{2+} converts H_2O_2 to $^{\bullet}OH$ radical. Among the ROS, H_2O_2 is the most stable and exhibits significantly physiological and the toxicological effects

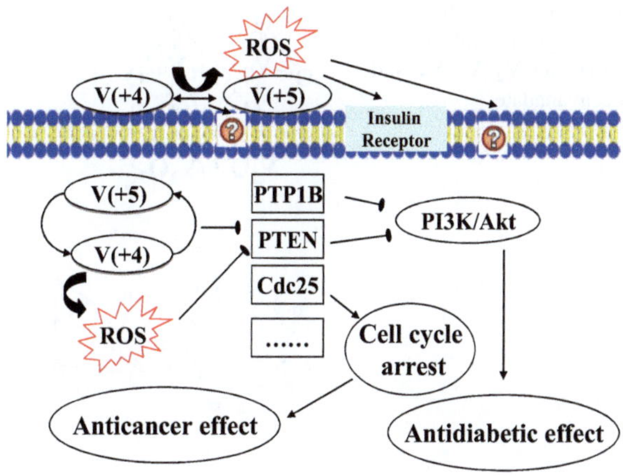

Fig. 1.3 Induction of ROS generation by vanadate and its derivatives may exert diverse biological effects. Vanadate can also inhibit the activities of PTP1B and PTEN directly as a phosphate analogue

(Shi and Dalal 1991, 1993; Sreedhara et al. 1997). The vanadate-induced H_2O_2 can act as a second messenger to activate different signaling pathways.

Considering that all the PTP superfamily of enzymes has a conserved cysteine residue in their catalytic domain, which must be in the reduced form for full activity (Savitsky and Finkel 2002), these enzymes might be direct targets of ROS. Therefore, it is conceivable that ROS-based inhibition of phosphatase should also be taken into account to examine the biological effects of vanadate rather than simply treat it as a phosphate analogue. For instance, as shown in Fig. 1.3, after a vanadium compound has penetrated the membrane, it may either inhibit the activities of PTP1B, PTEN, and CDC25 directly as a phosphate analogue or undergo one-electron electron transfer reactions by ROS formation to cause insulin-enhancing effects or cell cycle arrest. Besides the direct action of H_2O_2, the vanadate-induced H_2O_2 may turn back to coordinate with vanadate forming a series of monomeric and dimeric peroxovanadium species, V*m*X*n* ($m = 1, 2$; $n = 1, 2, 3$, with X stands for peroxo group) (Bortolini and Conte 2005; Pettersson et al. 2003). The peroxo species are more potent inhibitors of phosphatases than vanadate.

1.3 Chemical and Biochemical Basis Underlying the Biological Effects of Vanadate Oligomers

Theoretically, only the monomeric VO_4^{3-} is able to substitute PO_4^{3-} and inhibits phosphatases. In fact, the oligomeric species can also exert inhibitory effects, but probably by differential mode of action, some of which may stem from their different structures.

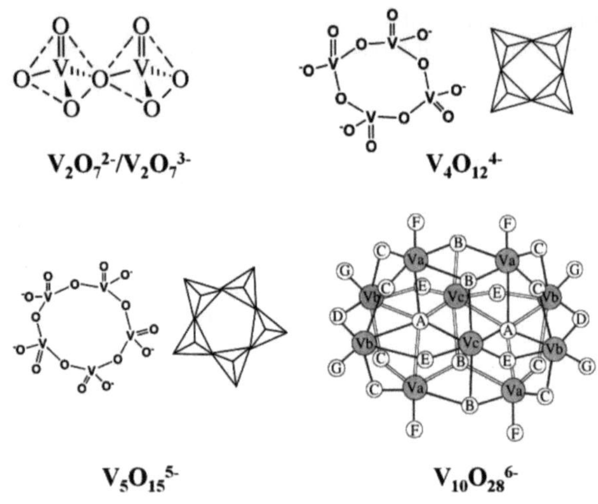

Fig. 1.4 Schematic drawings of tetrahedral vanadates V_2, V_4, V_5, and decavanadate

1.3.1 Structures and Properties of Oligovanadates

As shown in Fig. 1.4, lower oligovanadates are constructed from varied number of VO_4 units by sharing oxygen atoms, forming linear ($V_2O_7^{4-}$, V_2) and cyclic oligomers (V_4, V_5). Instead, V_{10} adopted a cage structure with octahedral VO_6 units. This structure renders it different from V_1, V_2, V_4, and V_5.

Firstly, this compact cage structure constructed from ten distorted octahedral VO_6 units with 28 shared oxygen atoms makes the polyanion $[V_{10}O_{28}]^{6-}$ relatively stable in solution. Although it tends to dissociate in neutral solution, but the reaction is rather slow.

Secondly, the basic oxygen surfaces and high negative charge density endow decavanadate with higher water solubility and nucleophilicity. These properties determined its high affinity to the positively charged sites.

Thirdly, the V(V) atoms in V_{10} are reducible by accepting one electron to yield V(IV). Since the electron transfer is affected by one or two vanadium atoms and only by the low-lying orbital, the V_{10} cage structure is basically unchanged in redox process. Thus, by V(V)/V(IV) shuttle, V_{10} can act as an electron transporter (Long et al. 2010).

1.3.2 Chemical and Biochemical Basis Underlying the Biological Effects of Decavanadate

Decavanadate is the only one among the oligomers isolated from the vanadate solution. A series of studies have been performed using solution of solid decavanadate or acidified vanadate solutions. In a recent review (Aureliano and

Crans 2009), Aureliano listed the effects of V_{10} on several enzymes and proteins. Most of the effects of V_{10} can be attributed to the strong affinity to multipositively charged sites of proteins and in a few cases accompanied with oxidation of thiol groups of proteins.

1.3.2.1 Binding Properties of Decavanadate

It is presumed that the nanosize clusters of V_{10} carrying a number of oxygen atoms in the surface have high affinity to bind with the positively charged domain of a protein (Soman et al. 1983). The binding causes conformation change or/and blocks the substrate binding of proteins.

V_{10} inhibits the nucleotide-dependent enzyme, adenylate kinase, which is not affected by monovanadate (Boyd et al. 1985; DeMaster and Mitchell 1973). The inhibition is explained by that V_{10} binds to adenylate kinase and fits across the phosphate-binding sites of both AMP and ATP sites and thus blocks nucleotide binding and inhibits phosphate transfer (Pai et al. 1977).

V_{10} binding may also result in the blocking of substrate binding to the enzyme. For example, it is able to inhibit rabbit skeletal muscle phosphorylases a and b by this way (Soman et al. 1983), while monovanadate not. Similarly, based on kinetic study and electrostatic potential maps of the surface of wild-type bovine pancreatic ribonuclease A (RNase A), Messmore et al. suggested that decavanadate inhibits RNase A by binding to the active site via electrostatic interaction (Messmore and Raines 2000).

Moreover, the protein-binding associated effect of decavanadate is unique in case of stabilizing the heat shock protein (Hsp90)–protein complex. Hsp90 is a chaperone protein, containing three domains: the ATP-binding, protein-binding, and dimerizing domains. ATP modulates the formation of Hsp90–protein complex, which stabilizes the protein. Both vanadate and V_{10} bind to Hsp90 and both stabilize Hsp90–protein complex but by different ways. The binding site of monovanadate was suggested to be the ADP/ATP-binding site of the Hsp90 N terminus, while V_{10} binds to the highly positively charged middle region and maintains Hsp90–protein interaction (Hou et al. 1992; Soti et al. 1998).

In some cases, decavanadate binds to the proteins in the form of nucleotide-bivalent ion (Mg^{2+})-V_{10} ternary complexes. MutS, a member of the ABC ATPases superfamily, plays an important role in repairing DNA biosynthetic errors. The inhibition by vanadate is attributed to a similar mechanism described above for other ATPases. But V_{10} binds to the ATP-binding region of MutS in forms of ADP–Mg–V_{10} and induces a steric impediment of the protein ATP/ADP exchange (Pezza et al. 2002).

Decavanadate binding to G-actin was found accompanied by thiol oxidation. The effect on G-actin polymerization is important in maintaining the integrity of cytoskeleton and also in muscle contraction. The interaction of decavanadate with G-actin was described as follows: decavanadate oxidizes cysteine thiol group with concurrent formation of vanadyl ions at first, and then the vanadyl ions bind to

actin, causes conformation change, and finally inhibition of G-actin polymerization (Aureliano 2011; Ramos et al. 2006, 2012). Differing from V_{10}, monovanadate does not interact with the cysteine thiol and acts merely as a phosphate analogue.

In summary, the actions of decavanadate might be ascribed mainly to its high affinity to specific positively charged sites of proteins, rather than as a phosphate analogue. Based on this property, decavanadate exhibits various effects on different proteins. Anyhow, the current experimental results are mostly obtained from biochemical studies, and the concentration is higher than expected in the organisms and cells. Thus further studies are necessary before ascertaining its biological significance. Other than this, the results as mentioned above are oriented to the phosphorylation-related proteins. It is expectable that the highly negatively charged decavanadate ions will bind to many proteins nonspecifically.

1.3.2.2 Redox Properties of Decavanadate

At cell level, mitochondrial pathway becomes more important in vanadate-associated redox processes. In vitro studies showed that decavanadate is stronger than vanadate in inducing membrane depolarization and inhibiting oxygen consumption to isolated hepatic and cardiac mitochondria (Soares et al. 2007a).

Aureliano's group had performed a series of studies on the impact of decavanadate on toadfish (Aureliano et al. 2002) and found it caused stronger oxidative stress and oxidative damages (Aureliano and Gândara 2005). It is interesting that differing from metavanadate, decavanadate did not induce cardiac mitochondrial ROS production and SOD activity, but reduce catalase activity. Similarly, by comparing intravenous exposure of decavanadate and vanadate (V_1–V_5) to marine teleost *Halobatrachus didactylus* (Lusitanian toadfish), they found that more vanadium was accumulated by mitochondria in case of decavanadate administration and decavanadate promoted stronger mitochondrial antioxidant enzymes activities than vanadate.

Putting the results together, we can see the importance of protein binding in V_{10}'s biological effects. However, most of the studies were limited to its biochemical actions; the biological effects and the outcomes were seldom reported. Based on current results, V_{10} is more toxic than vanadate, at least in the level of mitochondria. As to the contribution of V_{10} in the effects of vanadate, further studies are needed. Whether V_{10} is more active in redox reactions in biological systems than vanadate and how the in-cage electron transfer acts remains to be clarified.

1.3.3 *Biological Effects of Vanadate Dimer and Tetramer*

Since the lower oligomers are evidently present in solution as minor constituents and not isolated, their specific effects are hardly defined. Crans' group tried to clarify the different activities among V_1, V_2, V_4, and V_5 based on oligomer analysis

with ^{51}V NMR and the inhibition effect of vanadate solution. By this way, they claimed that in several cases, the observed effects of vanadate are in fact contributed by the oligomers, V_2 and V_4. They claimed that V_2 and V_4 are the acting species in vanadate solution to inhibit glucose-6-phosphate dehydrogenase (G6PD), but V_2 was the major inhibiting species with respect to NADP, and V_4, to G6P and to NAD (Crans and Schelble 1990). The inhibition of glycerol-3-phosphate dehydrogenase (G3PDH) was also suggested mainly by V_4, while dimer exhibits weaker effect (Crans and Simone 1991). In addition, in the inhibition of fructose-1,6-bisphosphate aldolase by vanadate, the active species was identified to be V_4. It is the tetramer that oxidizes the thiol group of aldose and thus inhibits aldolase irreversibly, whereas vanadate dimer is a reversible inhibitor (Crans et al. 1992). A similar case is that the contributor in vanadate-mediated specific photocleavage of myosin subfragment 1 was suggested to be V_4 (Cremo et al. 1990; Ringel et al. 1990).

Divanadate, $[V_2O_7]^{4-}$ may be envisaged as a pyrophosphate analogue. It presents in the solution of vanadate, but due to its low level and rapid conversion, it is difficult to define its specific effect. So far, the only supporting evidence was reported on the pyrovanadolysis as pyrophosphorolysis-like reaction (Akabayov et al. 2011). The role of pyrophosphate in biological effects is different from phosphate; thus the pertinent action of divanadate is intriguing.

1.4 Summary

The existing experimental results indicate that the effects of vanadate are the integrated manifestation of the species formed in biological systems. Among the species, the peroxocomplexes and the oligomeric vanadates, especially the decamer, are shown to bring about several effects different from monomeric vanadate. On account of species transformation, the various effects of monovanadate are discussed by its analogy with phosphate and its redox behaviors, including the formation and action of peroxocomplexes. However, due to the condensed cage structure and high negative charge density, decavanadate is unlikely to be a phosphate analogue but with a special multipoint binding ability to positively charged sites. Although the contributions and the behaviors of the dimeric and tetrameric vanadate species are still ill-defined, the studies have shown that they may exert differential modes of action on the enzymes or other biomolecules. Therefore, different oligomers may exert a specific biological effect or even the similar effect but via a different mechanism.

In this commentary, we only focus on the discussions on the interactions between vanadate and its oligomers with specific biomolecules without involving the scenario of certain biological systems. However, in order to elucidate the underlying mechanism of their biological effects, a detailed knowledge of genetic background of the cell types such as the expressions of tumor suppressors or oncogenes and redox state of various cells is required, which may be as biological

determinants to allow vanadate and its derivatives exert cell-specific effects. It is reported that vanadate induced cell cycle arrest at the G2/M phase by H_2O_2-activated ERK and p38 pathways (Zhang et al. 2003). But, there might be another mechanism leading to vanadate-induced cytotoxicity independent on H_2O_2 generation (Capella et al. 2002, 2007). Our group also demonstrated that vanadate can induce cell cycle arrest in HepG2 cells via an ROS-independent pathway (Fu et al. 2008), and by using antioxidants as synergistic agents, the damage to normal liver cells may be avoided (Wang et al. 2010). The study of antiproliferative effects of vanadate, tungstate, and molybdate on human prostate cancer cell line PC-3 demonstrated that all of the three oxoanions can cause G2/M cell cycle arrest but vanadate exerted much more potent effect in PC-3 cells than the other two oxoanions (Liu et al. 2012). The results reveal that ROS-mediated degradation of CDC25C is responsible for vanadate-induced G2/M cell cycle arrest. The study proposes a possible mechanism to clarify the differential effect of three oxoanions in biological systems beyond just considering that they are structural analogues of phosphate. The redox properties of vanadium may be important factors to exert pharmacological effects in human prostate cancer cells.

Therefore, the effects of vanadate and its derivatives might be cell specific, which means a result in one type of cell lines cannot be simply extended to another situation. However, if details of the specific conditions in experiment have been known and categorized including vanadium speciation, redox status and even the expressions of redox-sensitive transcription factors in the biological systems, we can optimize the performance of vanadate and its derivatives.

References

Adrogue HJ (2006) Metabolic acidosis: pathophysiology, diagnosis and management. J Nephrol 19(Suppl 9):S62–S69

Ahmed-Ouameur A, Diamantoglou S, Sedaghat-Herati MR, Nafisi S, Carpentier R, Tajmir-Riahi HA (2006) The effects of drug complexation on the stability and conformation of human serum albumin: protein unfolding. Cell Biochem Biophys 45:203–213

Akabayov B, Kulczyk AW, Akabayov SR, Theile C, McLaughlin LW, Beauchamp B, van Oijen AM, Richardson CC (2011) Pyrovanadolysis, a pyrophosphorolysis-like reaction mediated by pyrovanadate, Mn^{2+}, and DNA polymerase of bacteriophage T7. J Biol Chem 286:29146–29157

Andersson I, Pettersson L, Hastings JJ, Howarth OW (1996) Oxygen and vanadium exchange processes in linear vanadate oligomers. J Chem Soc Dalton Trans: 3357–3361

Ashraf SM, RajeshKaleem S (1995) Interaction of decavanadate polyanions with proteins. Anal Biochem 230: 68–74

Aureliano M, Crans DC (2009) Decavanadate ($V_{10}O_{28}^{6-}$) and oxovanadates: oxometalates with many biological activities. J Inorg Biochem 103:536–546

Aureliano M, Gândara R (2005) Decavanadate effects in biological systems. J Inorg Biochem 99:979–985

Aureliano M, Joaquim N, Sousa A, Martins H, Coucelo JM (2002) Oxidative stress in toadfish (Halobactrachus didactylus) cardiac muscle. Acute exposure to vanadate oligomers. J Inorg Biochem 90:159–165

Aureliano M (2011) Recent perspectives into biochemistry of decavanadate. World J Biol Chem 2:215

Baran EJ (2008) Vanadium detoxification: chemical and biochemical aspects. Chem Biodivers 5:1475–1484

Bortolini O, Conte V (2005) Vanadium (V) peroxocomplexes: structure, chemistry and biological implications. J Inorg Biochem 99:1549–1557

Boyd DW, Kustin K, Niwa M (1985) Do vanadate polyanions inhibit phosphotransferase enzymes? Biochim Biophys Acta 827:472–475

Brandão TAS, Hengge AC, Johnson SJ (2010) Insights into the reaction of protein-tyrosine phosphatase 1B. J Biol Chem 285:15874

Cannon JC, Chasteen ND (1975) Nonequivalence of the metal binding sites in vanadyl-labeled human serum transferrin. Biochemistry 14:4573–4577

Capella LS, Gefe MR, Silva EF, Affonso-Mitidieri O, Lopes AG, Rumjanek VM, Capella MAM (2002) Mechanisms of vanadate-induced cellular toxicity: role of cellular glutathione and NADPH. Arch Biochem Biophys 406:65–72

Capella MAM, Capella LS, Valente RC, Gefé M, Lopes AG (2007) Vanadate-induced cell death is dissociated from H_2O_2 generation. Cell Biol Toxicol 23:413–420

Casey JR, Grinstein S, Orlowski J (2010) Sensors and regulators of intracellular pH. Nat Rev Mol Cell Biol 11:50–61

Chasteen ND, Grady JK, Holloway CE (1986) Characterization of the binding, kinetics, and redox stability of vanadium (IV) and vanadium (V) protein complexes in serum. Inorg Chem 25:2754–2760

Chen ZL, Owens G (2008) Trends in speciation analysis of vanadium in environmental samples and biological fluids-a review. Anal Chim Acta 607:1–14

Crans DC, Schelble SM (1990) Vanadate dimer and tetramer both inhibit glucose-6-phosphate dehydrogenase from Leuconostoc mesenteroides. Biochemistry 29:6698–6706

Crans DC, Simone CM (1991) Nonreductive interaction of vanadate with an enzyme containing a thiol group in the active site: glycerol-3-phosphate dehydrogenase. Biochemistry 30:6734–6741

Crans DC, Tracey AS (1998) The chemistry of vanadium in aqueous and nonaqueous solution. ACS Publications, Washington, DC

Crans DC, Rithner CD, Theisen LA (1990) Application of time-resolved vanadium-51 2D NMR for quantitation of kinetic exchange pathways between vanadate monomer, dimer, tetramer, and pentamer. J Am Chem Soc 112:2901–2908

Crans DC, Sudhakar K, Zamborelli TJ (1992) Interaction of rabbit muscle aldolase at high ionic strengths with vanadate and other oxoanions. Biochemistry 31:6812–6821

Crans DC, Smee JJ, Gaidamauskas E, Yang L (2004) The chemistry and biochemistry of vanadium and the biological activities exerted by vanadium compounds. Chem Rev 104:849–902

Crans DC, Zhang B, Gaidamauskas E, Keramidas AD, Willsky GR, Roberts CR (2010) Is vanadate reduced by thiols under biological conditions? changing the redox potential of V (V)/V(IV) by complexation in aqueous solution. Inorg Chem 49:4245–4256

Crans DC, Trujillo AM, Pharazyn PS, Cohen MD (2011) How environment affects drug activity: localization, compartmentalization and reactions of a vanadium insulin-enhancing compound, dipicolinatooxovanadium (V). Coord Chem Rev 255:2178–2192

Cremo CR, Long GT, Grammer JC (1990) Photocleavage of myosin subfragment 1 by vanadate. Biochemistry 29:7982–7990

Davies DR, Hol WG (2004) The power of vanadate in crystallographic investigations of phosphoryl transfer enzymes. FEBS Lett 577:315–321

DeMaster EG, Mitchell A (1973) A comparison of arsenate and vanadate as inhibitors or uncouplers of mitochondrial and glycolytic energy metabolism. Biochemistry 12:3616–3621

Druskovich DM, Kepert DL (1975) Base decomposition of decavanadate. J Chem Soc Dalton Trans: 947–951

Etcheverry SB, Apella MC, Baran EJ (1984) A model study of the incorporation of vanadium in bone. J Inorg Biochem 20:269–274
Evangelou AM (2002) Vanadium in cancer treatment. Crit Rev Oncol Hematol 42:249–265
Faneca H, Figueiredo VA, Tomaz I, Goncalves G, Avecilla F, Pedroso de Lima MC, Geraldes CF, Pessoa JC, Castro MM (2009) Vanadium compounds as therapeutic agents: some chemical and biochemical studies. J Inorg Biochem 103(4):601–608
Felts RL, Reilly TJ, Tanner JJ (2006) Structure of Francisella tularensis AcpA: prototype of a unique superfamily of acid phosphatases and phospholipases C. J Biol Chem 281:30289–30298
Ferrer EG, Bosch A, Yantorno O, Baran EJ (2008) A spectroscopy approach for the study of the interactions of bioactive vanadium species with bovine serum albumin. Bioorg Med Chem 16:3878–3886
Fu Y, Wang Q, Yang XG, Yang XD, Wang K (2008) Vanadyl bisacetylacetonate induced G1/S cell cycle arrest via high-intensity ERK phosphorylation in HepG2 cells. J Biol Inorg Chem 13:1001–1009
Goc A (2006) Biological activity of vanadium compounds. Cent Eur J Biol 1:314–332
Gordon JA (1991) Use of vanadate as protein-phosphotyrosine phosphatase inhibitor. Methods Enzymol 201:477–482
Gorzsás A, Andersson I, Pettersson L (2006) On the fate of vanadate in human blood. Eur J Inorg Chem 2006:3559–3565
Grabe M, Oster G (2001) Regulation of organelle acidity. J Gen Physiol 117:329–344
Harris WR, Carrano CJ (1984) Binding of vanadate to human serum transferrin. J Inorg Biochem 22:201–218
Heath E, Howarth OW (1981) Vanadium-51 and oxygen-17 nuclear magnetic resonance study of vanadate (V) equilibria and kinetics. J Chem Soc Dalton Trans: 1105–1110
Heinemann G, Fichtl B, Mentler M, Vogt W (2002) Binding of vanadate to human albumin in infusion solutions, to proteins in human fresh frozen plasma, and to transferrin. J Inorg Biochem 90:38–42
Holtz KM, Stec B, Kantrowitz ER (1999) A model of the transition state in the alkaline phosphatase reaction. J Biol Chem 274:8351–8354
Hou D, Hagen KS, Hill CL (1992) Tridecavanadate,$[V_{13}O_{34}]^{3-}$, a new high-potential isopolyvanadate. J Am Chem Soc 114:5864–5866
Huyer G, Liu S, Kelly J, Moffat J, Payette P, Kennedy B, Tsaprailis G, Gresser MJ, Ramachandran C (1997) Mechanism of inhibition of protein-tyrosine phosphatases by vanadate and pervanadate. J Biol Chem 272:843–851
Iannuzzi M, Young T, Frankel GS (2006) Aluminum alloy corrosion inhibition by vanadates. J Electrochem Soc 153:B533
Jakusch T, Dean A, Oncsik T, Bényei AC, Di Marco V, Kiss T (2009) Vanadate complexes in serum: a speciation modeling study. Dalton Trans 39:212–220
Jakusch T, Pessoa JC, Kiss T (2011) The speciation of vanadium in human serum. Coord Chem Rev 255:2218–2226
Kalyani P, Ramasarma T (1992) Polyvanadate-stimulated NADH oxidation by plasma membranes–the need for a mixture of deca and meta forms of vanadate. Arch Biochem Biophys 297:244–252
Kalyani P, Vijaya S, Ramasarma T (1992) Characterization of oxygen free radicals generated during vanadate-stimulated NADH oxidation. Mol Cell Biochem 111:33–40
Khandke L, Gullapalli S, Patole MS, Ramasarma T (1986) Vanadate-stimulated NADH oxidation by xanthine oxidase: an intrinsic property. Arch Biochem Biophys 244:742–749
Kiss T, Jakusch T, Hollender D, Ágnes D, Enyedy Éva A, Pessoa JC, Sakurai H, Sanz-Medel A (2008) Biospeciation of antidiabetic VO (IV) complexes. Coord Chem Rev 252:1153–1162
Kiss T, Jakusch T, Gyurcsik B, Lakatos A, Enyedy ÉA, Sija É (2012) Application of modeling calculations in the description of metal ion distribution of bioactive compounds in biological systems. Coord Chem Rev 256:125–132

Kraut JA, Madias NE (2010) Metabolic acidosis: pathophysiology, diagnosis and management. Nat Rev Nephrol 6:274–285

Larson JW (1995) Thermochemistry of vanadium (5+) in aqueous solutions. J Chem Eng Data 40:1276–1280

Liochev S, Fridovich I (1986) The vanadate-stimulated oxidation of NAD(P)H by biomembranes is a superoxide-initiated free radical chain reaction. Arch Biochem Biophys 250:139–145

Liu TT, Liu YJ, Wang Q, Yang XG, Wang K (2012) Reactive-oxygen-species-mediated Cdc25C degradation results in differential antiproliferative activities of vanadate, tungstate, and molybdate in the PC-3 human prostate cancer cell line. J Biol Inorg Chem 17:311–320

Long DL, Tsunashima R, Cronin L (2010) Polyoxometalates: building blocks for functional nanoscale systems. Angew Chem Int Ed Engl 49:1736–1758

Macara IG, Kustin K, Cantley LC Jr (1980) Glutathione reduces cytoplasmic vanadate. Mechanism and physiological implications. Biochim Biophys Acta 629:95–106

Messmore JM, Raines RT (2000) Decavanadate inhibits catalysis by ribonuclease A. Arch Biochem Biophys 381:25–30

Pai EF, Sachsenheimer W, Schirmer RH, Schulz GE (1977) Substrate positions and induced-fit in crystalline adenylate kinase. J Mol Biol 114:37–45

Pettersson L, Andersson I, Gorzsás A (2003) Speciation in peroxovanadate systems. Coord Chem Rev 237:77–87

Pezza RJ, Villarreal MA, Montich GG, Argarana CE (2002) Vanadate inhibits the ATPase activity and DNA binding capability of bacterial MutS. A structural model for the vanadate-MutS interaction at the Walker A motif. Nucleic Acids Res 30:4700–4708

Porcelli AM, Ghelli A, Zanna C, Pinton P, Rizzuto R, Rugolo M (2005) pH difference across the outer mitochondrial membrane measured with a green fluorescent protein mutant. Biochem Biophys Res Commun 326:799–804

Purcell M, Neault JF, Malonga H, Arakawa H, Tajmir-Riahi HA (2001) Interaction of human serum albumin with oxovanadium ions studied by FT-IR spectroscopy and gel and capillary electrophoresis. Can J Chem 79:1415–1421

Ramos S, Manuel M, Tiago T, Duarte R, Martins J, Gutierrez-Merino C, Moura JJG, Aureliano M (2006) Decavanadate interactions with actin: inhibition of G-actin polymerization and stabilization of decameric vanadate. J Inorg Biochem 100:1734–1743

Ramos S, Moura JJG, Aureliano M (2012) Recent advances into vanadyl, vanadate and decavanadate interactions with actin. Metallomics 4:16–22

Rehder D (2008) Bioinorganic vanadium chemistry. Wiley, Chichester

Ringel I, Peyser YM, Muhlrad A (1990) ^{51}V NMR study of vanadate binding to myosin and its subfragment 1. Biochemistry 29:9091–9096

Rubinson KA (1981) Concerning the form of biochemically active vanadium. Proc R Soc Lond B Biol Sci 212:65–84

Sanna D, Micera G, Garribba E (2009) New developments in the comprehension of the biotransformation and transport of insulin-enhancing vanadium compounds in the blood serum. Inorg Chem 49:174–187

Saponja JA, Vogel HJ (1996) Metal-ion binding properties of the transferrins: a vanadium-51 NMR study. J Inorg Biochem 62:253–270

Savitsky PA, Finkel T (2002) Redox regulation of Cdc25C. J Biol Chem 277:20535–20540

Scior T, Guevara-Garcia A, Bernard P, Do QT, Domeyer D, Laufer S (2005) Are vanadium compounds drugable? structures and effects of antidiabetic vanadium compounds: a critical review. Mini Rev Med Chem 5:995–1008

Shi XL, Dalal NS (1991) Flavoenzymes reduce vanadium(V) and molecular oxygen and generate hydroxyl radical. Arch Biochem Biophys 289:355–361

Shi XL, Dalal NS (1993) Vanadate-mediated hydroxyl radical generation from superoxide radical in the presence of NADH: Haber-Weiss vs Fenton mechanism. Arch Biochem Biophys 307:336–341

Soares SS, Aureliano M, Joaquim N, Coucelo JM (2003) Cadmium and vanadate oligomers effects on methaemoglobin reductase activity from Lusitanian toadfish: in vivo and in vitro studies. J Inorg Biochem 94:285–290

Soares SS, Martins H, Aureliano M (2006) Vanadium distribution following decavanadate administration. Arch Environ Contam Toxicol 50:60–64

Soares SS, Gutiérrez-Merino C, Aureliano M (2007a) Decavanadate induces mitochondrial membrane depolarization and inhibits oxygen consumption. J Inorg Biochem 101:789

Soares SS, Martins H, Duarte RO, Moura JJ, Coucelo J, Gutierrez-Merino C, Aureliano M (2007b) Vanadium distribution, lipid peroxidation and oxidative stress markers upon decavanadate in vivo administration. J Inorg Biochem 101:80–88

Soman G, Chang YC, Graves DJ (1983) Effect of oxyanions of the early transition metals on rabbit skeletal muscle phosphorylase. Biochemistry 22:4994–5000

Soti C, Radics L, Yahara I, Csermely P (1998) Interaction of vanadate oligomers and permolybdate with the 90-kDa heat-shock protein, Hsp90. Eur J Biochem 255:611–617

Sreedhara A, Susa N, Rao CP (1997) Vanadate and chromate reduction by saccharides and L-ascorbic acid: effect of the isolated V (IV) and Cr (III) products on DNA nicking, lipid peroxidation, cytotoxicity and on enzymatic and non-enzymatic antioxidants. Inorg Chim Acta 263:189–194

Thompson KH, Orvig C (2006) Vanadium in diabetes: 100 years from Phase 0 to Phase I. J Inorg Biochem 100:1925–1935

Tracey AS, Jaswal JS, Angus-Dunne SJ (1995) Influences of pH and ionic strength on aqueous vanadate equilibria. Inorg Chem 34:5680–5685

Tracey AS, Willsky GR, Takeuchi E (2007) Vanadium: chemistry, biochemistry, pharmacology, and practical applications. CRC, Boca Raton, FL

Turner TL, Nguyen VH, McLauchlan CC, Dymon Z, Dorsey BM, Hooker JD, Jones MA (2011) Inhibitory effects of decavanadate on several enzymes and Leishmania tarentolae In Vitro. J Inorg Biochem 108:96–104

Wang K (1997) The analogy in chemical and biological behavior between non-essential ions compared with essential ions. South Afr J Chem 50:232–239

Wang Q, Liu TT, Fu Y, Wang K, Yang XG (2010) Vanadium compounds discriminate hepatoma and normal hepatic cells by differential regulation of reactive oxygen species. J Biol Inorg Chem 15:1087–1097

Weisz OA (2003) Organelle acidification and disease. Traffic 4:57–64

Yang X, Wang K, Lu J, Crans DC (2003) Membrane transport of vanadium compounds and the interaction with the erythrocyte membrane. Coord Chem Rev 237:103–111

Zhang B, Zhang S, Wang K (1996) Synthesis, characterization and crystal structure of cyclic vanadate complexes with monosaccharide derivatives having a free adjacent diol system. J Chem Soc Dalton Trans: 3257–3263

Zhang B, Ruan L, Chen B, Lu J, Wang K (1997a) Binding of vanadate to human erythrocyte ghosts and subsequent events. Biometals 10:291–298

Zhang M, Zhou M, Van Etten RL, Stauffacher CV (1997b) Crystal structure of bovine low molecular weight phosphotyrosyl phosphatase complexed with the transition state analog vanadate. Biochemistry 36:15–23

Zhang Z, Leonard SS, Huang C, Vallyathan V, Castranova V, Shi X (2003) Role of reactive oxygen species and MAPKs in vanadate-induced G(2)/M phase arrest. Free Radic Biol Med 34:1333–1342

Zorzano A, Palacín M, Marti L, García -Vicente S (2009) Arylalkylamine vanadium salts as new anti-diabetic compounds. J Inorg Biochem 103:559–566

Chapter 2
Structural Characterization of Inorganic Biomaterials

Irene M. Mavridis

Contents

Abstract Composite materials with unique architectures are ubiquitous in nature, e.g., marine shells, sponge spicules, bones, and dentine. These structured organic–inorganic systems are generated through self-assembly of organic matter (usually proteins or lipids) into scaffolds, onto which the inorganic component is deposited in organized hierarchical structures of sizes spanning several orders of magnitude. The development of bio-inspired materials is possible through the

I.M. Mavridis (✉)
National Center for Scientific Research Demokritos, Aghia Paraskevi, P. O. Box 60228, Athens, Greece
e-mail: mavridi@chem.demokritos.gr

W.E.G. Müller et al. (eds.), *Biomedical Inorganic Polymers*, Progress in Molecular and Subcellular Biology 54, DOI 10.1007/978-3-642-41004-8_2,

design of synthetic bottom-up self-assembly methods. Knowledge of the structure is required in order to assess the efficiency of their design and evaluate their properties. This chapter reviews the main methods used for structure determination of natural and synthetic inorganic biomaterials, namely, X-ray diffraction and scattering and electron diffraction and microscopy (TEM, SEM), as well as the AFM and CSLM microscopy methods. Moreover, spectroscopic (IR, NMR, and Raman) and thermal methods are presented. Examples of biomimetic synthetic materials are used to show the contribution of single or multiple techniques in the elucidation of their structure.

2.1 Introduction

Molecular recognition and self-assembly are well-known chemical and biological processes leading to selectivity, binding, and special functions, e.g., transport of molecules. Molecular recognition involves geometrical and energetic complementarities between molecular species. The geometric complementarity in chemistry has been recognized early by Fischer (1894) as the "lock and key" principle, but it was only in the late 1970s that the concept gained momentum in connection with supramolecular chemistry (Lehn 1995). It is recognized that high specificity of a molecular system A towards another B in the presence of other related species is achieved by complementarity in shape and size between A and B and multiple intermolecular nonbonding interactions, such as H-bonds and dipole–dipole, resulting in large contact areas and large difference in the binding free energy between A and B (Lehn 1995). Self-assembly by molecular recognition leads to unique architectures in nature that most of the time serve the purpose of specific physical and chemical functions. These structures are organic–inorganic composite materials generated through the self-assembly of organic matter (usually proteins or lipids, 1–100 nm) acting as frameworks that through molecular recognition generate orientation and deposition of the inorganic component in an organized manner. The overall structure is hierarchical with size scales ranging from the nano (nm) to the macro (cm) scale. Some examples relevant to this chapter are (a) sponge spicules comprising a proteinaceous axial filament and amorphous hydrated silica deposited concentrically around it; (b) marine shells, such as nacre of abalone, composed of films (<10 nm) of silk-like proteins, β-chitin, and inorganic aragonite (CaCO3) platelets (<500 nm); (c) bone, a very diverse and complex tissue, which can be described in simplified terms as comprising an organic component, mainly of type 1 collagen, and a mineralized inorganic component of carbonated apatite platelets; and (d) dentin, the bone-like matrix between the exterior enamel and the pulp of the tooth, composed also of inorganic and organic materials (mainly hydroxyapatite and collagen) structured in closely packed "dentinal tubules."

By mimicking nature's way, it is possible to design and synthesize hierarchically structured materials for various uses, especially for biomedical purposes, as, for example, the much-needed bone tissue regeneration. Detailed structural determination is required in order to monitor how close any synthetic biomaterial resembles its prototype. Moreover, it permits to assess the design of their synthesis, usually by synthetic bottom-up self-assembly methods, to get insight in the mechanism of formation and evaluate their properties, so that the biomimetic approach converges to the natural one in a self-consistent way. The modern methods available for analysis of the structure are very powerful, which has contributed greatly to the advancement of material science and the development of biomimetic materials.

2.2 Diffraction and Scattering Methods

2.2.1 Single-Crystal X-Ray Diffraction

The scattering of X-rays by single crystals (continuous and infinite 3-dimensional lattice) generates diffraction, which has been used to determine the structure of matter, since the beginning of the twentieth century. The method, known as X-ray crystallography, provides the detailed atomic positions of any examined substance in its crystal lattice (Blow 2002). However, in order to obtain this detailed information, single crystals are required. The inorganic biomaterials, as composite materials of more than one species and/or phases, cannot form single crystals, although parts of them can be polycrystalline and some parts frequently exhibit order of some degree. The presentation of the single crystal diffraction that follows will help the discussion of the diffraction methods used for inorganic biomaterials, which is that of polycrystalline or low-order materials. Figure 2.1 shows a diffraction image by a single crystal, each single spot corresponding to a unique diffraction direction hkl in space. Bragg gave a simplified picture of diffraction describing it as reflection of X-rays by infinite sets of equivalent planes P, A, B, etc. (infinite number in every set) of the crystal lattice. Each set of planes is characterized by a unique direction and interplanar spacing d_{hkl}. If a parallel X-ray beam (1, 2 in Fig. 2.1) falls on any set of planes P, diffraction occurs (1′, 2′ in Fig. 2.1) only if the planes P form the same angle θ_{hkl} with the incident beam and the diffracted beam (reflection condition). Thus, each diffraction direction hkl (called "reflection") corresponding to the unique crystal spacing d_{hkl} is related to the diffraction angle θ_{hkl} according to the well-known Bragg's law:

$$2d_{\mathrm{hkl}} \sin\theta_{\mathrm{hkl}} = n_{\mathrm{hkl}}\lambda \tag{2.1}$$

where λ is the wavelength of the radiation and n is an integer corresponding the order of the "reflection."

X-ray diffraction, the most commonly used method to examine the molecular structure, depends on the scattering of X-rays by the electron density of atoms and molecules. Neutron radiation can also be used to determine the molecular structure by the same techniques. The neutron beam, depending on scattering by the nuclei,

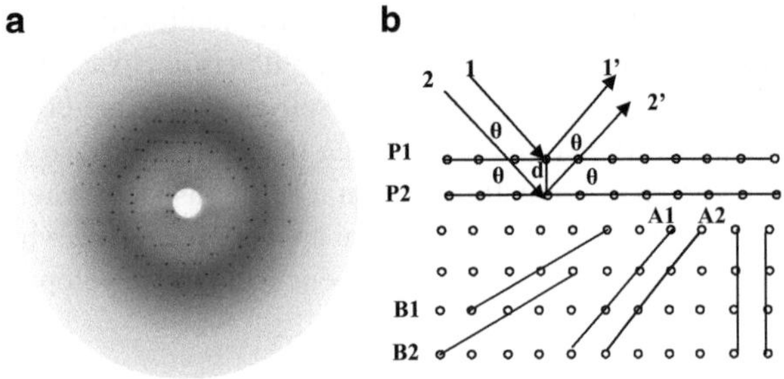

Fig. 2.1 Single-crystal X-ray diffraction image (**a**); schematic diagram of Bragg's law (**b**)

has scattering cross section approximately equal for most atoms, in contrast to this of X-rays that increases with atomic number. Therefore, neutron diffraction techniques are more sensitive to lighter atoms, e.g., it is ideal to detect hydrogen bonding in biological systems. However, neutron diffraction generally requires large crystals, which are not readily available.

For substances for which it is difficult to obtain single crystals, notably proteins and other biological systems, disordered or low-order materials, ceramics, or macromolecules, other methods are used, such as small-angle X-ray scattering, fiber diffraction, and powder X-ray diffraction (XRD).

2.2.2 *Powder X-Ray Diffraction*

A polycrystalline powder comprises grains of single crystals in all orientations; thus, the diffraction reduces to concentric circles, each characterized by a specific θ angle (Fig. 2.2) and a unique set of spacing according to Bragg's law (Klug and Alexander 1974). The pattern provides the fingerprint of polycrystalline inorganic materials, due to the fact that each mineral has a unique lattice and set of d-spacings. Determination of all d-spacings from the powder XRD data (all peaks in the diagram of Fig. 2.2) by Bragg's law provides identification of the inorganic materials by comparison with known standards or with the help of databases, as the International Centre for Diffraction Data, which stores a huge amount of XRD patterns. The use of the latter is facilitated by automated search/match routines.

Powder XRD is used (a) to determine the lattice parameters (unit cell of crystal) and the symmetry of the lattice, thus to determine the phase of the material; (b) to detect if a sample is a mixture of more than one crystalline compound; (c) to detect and monitor phase transitions, as these are characterized by change of the lattice, therefore change of the diffraction pattern; (d) to calculate thermodynamic parameters as thermal expansion, which can be observed from change of the 2θ angle of diffraction peaks with the change of temperature, resulting from expansion or contraction of lattice parameters; (e) to give information on the degree of crystallinity or the crystallite size for fine-grained materials, as composites and clays; and (f) to detect doping by certain ions, since the latter influences the unit cell

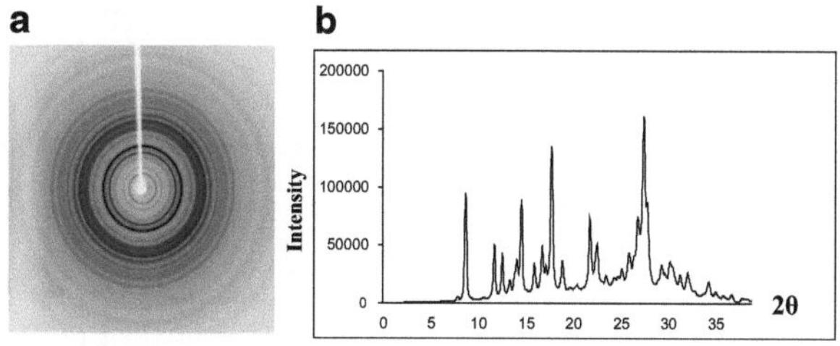

Fig. 2.2 X-ray powder diffraction image (**a**); intensity of diffraction along any radius of the diffraction image (**b**)

parameters of the studied samples or induces change of crystal phase (Kapoor and Batra 2010). In recent years, modern powder diffraction is applied to determine the molecular structure of the material by the Rietveld method (Rietveld 1969). High-resolution synchrotron radiation data and new analysis procedures through suites of software (Toby 2001) have enhanced greatly the abilities of the method so that it has even entered in the field of molecular structure of proteins (Margiolaki and Wright 2008; Margiolaki et al. 2007).

Powder XRD is routinely used for characterization of materials, e.g., synthetic bio-inspired materials, since it gives a quick picture of the degree of resemblance to the prototype. The following is a useful example of its use: The size-controlled growth of hydroxyapatite (HA, $Ca_{10}(PO_4)_6(OH)_2$) nanocrystals, dispersed into a biopolymer matrix of chitosan (CTS, biocompatible, biodegradable, and bioactive, easily obtained from chitin) by coprecipitation, depends on the CTS/HA ratio. This is a biomimetic approach to engineer bone tissue (Rusu et al. 2005), since it is known that bone (such as long bone and jaw bone) is a biologically and chemically bonded composite between HA nanocrystals and type I collagen. Figure 2.3 shows how XRD has been used to determine the crystallinity and crystallite size of samples.

Crystallites size can be calculated by Scherrer's equation (Klug and Alexander 1974) [Eq. (2.1)]:

$$L = K \cdot \lambda / \beta_{\mathrm{m}} \cdot \cos\theta \tag{2.2}$$

where L is the average crystallite size, β_m the full width of a diffraction order at half of the maximum intensity (in radians), θ the Bragg angle of this diffraction order, λ the wavelength of X-ray radiation, and K is a constant related to the crystallite shape, approximately equal to unity. Note that $\beta_m(\mathrm{rad}) = \beta_m(\mathrm{deg}) \cdot \pi/180$. Crystallite size is used to compare the synthesized products with each other and compare them with the natural prototype.

The crystallinity degree, X_c, is expressed by the fraction of crystalline inorganic material in a sample. Since the amorphous material is formed at the expense of the crystalline phase, the intensity of a diffraction peak will be proportionately reduced upon loss of crystallinity. Therefore, X_c can be extracted from the XRD pattern of the partially crystalline sample as

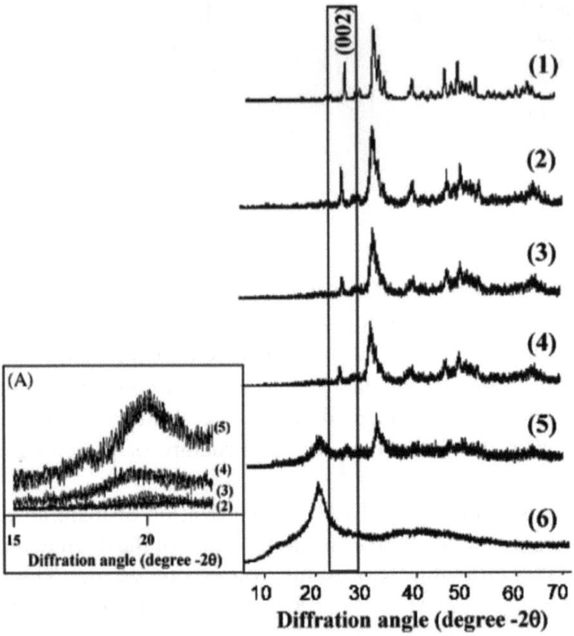

Fig. 2.3 Powder XRD patterns of a series of composite samples of increasing CTS/HA ratio (2)–(5), HA (1), and CST (6). Based on them, it was shown that the higher the amount of CTS in a composite sample, the smaller the average crystallite size, the larger the percentage of nano-sized hydroxyapatite, and the more the composites' structural features resemble those of biological apatites. The *inset* (A) shows an enlarged selection area used to determine the XRD peak domain of CTS. The broad diffraction peak around 2θ 20 is observed because CTS is a semicrystalline material. Reprinted from Biomaterials (Rusu et al. 2005). Copyright 2005 with permission from Elsevier

$$X_c = 1\text{-}V_{mn}/I_n \tag{2.3}$$

where I_n is the intensity of an XRD peak and V_{mn} the intensity of the hollow between peak n and a neighboring peak m of close 2θ value (Kapoor and Batra 2010).

2.2.3 Small-Angle Scattering

Being a scattering and not a diffraction method, small-angle scattering (SAS) is used for low-order condensed matter, metal alloys, liquid crystals, and porous materials, powders, ceramics, synthetic polymers in solution and in bulk, and biological macromolecules in solution. It has been developed after the discovery by A. Guinier (Guinier and Fournet 1955) that scattering at very small angles θ (close to the incident beam) is related to electron-density fluctuations of the examined noncrystalline matter. Modern small-angle scattering by X-rays (SAXS) and neutrons (SANS) provide structural information not only (resolution one to a few hundred nm) on particle size and size distributions, shape, and orientation distributions but also on the internal low-resolution molecular structure of biological systems in the absence of single-crystals or structure of disordered and partially ordered systems (Svergun 2007). As SAXS has short response times, it can follow biological processes, which make it an ideal complement to time-consuming single-crystal diffraction method.

SAXS is used for structural studies of mesoporous materials, as zeolites and mesoporous silicas. These are products of self-assembly between organic surfactants and inorganic molecules that are organized like the liquid crystalline hexagonal, lamellar, or cubic phases. The structure-directing surfactant species are subsequently removed by calcination, leaving behind mesoporous rigid structures. The latter may exhibit uniform pores forming lattices with long-range organization, sometimes of high degree meaning that several Bragg reflections of the lattice can be identified, e.g., silica mesophases with highly ordered 2D hexagonal symmetry and a wide range of d-spacing (Zhao et al. 1998). The lattice spacing and the symmetry can be determined from the 2θ angles. Samples of low-degree order do not exhibit multi-peak patterns but single peaks, from which the pore size can be determined. Figure 2.4 shows the SAXS pattern of mesoporous functionalized aluminosilica films forming a long-range ordered cubic lattice (Athens et al. 2011). Note that the smaller 2θ angle, the larger the d-spacing.

2.2.4 Grazing Incidence X-Ray Diffraction

It is sometimes very difficult to analyze thin films due to their small diffracting volumes, which result in low diffracted intensities. The synchrotron technique of grazing incidence X-ray diffraction (GID) is a powerful tool for studying surfaces (Tanner et al. 2004). GID measurements are performed at very low incident angles to maximize the observed signal from the thin layers (Fig. 2.5). The intense synchrotron radiation and increased path length of the incident X-ray beam through the film result in increased intensity of the diffracted (or scattered) beam so that conventional analysis can be obtained.

An interesting example shows how, by SAXS grazing incidence X-ray diffraction (GID) experiment, it was possible to follow in real time the formation of highly oriented mesostructured silica films obtained by dip coating using cetyltrimethylammonium bromide (CTAB) as a templating agent and tetraethoxysilane (TEOS). This self-assembly complex process gives films with three different symmetries of the porous networks depending on the CTAB/TEOS molar ratio: 3D hexagonal (P63/mmc), cubic (Pm3n), and 2D hexagonal (p6m) (Grosso et al. 2002). The results were verified with transmission electron microscopy (see below).

A more complex application of GID is the study of the ordering influence of a freshly cleaved mica surface on surfactant tubular molecules (up to two layers)

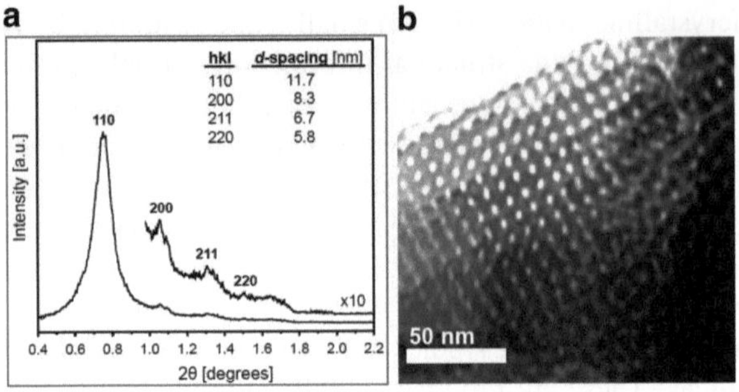

Fig. 2.4 Small-angle X-ray scattering pattern for perfluorosulfonic acid-grafted cubic mesoporous aluminosilica film (**a**). The diffraction pattern is indexed to the body-centered-cubic (Im3m) structure. Transmission electron microscopy image (**b**). Reproduced by permission from Am. Chem. Soc. (Athens et al. 2011). Copyright 2011 Am. Chem. Soc.

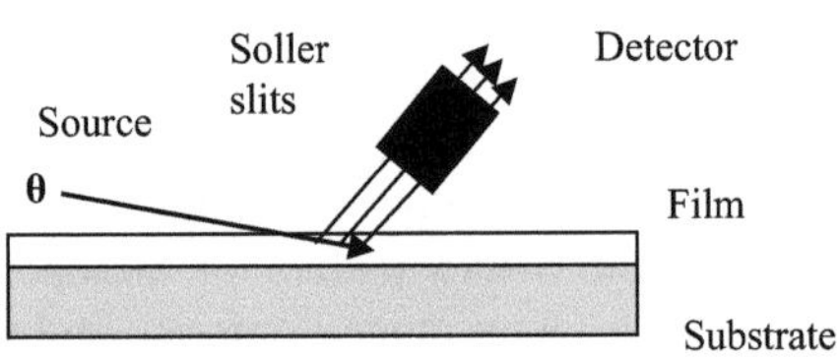

Fig. 2.5 Schematic diagram of the geometry of grazing incidence X-ray diffraction

and subsequently on mesoscopic silica films grown on top from TEOS precursor (Aksay et al. 1996). Grazing incidence X-ray diffraction during growth revealed epitaxial alignment of the surfactant tubules with crystalline mica and significant strain on the silica overlayer. That is, mica forces the surfactant molecules to align parallel to the surface (by van der Waals or ionic forces) but exerts strain on the silica layer formed on top in the direction perpendicular to the surface and this strain is measured by GID as a 4 % distortion of the hexagonal packing of the mesophase silica. The strain on the surfaces of films seems to be released only if they are left to grow for a long period of time, after several surfactant-silica layers have been deposited.

2.2.5 Fiber Diffraction

The diffraction image of oriented fibers of hydrated B form of DNA (the famous Photo 51) taken by Rosalind Franklin in 1952, which has been a critical evidence in identifying the structure of DNA, is a well-known and characteristic fiber diffraction pattern. Fiber diffraction is obtained by other systems with one-dimensional periodicity, e.g., components of the cytoskeleton. The patterns

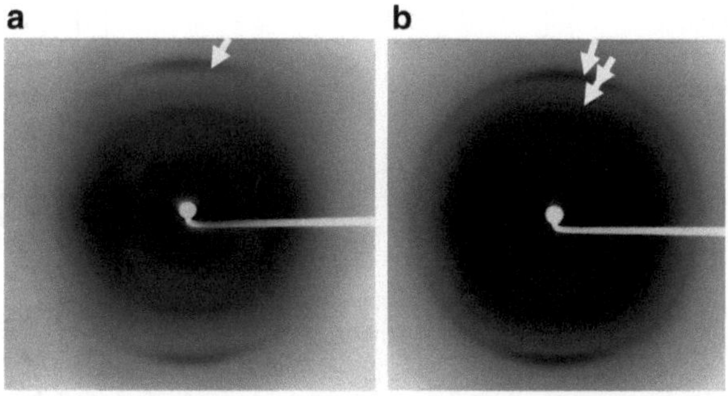

Fig. 2.6 (**a**) X-ray diffraction pattern of an air-dried gelatin film under constant elongation (80 %). The vertical direction of the pattern coincides with the direction of elongation. The arrow indicates the collagen reflection corresponding to the periodicity of 0.29 nm. The 1.1 nm collagen reflection is oriented in the direction orthogonal to the stretching. (**b**) Under the same conditions, X-ray pattern from HA–gelatin film containing 9 % of inorganic phase under constant elongation (100 %). The arrow indicates the collagen reflection corresponding to the periodicity of 0.29 nm. The crooked arrow indicates the (0 0 2) HA reflection, corresponding to the periodicity of 0.344 nm. Reprinted from Biomaterials (Bigi et al. 1998). Copyright 1998, with permission from Elsevier

are obtained if oriented fibers with their axes parallel to each other are placed at right angles with respect to an X-ray beam. As in the case of DNA, a model of the fiber is constructed and its calculated fiber diffraction is compared to the observed.

Sometimes fiber diffraction is induced by external means on the examined samples, e.g., mechanically. Figure 2.6 shows diffraction images of films of gelatin–hydroxyapatite (HA) composites under mechanical stress (Bigi et al. 1998). Pure unstretched gelatin films exhibit an X-ray diffraction pattern of two complete rings corresponding to the characteristic periodicities of 0.29 and 1.1 nm. Upon stretching, the collagen molecules align their long axis parallel to the direction of deformation and the gelatin-layered structure becomes more ordered exhibiting fiber diffraction (Fig. 2.6, Left). The patterns from unstretched composite HA–gelatin films show the diffraction rings characteristic of gelatin (0.29 and 1.1 nm), as well as the rings of crystalline HA in agreement with a random distribution of the inorganic crystallites inside the disoriented gelatin. Under constant deformation, the fiber pattern (Fig. 2.6, right) displays the 0.29 nm collagen reflection along with the (0 0 2) HA reflection indicating that the orientations of the *c*-axis of the HA crystallites are preferentially oriented along the direction of elongation. Therefore, under deformation, the inorganic crystals, which are embedded in the gelatin layers, seem to squeeze out in the interlayer spaces and assume a preferred orientation parallel to the force trajectories.

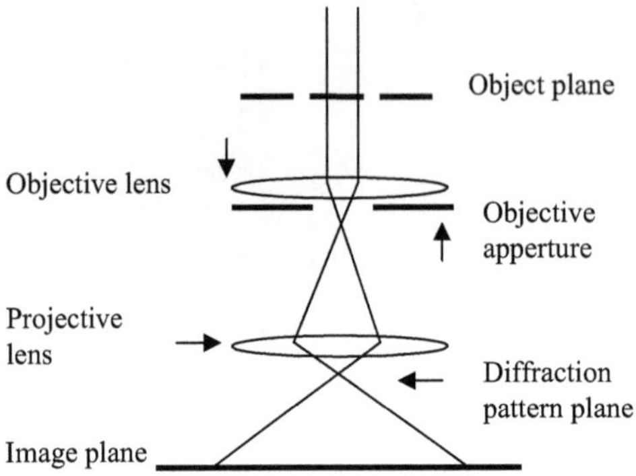

Fig. 2.7 Schematic diagram of the electron diffraction geometry

2.2.6 Electron Diffraction

Electron diffraction is similar to X-ray and neutron diffraction (Cowley 1992). Commonly, it is used for phase identification, lattice parameters and symmetry determination, and disorder and defect identification. Experiments are performed in a transmission electron microscope (TEM). Figure 2.7 shows a schematic diagram of TEM. Electrons are accelerated by an electrostatic potential before they interact with the object to be studied. By the interaction of the atomic potential of the object and the incident electron beam transmitted through it, diffracted electrons are generated that are focused into a regular arrangement of diffraction spots by an electromagnetic objective lens. The experimental setup allows for projection and collection of the electron diffraction pattern. Alternatively, if the diffracted waves and the transmitted beam interfere on the image plane, they form an enlarged image of the sample (TEM micrograph).

A great advantage of the transmission electron microscope is that by adjusting the electron lenses, it is possible to observe for the same region of a polycrystalline material both its electron microscope image (real space) and diffraction patterns (reciprocal space) as shown in Fig. 2.8. More importantly, it is possible to focus on very small areas, the size of a grain of the material, which can be a single crystal at specific orientation. Thus, the crystal lattices and orientation for all the crystalline grains can be determined. Therefore, it is possible to obtain the structural organization of a macroscopic sample by electron diffraction and TEM. This important advantage of electron diffraction derives from the fact that the scattering cross section of matter for electrons is 10^3–10^4 times larger than that of X-rays and neutrons, with typical wavelengths (~2×10^{-12}) one hundredth of the X-ray or neutron wavelength. Due to the large cross section, the electron beam can produce quite intense diffraction from extremely fine crystalline probe sizes (nm). By applying special geometry of the electron diffraction experiment, it is possible to collect 3D single-crystal diffraction data from a very small crystalline specimen and determine its molecular structure. Thus, recent highly improved electron diffraction techniques permit structure determination

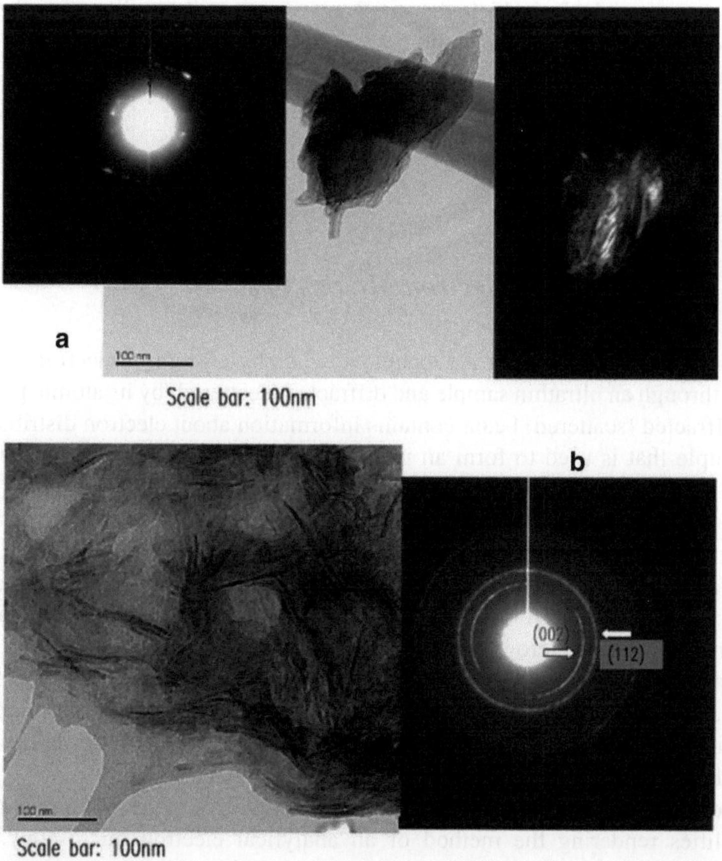

Fig. 2.8 Electron diffraction patterns for organized gelatin (GEL)–HA nanocomposites (Chang et al. 2003) by a biomimetic chemical coprecipitation process strongly indicate the self-organization of well-developed HA nanocrystallites, embedded in GEL molecules (**a**). The electron diffraction of samples containing small amount of GEL shows discrete strong spots, almost like a single HA crystal, indicating a strongly preferred (0 0 2) orientation of HA crystals (and the higher orders (0 0 4), (0 0 6), and so on). The size of the tiny crystallites is quite big (30 nm × 70 nm, micrograph in (**a**)). This is suggested to result from a specific arrangement of reactive carboxylic groups of GEL that mimic the (0 0 2) planes of the HA crystal and induce its nucleation in nm crystallites from the stable aqueous solution that subsequently grow and fuse into single crystals. It is known that the (0 0 2) axis of HA nanocrystals in bone aligns with the collagen matrix (*c*-axis parallel with the COL fibers) (Chang et al. 2003). In contrast, as the amount of GEL added to the reaction is increased (**b**), the electron diffraction pattern is differentiated from the single-crystal diffraction spots to that of a polycrystalline sample with very sharp rings indicating growth of tiny nanocrystals. The crystallite size (5 nm × 14 nm micrograph in (**b**)) is smaller with relatively longer particles suggesting a preferred orientation quite unlike the crystal habit of pure HA. Reprinted from Biomaterials (Chang et al. 2003), Copyright 2003 with permission from Elsevier

for many materials for which the sizes are very small to allow for the usual single-crystal X-ray data collection, i.e., large complexes of macromolecules, nano-sized particles, and low-dimensional objects.

2.3 Microscopy

2.3.1 Transmission Electron Microscopy

As described above, TEM is the technique whereby a beam of electrons is transmitted through an ultrathin sample and diffracted (scattered) by its atomic potential. The diffracted (scattered) beam contains information about electron distribution in the sample that is used to form an image (Fig. 2.8). Thus, TEM gives the direct image of the sample, which frequently includes high-resolution imaging of the crystal lattice (Fig. 2.4). TEM's common uses have been to provide information (Wang et al. 2011) on shape and size of crystalline samples, e.g., nanocrystals and/or size of crystallites in amorphous samples, thus complementing the XRD results: For the CST/HA samples presented in Fig. 2.3 (Rusu et al. 2005), area domains from TEM micrographs were examined in order to find how many groups of crystallites with the same characteristic features existed (either particle length or width), and the percentage of crystallites in each population was determined. The analysis clearly confirmed the bimodal distribution of the HA crystallite length found by XRD, whose sizes are in the range of biological HA.

Modern transmission electron microscopes provide a range of complementary capabilities rendering the method of an analytical electron microscopy. There exist detectors that analyze inelastically scattered electrons (electron energy-loss spectroscopy, or EELS), excited electromagnetic waves (energy dispersion spectroscopy, EDS), and Z-contrast that provide information on chemical compositions and local atomic environments (Andre et al. 2011). As described in the electron diffraction section, these studies can be combined with elastic electron diffraction on extremely fine crystalline probes (nm). It is possible to generate 3D single-crystal diffraction data and hence the molecular structure of the material, a very important tool in developing structural models of multiple phase materials.

Electron diffraction and TEM are used to study the short-range order of amorphous solids. TEM is used to study transformations from an amorphous state to phases of increased structural organization of various materials or vice versa. Amorphous samples do not show distinct features in TEM because of lack of diffraction; thus, TEM is used to detect polycrystalline particles in amorphous matrices (Karavas et al. 2007). They are also used to see long-range orders, e.g., mesoporous silica (Grosso et al. 2002).

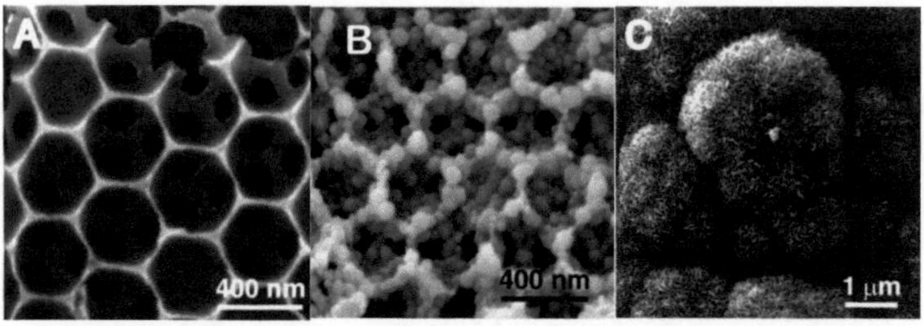

Fig. 2.9 SEM micrograph of 3D structured macroporous bioglass (**A**), after immersion in simulated biofluids for 3 h (**B**) showing overgrowth of amorphous phosphate on the pore walls and after 4 days (**C**). Reproduced by permission from Chem. Mat. (Yan et al. 2001), Copyright 2001 Am. Chem. Soc.

2.3.2 Scanning Electron Microscopy

SEM images a sample by scanning it with a beam of electrons which interact with atoms at or near the surface of a sample producing signals that contain information about the sample's surface structure, composition, and electrical conductivity. In contrast to TEM, it is using detectors placed before the sample on the same side of the incident electron beam (in Fig. 2.7) and does not require thin samples. Due to the very narrow electron beam, SEM micrographs have a large depth of field yielding a characteristic three-dimensional appearance (Fig. 2.9) that provides information on crystal shape, particle size, and the surface topography (Yan et al. 2001).

For conventional SEM imaging, the specimens must be electrically conductive in order to prevent the accumulation of electrostatic charges; therefore; they are usually coated with a thin layer of electrically conducting material (gold, platinum, gold/palladium alloy, osmium, chromium, graphite, and others). Alternatively, environmental SEM (ESEM) can be used (low-voltage mode), where the working distance is short and the specimen is placed in a relatively high-pressure chamber that helps to neutralize charges. In the standard detection mode by the method of secondary electron imaging (SEI, resolution less than 1 nm), steep surfaces and edges appear brighter than flat surfaces, which results in images with a well-defined, 3D appearance. The backscattered electrons (BSE) mode (electrons are reflected out of the specimen) is used to detect contrast between areas with different chemical compositions, because the intensity of electron backscattering depends on the atomic number of the atoms.

2.3.3 Confocal Scanning Laser Microscopy

Confocal scanning laser microscopy (CSLM) reconstructs 3D images of the specimen by assembling a series of thin slices, up to 100 μm, taken along an axis perpendicular to the slices (Semwogerere and Weeks 2005). CSLM scans point by point via 3–5 laser systems and combines the images of the slices by electronic means similar to those used in SEM to create the overall picture. Confocal microscopes image either by reflecting light off the specimen or by inducing fluorescence from fluorophores applied to the specimen (Andre et al. 2011).

2.3.4 *Atomic Force Microscopy*

Atomic force microscopy (AFM) is a very versatile method of nano-science used for (a) imaging (resolution of the order of nm), (b) measuring forces, and (c) manipulating molecules on surfaces. AFM operates by a fine tip (cantilever) that scans a surface and "feels" the forces from the sample. Depending on the mode, the measured forces include mechanical contact force, van der Waals forces, capillary forces, chemical bonding, electrostatic forces, and magnetic forces. In contrast to the electron microscopes (TEM, SEM), AFM does not require vacuum; it can be used at ambient conditions and in liquid samples; therefore, it is suitable for biological systems. Since AFM's resolution is very high, it is used complementary to TEM and SEM methods to provide the 3D profile of surfaces (Andre et al. 2011). Sometimes it can reveal details undetectable with high-resolution SEM, e.g., AFM revealed the presence of nanoparticulate silica organized around the central axial filament of freshly cleaved (un-etched) spicule cross sections (Weaver et al. 2003). AFM is commonly used to image roughness, epitaxial deposition, grain size, study the morphology (or symmetry) of a growing overlayer on substrates of various symmetries in order to examine the influence of the latter (Aksay et al. 1996), monitor phase changes, and so on.

2.4 Spectroscopy

2.4.1 *Infrared Spectroscopy (IR)*

IR is usually used as absorption spectroscopy at the 4,000–400 cm^{-1} range, where vibrational or rotational–vibrational excitation of covalently bonded atoms and groups takes place. The IR absorbance frequencies are characteristic of the chemical groups and their chemical environment; thus, the spectrum can be used as a fingerprint for identification of unknown compounds or to judge about close intermolecular interaction of specific groups. For example, changes in the absorption frequencies of the spectrum are correlated to changes in the environment of an atom and emergence of new peaks to new covalent bond formation. The following are some characteristic examples: (a) IR has been employed to examine the propensity of a three-dimensionally ordered macroporous bioactive glass for apatite formation by soaking in simulated body fluid at body temperature. During the soaking, the growing of the absorptions due to phosphate (mainly the low-intensity band at 567 cm^{-1} corresponding to the antisymmetric vibrational mode of P-O in amorphous calcium phosphate) and carbonate groups could be observed close to the Si peaks (typical bands at the low-wave number range the 1,080, 800, and 470 cm^{-1} corresponding to the asymmetric Si–O–Si stretch, the symmetric Si–O–Si stretch, and the Si–O–Si deformation mode, respectively). The growing of the peak corresponding to phosphates suggested that deposition of amorphous calcium phosphate was taking place (Yan et al. 2001). (b) In the gelatin (GEL)–hydroxyapatite (HA) composite (Chang et al. 2003), FTIR analysis showed a red shift of the band at 1,339 cm^{-1} for a series of HA–GEL composites attributed to the covalent bond formation of gelatin with Ca^{2+} ions of HA nanocrystals. This band characterizes the wagging vibration of proline side chains of GEL (as in collagen). (c) In the micro-FTIR study of the chemical structure of synthesized silk/silica hybrids by the sol–gel process in the presence of a cross-linking agent, two new peaks at 1,469 cm^{-1} and 958 cm^{-1} emerge that are ascribed to the group –N=C– of the cross-linking agent showing that the latter is able to form covalent bonds between silk and sol–gel dispersion particles (Hou and Chen 2010).

2.4.2 *Raman Spectroscopy*

As IR spectroscopy, Raman spectroscopy is used to study low-frequency modes of a molecular system, as vibrational or rotational, in the solid state. It is based on Raman scattering, a scattering that results from interaction of IR radiation with vibrational modes of the examined molecular system. A vibrational spectrum may be obtained from Raman scattering that yields similar, but complementary, information to IR spectroscopy. In materials science the 2D micro-Raman spectroscopy is very helpful, because it maps the surface of the sample and provides a chemical characterization at every point: In a study that examined whether a poorly water-soluble drug dispersed in a polymeric matrix exist in the form of amorphous nano-dispersions or is molecularly dispersed, it was revealed by SEM and TEM that the drug forms amorphous nano-dispersions into the polymer matrix. However, it was found by micro-Raman mapping of the whole surface that some portion of the drug exists also as molecular dispersion inside the amorphous polymer matrix (Karavas et al. 2007).

2.4.3 *Nuclear Magnetic Resonance Spectroscopy*

Solid-state nuclear magnetic resonance (NMR) provides the same type of information as NMR spectroscopy in solution, but special equipment and methods (including magic angle spinning and cross polarization) are needed due to anisotropic or orientation-dependent interactions in the solid state (Laws et al. 2002). Since silicon and phosphorous, as well as hydrogen and carbon, have isotopes with a spin of ½, solid-state NMR can provide information on the intermolecular interactions of these atoms that are common composite inorganic biomaterials. Therefore, it provides information on covalent and intermolecular interactions and contributes to build indirectly a model of the structure of the material, as in the case of IR spectroscopy. Thus, single-pulse ^{29}Si magic angle spinning NMR spectra of siliceous spicules isolated from *T. aurantia* heated to 300 and 600 °C revealed different degrees of heat-induced silica condensation. The results show that the inorganic silica framework is moderately condensed, even in the non-heat-treated sample (Weaver et al. 2003). Another application, by ^{29}Si NMR spectra, is the determination of the proportions of Qn species (Qn representing a Si atom bonded to n other Si atoms via O-bridges, e.g., Q3 and Q4) allowing the quantification of the cross-linking degree of silica (Hou and Chen 2010).

2.5 Thermal Analysis

Thermal analysis monitors the thermal response of the systems on heating and measures the energy released or absorbed (exothermic or endothermic); thus, it can investigate the mechanisms of thermal processes of a system. Thermal behavior gives indirect information on the examined structure and complements results from other methods. Both thermal analysis techniques, *differential thermal/thermogravimetric analysis* (DT/TGA) and *differential scanning calorimetry* (DSC), are usually applied in the thermal analysis. DT/TGA measures the weight loss of the samples upon heating, which it is usually interpreted as due to either elimination of water or to degradation and loss of the organic matter. DSC can detect the above, as well as monitor possible structural modifications and phase changes and measure the energetics involved.

Examples are as follows: (a) DT/TGA and DSC measurements in sponge spicules (Croce et al. 2004) indicate the presence of organic matter and its degradation, which takes place with weight loss at high temperature. At low temperature endothermic effects in DSC are either due to loss of coordinated water to protein molecules in the spicules or structural changes of the proteins. The kinetics of DSC provide information on the efficiency of the energy transfer from the instrument furnace to the organic matter located within the spicule cavity, therefore on the thickness of the walls of the silica matrix, which is different for the various species studied. (b) DT/TGA was employed to determine the phase stability and the reaction temperatures during the fabrication of Mg-substituted tricalcium phosphate scaffolds by solid-state reaction processes in various studies for biomimetic bone regeneration (Zhang et al. 2009). (c) The study of organized gelatin [GEL]/HA nanocomposites (Chang et al. 2003) strongly supports the existence of a chemical bond between HA and GEL in the composite. DT/TGA data for certain [GEL]/HA ratios showed similar results to real bone.

2.6 Concluding Remarks

The analysis of such complex structures as inorganic biomaterials requires the use of various complementary techniques. As a concluding example, the structural study of intact spicules from various species of sponges is presented. It was carried out with a selected combination of methods, SEM, thermogravimetric and calorimetric analysis, IR spectroscopy, fiber diffraction, and theoretical molecular modeling (Croce et al. 2004). The axial filament composed of proteins and positioned at the spicule axis is crucial to the spicule formation because it functions as template for silica deposition. Both mega- and microscleres spicules of hexactinellid *S. joubini* and the demosponges *G. cydonium*, *P. ficiformis*, and *T. aurantium* were used in the study. The above combination of techniques allowed the analysis of the axial filaments inside their natural siliceous case, which give a more realistic picture of their organization, in contrast to structural characterization carried out on filaments extracted by strong treatment (Shimizu et al. 1998) that may have caused denaturation or structural changes in the proteins.

SEM with energy dispersive spectrometry (EDS) analysis indicated that in all samples, the inorganic envelope is composed almost exclusively of silica (Si and O). In megascleres, the presence of cavities was revealed. These cavities are the sites where the proteins responsible for the spicule growth are hosted (*vide infra*). The thermal analysis showed that for *S. joubini* (a) the overall energy involved during the heating is significantly larger than in the other spicules and (b) the onset of the temperature process is at lower temperature. These observations have been interpreted as deriving from greater amount of biomolecules in the axial filament of the hexactinellid species and that the biomolecules are less tightly bound. FTIR spectroscopy was employed to characterize the surface of siliceous materials, in particular to evaluate the presence of both surface hydroxyls and organic molecules, with particular interest to the spicule proteins. The fact that the band at 3,680 cm^{-1}, attributable to the stretching modes of Si–OH groups of the inorganic matrix, becomes evident upon evacuation at 250 °C indicates that these –OH groups at room temperature must be strongly involved in hydrogen bond interactions with the organic matter. FTIR spectroscopy showed also evidences of beta-sheet structures in the analyzed spicule proteins. Synchrotron radiation fiber diffraction experiments on intact spicules gave a more realistic picture of the protein organization. The sharp diffraction spots obtained from bundles of well-aligned fibers indicated that the protein units forming the axial filaments inside the spicules must be highly organized. In the planes perpendicular to the spicule axis, the protein units appear to be packed in a compact hexagonal way (Fig. 2.10a) with repeat distances 5.8 nm for demosponge spicules (*T. aurantium*, *G. cydonium*, and *P. ficiformis*) and 8.4 nm for the hexactinellid *S. joubini*. These distances were calculated from the

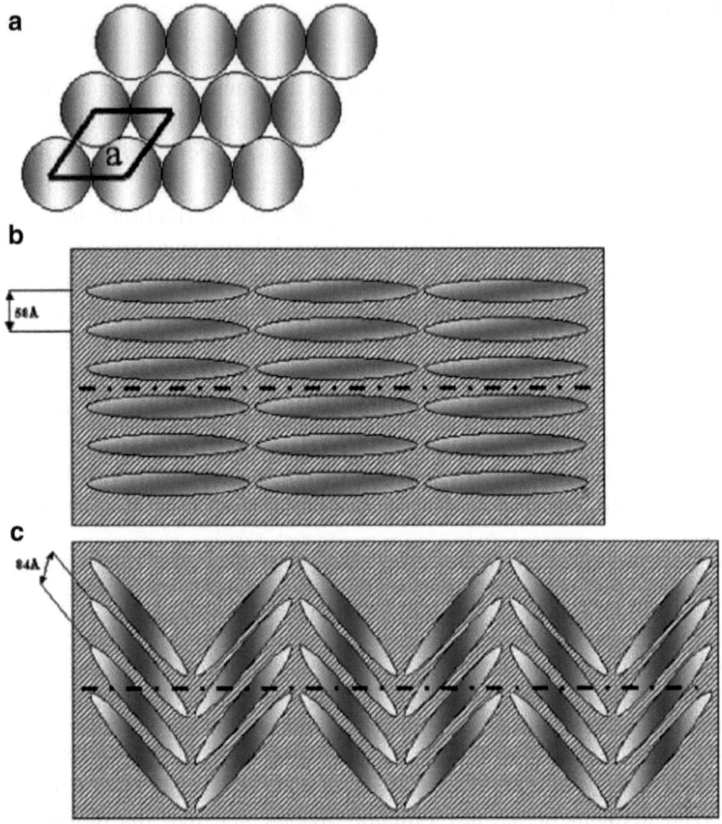

Fig. 2.10 2D hexagonal packing (**a**); structural model of the organization of the filaments in *G. cydonium* spicules (**b**); structural model of the organization of the filaments in *S. joubini* spicules (**c**). The *dot* and *dash lines* represent the longitudinal axis of the filament. Reprinted from Biophys J (Croce et al. 2004), Copyright 2004 with permission from Elsevier

position and distribution of the spots in the fiber diffraction (Croce et al. 2003). Equatorial spots up to the third order of *G. cydonium* spicules are consistent with a very regular hexagonal arrangement of protein units aligned along the spicule axis (Fig. 2.10b). The presence of nonequatorial spots in the diffraction patterns of *S. joubini* suggests also a hexagonal packing but of spirally oriented protein units along the spicule axis, according to the 2D model of (Fig. 2.10c). The diffraction spots in all patterns are much sharper than one would normally expect from a fiber, and this feature has been explained by the authors by the assumption that the silica present in the axial filament is organized around the protein units as a highly ordered mesoporous material and is embodying the protein units in regular mesoporous scaffolding.

References

Aksay IA, Trau M, Manne S, Honma I, Yao N, Zhou L, Fenter P, Eisenberger PM, Gruner SM (1996) Biomimetic pathways for assembling inorganic thin films. Science 273:892–898

Andre R, Tahir MN, Link T, Jochum FD, Kolb U, Theato P, Berger R, Wiens M, Schroeder H-C, Muller WEG et al (2011) Chemical mimicry: hierarchical 1D TiO2@zro2 core-shell structures reminiscent of sponge spicules by the synergistic effect of silicatein-r and silintaphin-1. Langmuir 27:5464–5471

Athens GL, Kim D, Epping JD, Cadars S, Ein-Eli Y, Bradley FC (2011) Molecular optimization of multiply-functionalized mesoporous films with ion conduction properties. J Am Chem Soc 133:16023–16036

Bigi A, Panzavolta S, Roveri N (1998) Hydroxyapatite-gelatin Þlms: a structural and mechanical characterization. Biomaterials 19:739–744

Blow DM (2002) Outline of crystallography for biologists. Oxford University Press, Oxford

Chang MC, Ko CC, Douglas WH (2003) Preparation of hydroxyapatite—gelatin nanocomposites. Biomaterials 24:2853–2862

Cowley JM (ed) (1992) Electron diffraction techniques. International Union of Crystallography, Oxford University Press, Oxford

Croce G, Frache A, Milanesio M, Viterbo D, Bavestrello G, Benatti U, Giovine M, Amenitsch H (2003) Fiber diffraction study of spicules from marine sponges. Microsc Res Tech 62:378–381

Croce G, Frache A, Milanesio M, Marchese L, Causá M, Viterbo D, Barbaglia A, Bolis V, Bavestrello G, Cerrano C et al (2004) Structural characterization of siliceous spicules from marine sponges. Biophys J 86:526–534

Fischer E (1894) Einfluss der configuration auf die Wirkung der enzyme. Ber Deutsch Chem Ges 27:2985–2993

Grosso D, Babonneau F, Albouy P-A, Amenitsch H, Balkenende AR, Brunet-Bruneau A, Rivory J (2002) An in situ study of mesostructured CTAB-silica film formation during dip coating using time-resolved SAXS and interferometry measurements. Chem Mater 14:931–939

Guinier A, Fournet G (1955) Small-angle scattering of X-rays. Wiley, New York

Hou A, Chen H (2010) Preparation and characterization of silk/silica hybrid biomaterials by sol–gel crosslinking process. Mat Sci Eng B 167:124–128

Karavas E, Georgarakis M, Docoslis A, Bikiaris D (2007) Combining SEM, TEM, and micro-Raman techniques to differentiate between the amorphous molecular level dispersions and nanodispersions of a poorly water-soluble drug within a polymer matrix. Int J Pharm 340:76–83

Kapoor S, Batra U (2010) Preparation and bioactivity evaluation of bone like hydroxyapatite - bioglass composite. Int J Chem Biol Engin 3:13, 24–28

Klug HP, Alexander LE (1974) X-ray diffraction procedures for polycrystallite and amorphous materials, 2nd edn. Wiley, New York

Laws DD, Bitter H-ML, Jerschow A (2002) Solid-state NMR spectroscopic methods in chemistry. Angew Chem Int Ed 41:3096–3129

Lehn J-M (1995) Supramolecular chemistry. VCH, Weinheim

Margiolaki I, Wright JP, Wilmanns M, Fitch AN, Pinotsis N (2007) Second SH3 domain of ponsin solved from powder diffraction. J Am Chem Soc 129:11865–11871

Margiolaki I, Wright JP (2008) Powder crystallography on macromolecules. Acta Cryst A64:169–180

Rietveld HM (1969) A profile refinement method for nuclear and magnetic structures. J Appl Cryst 2:65–71

Rusu VM, Ng C-H, Wilkec M, Tierscha B, Fratzld P, Peter MG (2005) Size-controlled hydroxy-apatite nanoparticles as self-organized organic–inorganic composite materials. Biomaterials 26:5414–5426

Semwogerere D, Weeks ER (2005) Confocal microscopy. In encyclopedia of biomaterials and biomedical engineering. Taylor & Francis, London

Shimizu K, Cha J, Stucky GD, Morse DE (1998) Silicatein a cathepsin L-like protein in sponge biosilica. Proc Natl Acad Sci USA 95:6234–6238

Svergun DI (2007) Small-angle scattering studies of macromolecular solutions. J Appl Crystallogr 40:s10–s17

Tanner BK, Hase TPA, Lafford TA, Goorsky MS (2004) Grazing incidence in-plane X-ray diffraction in the laboratory. Adv X-Ray Anal 47:309–314

Toby BH (2001) EXPGUI, a graphical user interface for GSAS. J Appl Cryst 34:210–213

Wang G, Peng Q, LI Y (2011) Lanthanide-doped nanocrystals: synthesis, optical-magnetic properties, and applications. Acc Chem Res 44:322–332

Weaver JC, Pietrasanta LI, Hedin N, Chmelka BF, Hansma PK, Morse DE (2003) Nanostructural features of demosponge biosilica. J Struct Biol 144:271–281

Yan H, Zhang K, Blanford CF, Francis LF, Stein A (2001) In vitro hydroxycarbonate apatite mineralization of CaO-SiO2 sol-gel glasses with a three-dimensionally ordered macroporous structure. Chem Mater 13:1374–1382

Zhang X, Takahashi T, Vecchio KS (2009) Development of bioresorbable Mg-substituted tricalcium phosphate scaffolds for bone tissue engineering. Mat Sci Eng C 29:2003–2010

Zhao D, Feng J, Huo Q, Melosh N, Fredrickson GH, Chmelka BF, Stucky GD (1998) triblock copolymer synthesis of mesoporous silica with periodic 50 to 300 Angsrom pores. Science 279:548–552

Chapter 3
Enzymes of Inorganic Polyphosphate Metabolism

Tatyana Kulakovskaya and Igor Kulaev

Contents

Abstract Inorganic polyphosphate (PolyP) is a linear polymer containing a few to several hundred orthophosphate residues linked by energy-rich phosphoanhydride bonds. Investigation of PolyP-metabolizing enzymes is important for medicine, because PolyPs perform numerous functions in the cells. In human organism, PolyPs are involved in the regulation of Ca^{2+} uptake in mitochondria, bone tissue development, and blood coagulation. The essentiality of polyphosphate kinases in the virulence of pathogenic bacteria is a basis for the discovery of new antibiotics. The properties of the major enzymes of PolyP metabolism, first of all polyphosphate kinases and exopolyphosphatases, are described in the review. The main differences between the enzymes of PolyP biosynthesis and utilization of

T. Kulakovskaya (✉) • I. Kulaev
Skryabin Institute of Biochemistry and Physiology of Microorganisms, Russian Academy of Sciences, Pushchino, Moscow Region 142290, Russia
e-mail: alla@ibpm.pushchino.ru

W.E.G. Müller et al. (eds.), *Biomedical Inorganic Polymers*, Progress in Molecular and Subcellular Biology 54, DOI 10.1007/978-3-642-41004-8_3,

prokaryotic and eukaryotic cells, as well as the multiple functions of some enzymes of PolyP metabolism, are considered.

3.1 Introduction

Inorganic polyphosphates (PolyPs) are linear polymers containing a few to several hundred orthophosphate residues linked by energy-rich phosphoanhydride bonds. They perform numerous functions in living cells (Kulaev and Vagabov 1983; Kornberg et al. 1999; Reusch 1992; Docampo and Moreno 2001; Kulaev et al. 2004; Omelon and Grynpas 2008; Rao et al. 2009):

- Phosphorus storage, participation in the phosphorus homeostasis in cells and in the phosphorus cycle of the biosphere
- Energy storage
- Detoxification of heavy metal cations
- Induction of the synthesis of RpoS, an RNA-polymerase subunit in bacteria, which is responsible for expression of the genes involved in the stationary phase and stress adaptation
- Involvement in bacterial cell motility, biofilm formation, and virulence
- Regulation of the level of the stringent response factor, guanosine 5′-diphosphate 3′-diphosphate (ppGpp), the second messenger in bacteria
- Formation of PolyP/poly-β-hydroxybutyrate/Ca^{2+} channels involved in membrane transport
- Regulation of enzyme activities
- Water retaining and antibacterial activity
- Involvement in biomineralization, including bone tissue development
- Participation in the blood coagulation cascade

In view of the role of PolyP in cell metabolism, bone tissue development (Schröder et al. 2000; Omelon et al. 2009; Morita et al. 2010; Müller et al. 2011), and blood coagulation (Caen and Wu 2010; Smith et al. 2006, 2010), the study into the biochemistry and cell biology of PolyP-metabolizing enzymes offers great prospects for medicine. The review presents data on the properties, functional significance, and localization of the major enzymes catalyzing PolyP biosynthesis and degradation. In addition, PolyP metabolism involves numerous pyrophosphatases and nonspecific (alkaline and acid) phosphatases. Description of these groups of enzymes is not a subject of the review.

3.2 Enzymes of PolyP Biosynthesis

3.2.1 Polyphosphate Kinase (Polyphosphate:ADP Phosphotransferase, EC 2.7.4.1)

The reaction of reverse transfer of energy-rich phosphate residues from ATP to PolyP and from PolyP to ADP, linking energy-rich pools, was discovered by Kornberg and coauthors (Kornberg et al. 1956).

$$\text{PolyP}_n + \text{ATP} \leftrightarrows \text{PolyP}_{n+1} + \text{ADP}$$

The polyphosphate kinase of *Escherichia coli* was the first one of the purified enzymes performing such reaction. Its properties were studied in detail, and the encoding gene *ppk1* was cloned and sequenced (Ahn and Kornberg 1990; Akiyama et al. 1992). The crystal structure of enzyme was determined (Zhu et al. 2003).

Later, the *ppk1* genes of many other bacteria (see http://www.expasy.org) were cloned, sequenced, and characterized. The deduced amino acid sequences of these enzymes show extensive homology in different bacterial species (Tzeng and Kornberg 1998). Some conserved amino acid residues are important for enzymatic activity. Replacement of conserved His-441 and His-460 by either glutamine or alanine by site-specific mutagenesis rendered an enzymatically inactive protein in *E. coli* (Kumble et al. 1996).

Another polyphosphate kinase activity different from ppk1 was revealed in the *ppk1*-lacking null mutant of *Pseudomonas aeruginosa* (Ishige et al. 2002; Zhang et al. 2002). It is considered that the main function of ppk2 is to utilize PolyP for GTP synthesis (Ishige et al. 2002; Rao et al. 2009; Gangaiah et al. 2010). The ppk2 catalyzes also the reaction:

$$\text{PolyP}_n + \text{AMP} \leftrightarrows \text{PolyP}_{n-1} + \text{ADP}$$

It was observed for the first time in *Corynebacterium xerosis* (Dirheimer and Ebel 1965), and the enzyme was partly purified from *Acinetobacter johnsonii* (Bonting et al. 1991). High PolyP:AMP phosphotransferase and PolyP:ADP phosphotransferase activities were found in *P. aeruginosa* (Ishige and Noguchi 2000, 2001). Partially purified PolyP:ADP phosphotransferase from this bacterium was independent of polyphosphate kinase encoded by the *ppk1* gene and could act as a PolyP-dependent nucleoside-diphosphate kinase, which preferred GDP as a phosphate acceptor (Ishige and Noguchi 2001). It was revealed that the activity in *E. coli* and *P. aeruginosa* is due to ppk2 (Ishige and Noguchi 2000; Ishige et al. 2002; Zhang et al. 2002). Some bacterial genomes possess 2–3 ppk2 paralogs (Nocek et al. 2008). The genome of *P. aeruginosa* possesses two one-domain ppk2 genes and one two-domain ppk2 gene. Both proteins were overexpressed in *E. coli* and purified. The one-domain proteins exhibited PolyP-ADP phosphorylation and generated ATP, while the two-domain protein catalyzed PolyP-dependent

Table 3.1 Some properties of bacterial polyphosphate kinases (Rao et al. 2009; Hooley et al. 2008)

Enzyme name	ppk1	ppk2
The encoding gene	*ppk1*	*ppk2*
Enzymatic properties in PolyP synthesis		
1. Nucleotide phosphate using for PolyP synthesis	ATP	ATP and GTP
2. PolyP product chain length	500–800	200–800
3. Metal ion	Mg^{2+}	Mn^{2+}
Molecular mass	320 kDa (a tetramer of 80 kDa subunits)	384 kDa (an octamer of 44 kDa subunits)
The domains similar to other proteins	C-terminal C1 and C2 domains are similar to the catalytic domain of phospholipase D	Thymidine monophosphate kinase domain
The properties of null mutants	Lesser survival at stationary growth stage defective in motility, quorum sensing, biofilm formation, and virulence enhanced sensitivity to some antibiotics, UV light, and oxidative stress	Δ*ppk2* mutant exhibited a significant survival defect under osmotic, nutrient, aerobic, and antimicrobial stresses

phosphorylation of AMP and produced ADP. Both groups of proteins were capable of phosphorylating GMP to GDP and GDP to GTP. Two-domain ppk2 enzymes are PolyP-dependent AMP kinases (Nocek et al. 2008).

Thus, prokaryotes have several genes encoding the enzymes with polyphosphate kinase activity. Some of the properties of ppk1 and ppk2 are compared in Table 3.1. The *ppk1* and *ppk2* genes are highly conserved and have been identified in many species (Kornberg et al. 1999; Rao et al. 2009). The *ppk1* homologs have been found in more than 100 prokaryotic species, including 20 major pathogens (Rao et al. 2009). Some bacterial species carry only *ppk1*, or only *ppk2*, or both genes (Rao et al. 2009). The data on their presence are summarized in Ishige et al. (2002), Zhang et al. (2002), and Rao et al. (2009).

In view of the high conservatism of the *ppk* genes, microbial associates and activated sludges are often assayed for the presence of similar genes. The fragments of putative *ppk1* genes have been found using PCR primers in bacteria from activated sludges containing high phosphate level (McMahon et al. 2002) and in marine oligotrophs living in phosphate-limited environment (Temperton et al. 2011).

In many bacteria, ppk1 is the major enzyme of PolyP synthesis. This fact is confirmed by a sharp decrease in PolyP content in the *ppk1* mutants of *E. coli* (Crooke et al. 1994; Rao and Kornberg 1996; Rao et al. 1998), *Neisseria meningitidis* (Tinsley and Gotschlich 1995), and *Vibrio cholerae* (Ogawa et al. 2000a). At the same time, the decrease in PolyP level has not been shown for *ppk2* null mutant of *Campylobacter jejuni* (Gangaiah et al. 2010).

The experiments with the mutants in the *ppkl* gene have shown that polyphosphate kinase ppk1 is essential for stationary phase responses and viability of many bacteria, including pathogens (Crooke et al. 1994; Rao and Kornberg 1996; Rao et al. 1998, 2009; Kornberg et al. 1999). The *Δppkl* mutants are characterized by growth defects in the stationary phase; defective responses to stress and starvation; higher sensitivity to stress factors including heat, antibiotics, antiserum, UV light, and other effectors; impairment in motility; biofilm formation; and virulence (Kornberg et al. 1999; Rao et al. 2009). The polyphosphate kinase ppk1 is a colonization factor of *Helicobacter pylori*. The mutant with the enhanced level of this enzyme exhibited better colonizing capacity in mice, while *ppkl*-deficient mutants could not colonize the gastric mucosa of mice (Ayraud et al. 2005).

The polyphosphate kinase ppk2 is involved in alginate synthesis and biofilm formation associated with virulence of some bacteria (Rao et al. 2009). Compromised survival of a ppk2 knockout of *Mycobacterium smegmatis* under heat or acid stress or hypoxia, and the ability of ppk2 of *M. tuberculosis* to complement this, confirmed that ppk2 plays a role in mycobacterial survival under stress (Sureka et al. 2009). The enzyme is critical for *C. jejuni* survival, adaptation, and persistence in the host environments (Gangaiah et al. 2010).

No *ppk* gene homologs have been found in mammals. A. Kornberg was the first to propose polyphosphate kinase as a new antimicrobial drug target (Kornberg et al. 1999). The methods and compositions reducing polyphosphate kinase activity, PolyP amounts, and virulence of many microbial pathogens are described in the patent (Kornberg and Portola 2003). Like ppk1, ppk2 may also be a target for designing novel antibiotics (Rao et al. 2009). DNA-based aptamers inhibiting *M. tuberculosis* and *V. cholerae* ppk2 catalytic activity were obtained and tested (Shum et al. 2011).

In respect to eukaryotes, *ppkl* homologs have been found in *Dictyostelium discoideum* (Zhang et al. 2005, 2007). The slime mold *D. discoideum* possesses polyphosphate kinase activity owing to the presence of a bacterial *ppkl* homolog, *DdPPK1*, which probably arose by horizontal gene transfer from bacteria. The deletion mutants still retain modest levels of PolyP indicating alternative pathways of its synthesis (Brown and Kornberg 2008). The enzyme responsible for sustaining the PolyP level, Dd PPK2, is an actin-related protein (Arp) complex, which is polymerized into an actin-like filament concurrently with the reversible synthesis of PolyP from ATP (Gómez-García and Kornberg 2004; Zhang et al. 2007; Rao et al. 2009).

Polyphosphate-synthesizing activity was characterized in extracts of the yeast *Candida humicola* (McGrath et al. 2005). Its properties were similar to those of bacterial polyphosphate kinase enzymes. PCR amplification of *C. humicola* genomic DNA using universal primers for bacterial polyphosphate kinase genes yielded a product, the translated sequence of which showed up to 34 % amino acid similarity to the bacterial enzyme.

BLAST searches have shown possible bacterial-type *ppkl* and *ppk2* homologs in some insects (Hooley et al. 2008). The BLAST search in human protein databases using the proteins that constitute DdPPK2 activity in *D. discoideum* (ERpA ArpD and ACT28) shows highly significant matches of actin-like proteins. The occurrence of PolyP-synthesizing activity in these proteins is still open to question (Hooley et al. 2008).

Polyphosphate kinase activity was observed in the yeast vacuolar membrane long ago (Shabalin et al. 1977). This activity proved to be exhibited by vacuolar membrane protein Vtc4 (Hothorn et al. 2009), a vacuolar transporter chaperon (Muller et al. 2002). X-ray crystallography showed that the VTC4p structure contained a long chain of electron-dense domain winding through the tunnel, suggesting that this module generated PolyP from ATP during dialysis or crystallization (Hothorn et al. 2009). In the presence of Mn^{2+}, Vtc4 synthesizes PolyP from ATP. The yeast cells with deletion in the *VTC4* gene lack the vacuolar PolyP pool (Ogawa et al. 2000b; Hothorn et al. 2009). The catalytic domain faces the cytoplasm and the polymer must pass the membrane (Hothorn et al. 2009). It is very likely that a similar protein provides polyphosphate kinase activity in mycorrhizal fungi where the polyphosphate-synthesizing activity using ATP is localized in the organelle fraction but not in the cytosol or at the plasma membrane (Tani et al. 2009).

The VTC complex seems not to be conserved in animals and plants, and other PolyP synthetases are still to be discovered in these organisms.

3.2.2 3-Phospho-D-Glyceroyl-Phosphate:Polyphosphate Phosphotransferase (EC 2.7.4.17)

The enzyme also known as 1,3-diphosphoglycerate-polyphosphate phosphotransferase (Kulaev and Bobyk 1971; Kulaev et al. 1971) catalyzes the following reaction:

$$\text{3-phospho-D-glyceroyl-1-phosphate} + \text{PolyP}_n \rightarrow \text{3-phosphoglycerate} + \text{PolyP}_{n+1}$$

The activity was found first in the *Neurospora crassa* mutant deficient in adenine, where the concentrations of ATP and other adenyl nucleotides were sharply reduced (Kulaev and Bobyk 1971). Later this activity was observed in the cell-free extract of the wild-type *N. crassa* strain and in other microorganisms including bacteria, but it was much lower. Probably, this pathway of PolyP synthesis occurs during glycolytic phosphorylation under a low ATP content in the cell and might actually be involved in the biosynthesis of some (presumably low-molecular-mass) PolyP fractions. Some authors, however, believed PolyP biosynthesis in the lower eukaryotes to be provided by 1,3-diphosphoglucerate:PolyP phosphotransferase (Schuddemat et al. 1989). The enzyme has not been purified and needs further investigations. This fossil reaction is believed to be evidence that PolyP preceded ATP at the earlier stages of evolution.

3.2.3 Dolichyl Diphosphate: Polyphosphate Phosphotransferase (EC 2.7.4.20)

The enzyme activity was found in the membrane fraction of yeast cells (Shabalin et al. 1979, 1984, 1985), where the PolyP synthesis using β-phosphate groups of dolichyl diphosphate took place:

$$\text{Dolichyl diphosphate} + \text{PolyP}_n \rightarrow \text{dolichyl phosphate} + \text{PolyP}_{n+1}$$

The enzyme was solubilized from the membrane fraction using Triton X-100 (Shabalin and Kulaev 1989). Dolichyl diphosphate:PolyP phosphotransferase activities of membrane preparation and solubilized fraction were metal dependent and stimulated by Mg^{2+} or Ca^{2+}. The same membrane fraction possessed dolichyl diphosphate:phosphohydrolase activity, which was however inhibited by divalent metal cations. The gene encoding the enzyme is yet unidentified.

The function of the enzyme is associated with mannan biosynthesis in the yeast cell wall.

The putative pathway of coordination of mannan and PolyP biosynthesis under cell wall formation (Kulaev 1994) explains the presence of PolyP outside the cytoplasmic membrane. Dolichol phosphates (Dol-P) act as transmembrane carriers of carbohydrate residues in glycoprotein biosynthesis. GDP-mannose at the cytoplasmic side of the endoplasmic reticulum interacts with the phosphate residue of Dol-P. The Dol-P-P-mannose is transported across the membrane so that the phosphomannose residue enters the lumen, where mannosyl transferase and Dol-P-P:PolyP phosphotransferase reactions occur. The formed Dol-P again crosses the membrane and interacts on its cytoplasmic side with a new molecule of GDP-mannose. The mannoproteins and PolyP are transported to the cell envelope by special vesicles.

3.3 Enzymes of PolyP Degradation

3.3.1 Polyphosphate-Glucose Phosphotransferase (EC 2.7.1.63)

The enzyme catalyzes glucose phosphorylation using PolyP or ATP as a phosphoryl donor:

$$\text{PolyP}_n + \text{D-glucose} \rightarrow \text{PolyP}_{n-1} + \text{D-glucose 6-phosphate}$$

The enzyme activity was first observed in *Mycobacterium phlei* (Szymona 1957) and then in other bacteria, including important pathogens such as Mycobacteria (Szymona and Szymona 1978) and *Corynebacterium diphtheriae* (Szymona and

Szymona 1961) and the bacteria of activated sludge accumulating high PolyP levels (Kuroda and Ohtake 2000).

PolyP glucokinase utilized PolyP via a quasiprocessive or nonprocessive mechanism (Pepin and Wood 1987; Hsieh et al. 1996). $PolyP_3$ and $PolyP_4$ were reported to be the end products formed from long-chain PolyP in glucose phosphorylation (Kowalczyk and Phillips 1993). The detailed characterization of purified enzyme and the cloning of the *ppgK* gene demonstrated that a single enzyme catalyzed both PolyP- and ATP-dependent glucokinase activities (Hsieh et al. 1993; Phillips et al. 1999). The evidence of existence of separate binding sites for inorganic (PolyP) and organic (ATP) phosphoryl donors were obtained (Phillips et al. 1999).

The enzyme responsible for the PolyP- and ATP-dependent mannokinase activities was purified to homogeneity from the cell extract of the bacterium *Arthrobacter* sp. (Mukai et al. 2003). This enzyme phosphorylated glucose and mannose with a high affinity for glucose, utilizing PolyP as well as ATP. The catalytic sites for PolyP-dependent phosphorylation and ATP-dependent phosphorylation of the enzyme were found to be shared, and the PolyP-utilizing mechanism of the enzyme was shown to be nonprocessive (Mukai et al. 2003). The deduced amino acid sequence of the polypeptide exhibited homology to the amino acid sequences of the PolyP/ATP-glucokinase of *M. tuberculosis* and the glucokinase of *Corynebacterium glutamicum* (Mukai et al. 2003). The *Corynebacterium glutamicum* gene *cg2091* encodes a PolyP-/ATP-dependent glucokinase (PPGK). The association of PPGK with PolyP granules was shown. The PPGK of *C. glutamicum* was purified from recombinant *E. coli*. PolyP was highly preferred over ATP and other NTPs; the protein was most active with PolyP of about 75 phosphate residues (Lindner et al. 2010). The highly thermostable polyphosphate glucokinase from *Thermobifida fusca* has a potential in glucose-6-phosphate generation based on low-cost polyphosphate (Liao et al. 2011).

The screening for polyphosphate glucokinase activities in a variety of different organisms showed its presence in the phylogenetically ancient bacteria (Kulaev and Vagabov 1983; Phillips et al. 1999). All these observations suggested a hypothesis that PolyP was a precursor of ATP in bioenergetic processes at the early stage of evolution (Kulaev 1994). There might have been a gradual transition from PolyP to ATP as a phosphoryl donor in glucose phosphorylation (Phillips et al. 1999). The comparison of kinetic features of PolyP- and ATP-dependent reactions for the enzymes from different sources supports the hypothesis that glucokinase in the earliest organisms may have predominantly been dependent on PolyP rather than ATP (Phillips et al. 1999). There is a progressive decrease in the efficiency of PolyP utilization by glucokinases, from older to newer organisms (Phillips et al. 1999). Some enzymes occurring among glucokinases of different bacteria are able to use for glucose phosphorylation mainly PolyP, like in *Microlunatus phosphovorus* (Tanaka et al. 2003); PolyP and ATP, like in *P. shermanii* and *M. tuberculosis* (Phillips et al. 1999); or ATP only, like in *E. coli*. The active center responsible for PolyP utilization was probably eliminated in the course of evolution, and the glucokinases of eukaryotes already cannot use this inorganic substrate.

3.3.2 NAD Kinase (ATP:NAD 2′-Phosphotransferase, EC 2.7.1.23)

In some prokaryotes, e.g., in *Brevibacterium ammoniagenes* (Murata et al. 1979), *Micrococcus luteus* and *Corynebacterium ammoniagenes* (Fillipovich et al. 2000), and *M. tuberculosis* (Kawai et al. 2000), the enzyme catalyzes phosphorylation of NAD using both ATP and PolyP as phosphate donors:

$$\mathrm{PolyP}_n + \mathrm{NAD} \rightarrow \mathrm{PolyP}_{n-1} + \mathrm{NADP}$$
$$\mathrm{ATP} + \mathrm{NAD} \rightarrow \mathrm{ADP} + \mathrm{NADP}$$

The NAD kinase of *E. coli* utilized ATP or other nucleoside triphosphates but not PolyP as phosphoryl donors for NAD phosphorylation (Kawai et al. 2001). The *ppnk* gene of *M. tuberculosis* was cloned and expressed in *E. coli* and showed an activity with PolyP (Kawai et al. 2000, 2001). The NAD kinase from *M. tuberculosis* overexpressed in *E. coli* was purified and crystallized in the presence of NAD (Mori et al. 2004). The *ppnK* gene product from the *Corynebacterium glutamicum* genome was purified from recombinant *E. coli*, and enzymatic characterization revealed its activity as a PolyP-/ATP-dependent NAD kinase (PPNK). PPNK from *C. glutamicum* was shown to be active as a homotetramer accepting PolyP, ATP, and even ADP for NAD phosphorylation. The catalytic efficiency with ATP as a phosphate donor for NAD phosphorylation was higher than with PolyP. With respect to PolyP chain length, PPNK was active with short-chain PolyPs. PPNK activity was independent of bivalent cations when using ATP but was enhanced by manganese and particularly by magnesium ions. When using PolyP, PPNK required bivalent cations, preferably manganese ions, for its activity (Lindner et al. 2010).

The amino acid sequence of the PolyP/ATP-NAD kinase of *M. tuberculosis* exhibited a homology with that of the ATP-NAD kinase of *E. coli* (Kawai et al. 2001).

The enzymes are active or not active with PolyP, depending on a microorganism under study. The evolutionally older *Mycobacteria* possess both activities, in contrast to the evolutionally younger *E. coli*. No data on the activity of eukaryotic NAD kinases with PolyP were reported. So, NAD kinases show the same feature as PolyP/ATP glucose kinases. This fact, together with the data on distribution of polyphosphate glucokinase in bacteria, confirms the idea of the greater role of PolyP in cell energetics at the early stages of evolution (Kulaev 1994).

The essentiality and conservation of prokaryotic NAD kinases among a wide variety of pathogenic microorganisms, as well as their differential structural features distinguishing the microbial and human cell types of the enzyme, are a basis for developing novel antibiotics being inhibitors of this enzyme (Magni et al. 2009; Petrelli et al. 2011).

3.3.3 Exopolyphosphatase (Polyphosphate Phosphohydrolase EC 3.6.1.11)

One of the important enzymes involved in PolyP metabolism is exopolyphosphatase, the enzyme that splits P_i from the end of the PolyP chain:

$$\text{PolyP}_n + \text{H}_2\text{O} \rightarrow \text{PolyP}_{n-1} + \text{P}_i$$

Bacteria possess two exopolyphosphatases: one encoded by the *ppx* gene (Akiyama et al. 1993) and the other encoded by the *gppA* gene (Keasling et al. 1993). The proteins demonstrate high sequence similarity: a 39 % identity (Reizer et al. 1993). Both enzymes have one hydrophobic region. The *ppx* and *gppA* possess five conserved boxes suggesting that both proteins belong to the sugar kinase/actin/heat-shock protein hsp70 superfamily (Reizer et al. 1993).

The exopolyphosphatase encoded by the *ppx* of *E. coli* is a dimer with a subunit molecular mass of about 58 kDa (Akiyama et al. 1993). Its affinity to high-molecular-mass PolyP was nearly 100-fold higher than that of yeast polyphosphatases ($K_m = 9$ nM PolyP_{500} as a polymer). This enzyme exhibits high demand for K^+ (21-fold stimulation by 175 mM of K^+) (Akiyama et al. 1993). Exopolyphosphatase ppx of *E. coli* is a highly processive enzyme demonstrating the ability to recognize long-chain PolyP. It is low active with short-chain PolyP. Multiple PolyP binding sites were identified in of the enzyme and shown to be responsible for the enzyme polymer length recognition (Bolesch and Keasling 2000).

The exopolyphosphatase encoded by the *ppx* from *A. johnsonii* has a K_m value of 5.9 μM for PolyP with the average chain length of 64 phosphate residues (Bonting et al. 1993). The activity is maximal in the presence of 2.5 mM Mg^{2+} and 0.1 mM K^+. No activity is observed in the absence of cations or in the presence of Mg^{2+} or K^+ alone. The enzyme of *A. johnsonii* was active with PolyP_3 and PolyP_4 in the presence of 300 mM NH_4 and 10 mM Mg^{2+}, but no activity with PolyP_3 was observed in the presence of 0.1 mM K^+ and 2 mM Mg^{2+}.

The purified bacterial exopolyphosphatases encoded by the *ppx* genes have a low specific activity: 1 μmol P_i/min per mg protein (*A. johnsonii*), 22 μmol P_i/min per mg of protein (*E. coli*) compared to the yeast enzymes, 200–400 μmol P_i/min per mg of protein and more. They are low active with PolyP_3 and short-chain PolyP and need K^+ for the maximal activity. The *ppx* genes were cloned and sequenced from *E. coli* (Akiyama et al. 1993), *P. aeruginosa* (Miyake et al. 1999), *V. cholerae* (Ogawa et al. 2000a), and other bacteria (see http://www.expasy.org). In *E. coli* (Akiyama et al. 1993), *V. cholerae* (Ogawa et al. 2000a), and *Serratia marcescens* (Lee et al. 2006), *ppk1* and *ppx* are in one operon, which suggests co-regulation of their transcription activities, while in *P. aeruginosa*, *ppx* is located in the opposite direction from the *ppk1* gene and they do not constitute an operon (Miyake et al. 1999). The overexpression of *ppx* in *Pseudomonas* mimicked some pleiotropic defects found in the *ppk1* mutants (Varela et al. 2010): they have defects in biofilm formation and in the shape of colonies similar to *Δppk1* mutants.

Another enzyme encoded by the *gppA* gene and possessing exopolyphosphatase activity was purified from *E. coli* (Keasling et al. 1993). This enzyme is a dimer with a monomer molecular mass of 50 kDa. K_m is 0.5 nM for $PolyP_{500}$. It has a preference for long-chain PolyP. The enzyme also performs the following reaction:

$$\text{Guanosine } 5'\text{-triphosphate, } 3'\text{-diphosphate} \rightarrow \text{guanosine } 5'\text{-diphosphate, } 3'\text{-diphosphate} + P_i$$

Guanosine 5′-triphosphate, 3′-diphosphate (pppGpp) and guanosine 5′-diphosphate, 3′-diphosphate (ppGpp), the so-called alarmones, play a central role in the bacterial cell response to stress conditions, particularly starvation (Magnusson et al. 2005; Kuroda 2006; Potrykus and Cashel 2008; Srivatsan and Wang 2008). The involvement of exopolyphosphatases and polyphosphate kinase in such response has been most thoroughly studied in *E. coli*.

The rRNA biosynthesis and ribosome formation in *E. coli* cells decrease under amino acid starvation; this phenomenon is known as "stringent response" (Magnusson et al. 2005; Potrykus and Cashel 2008; Srivatsan and Wang 2008). The ppGpp is produced not only in response to amino acid starvation but also to other nutrient limitations and to the conditions finally causing cessation of growth. *E. coli* has two proteins, RelA and SpoT, catalyzing the reaction of pppGpp synthesis:

$$\text{ATP} + \text{GTP} = \text{AMP} + \text{guanosine } 3'\text{-diphosphate} 5'\text{-triphosphate}$$

SpoT is also able to hydrolyze ppGpp to GTP and pyrophosphate:

$$\text{Guanosine } 3', 5'\text{-bis(diphosphate)} + H_2O = \text{GTP} + PP_i$$

The protein RelA is linked with ribosomes and produces pppGpp in response to the entry of free tRNA into the ribosomal A-site. Little is known about the signals regulating the activity of SpoT; however, this protein is responsible for pppGpp synthesis in response to stress conditions and nutrient limitations different from amino acid starvation (Magnusson et al. 2005; Potrykus and Cashel 2008; Srivatsan and Wang 2008).

The ppGpp enhances the expression of many stress-induced genes via transcription by RNA polymerase with σ^s, σ^e, and σ^N and forms a complex with the DksA protein (one of transcription factors) and directly inhibits the transcription of ribosomal RNA genes via destabilization of the RNA-polymerase-promoter complex (Potrykus and Cashel 2008; Srivatsan and Wang 2008). The ppGpp is also able to inhibit replication initiation, probably by interacting with primase, and to suppress translation via binding to the initiation factor IF2 and, probably, to the elongation factors. At the same time, the exact mechanisms of interaction between this alarmone and all of the above-mentioned proteins are not quite clear, and this compound involved in many bacterial processes is hitherto referred to as a "magic spot" (Magnusson et al. 2005; Potrykus and Cashel 2008).

PolyP inhibits (p)ppGpp hydrolysis by gppA enzyme; on the other hand, the elevated (p)ppGpp concentration reduces the rate of PolyP degradation both by this enzyme and by the major bacterial exopolyphosphatase ppx. As a result, the PolyP content in *E. coli* increases manifold under amino acid starvation. PolyP forms a complex with ATP-dependent Lon proteinase and increases its activity (Kuroda 2006). These PolyPs with an average chain length of 65–700 phosphate residues bind up to four Lon molecules per one PolyP chain, and such complex proves to be effective for the proteolysis of ribosomal proteins. The in vitro experiments showed that the proteins L1, L2, L6, and L24 were degraded by Lon only in the presence of PolyP (Kuroda 2006). Degradation of the proteins L9, L13, and L17 is stimulated in presence of PolyP. The Lon-PolyP complex does not attack intact ribosomes (Kuroda 2006). Free amino acids formed as a result of hydrolysis of ribosomal proteins are used for biosynthesis of inducible carriers and enzymes of amino acid biosynthesis under amino acid starvation.

The increase in PolyP level under amino acid starvation occurs even in P_i-deficient medium, because (p)ppGpp accumulating in these conditions suppresses PolyP hydrolysis by exopolyphosphatases. The *ppkl* mutant of *E. coli* is incapable of enhancing the level of (p)ppGpp under amino acid starvation, and the regulatory mechanism for the induction of enzymes synthesizing amino acids is not triggered (Kuroda 2006; Rao et al. 2009). The degradation of cell proteins under amino acid starvation intensifies in wild-type cells but remains at the initial level in the mutant. Consequently, PolyPs synthesized by polyphosphate kinase are necessary for this process. PolyPs regulate the expression of subunit σ^{38}, a product of the *rpoS* gene (Shiba et al. 1997), probably due to the suppression of (p)ppGpp hydrolysis by the *gppA* gene product.

So, both bacterial proteins with exopolyphosphatase activities are involved in signal transduction and stress response.

There are few data on the exopolyphosphatase activity in Archaea. It was very low in *Halobacterium salinarium* (Andreeva et al. 2000). A functionally active exopolyphosphatase gene was found in *Sulfolobus solfataricus,* cloned, and overexpressed. This gene showed the highest (25–45 %) similarity to the sequences of bacterial *ppx* and possessed all of their conserved motifs and split pppGpp in vitro (Cardona et al. 2002).

As for eukaryotes, the exopolyphosphatases of the yeast *Saccharomyces cerevisiae* are best studied. Exopolyphosphatase activity in the cell homogenate of *S. cerevisiae* is high (0.10–0.13 μmol P_i/min per mg of protein) (Andreeva et al. 1994; Wurst and Kornberg 1994) compared to bacteria (0.02–0.04 μmol P_i/ min per mg of protein) (Akiyama et al. 1993; Bonting et al. 1993). Exopolyphosphatases from the *S. cerevisiae* cell envelope (Andreeva et al. 1990; Andreeva and Okorokov 1993), cell homogenate (Wurst and Kornberg 1994), cytosol (Andreeva et al. 1998a, 2006), vacuolar sap (Andreeva et al. 1998b), and mitochondrial matrix (Lichko et al. 2000) were purified and characterized.

All of the studied exopolyphosphatases of *S. cerevisiae* exhibit some similar features: hydrolyze PolyP with different chain lengths followed by P_i release; do not hydrolyze *p*-nitrophenylphosphate (the substrate of phosphatases with a broad spectrum of action), ATP and other nucleoside triphosphates, and PP_i; have neutral pH optima; are insensitive to molybdate (the common phosphohydrolase inhibitor) and fluoride (the inhibitor of pyrophosphatases); and are competitively suppressed by heparin. The exopolyphosphatases localized in separate yeast cell compartments significantly differ in their molecular masses, substrate specificity, kinetic properties, and stimulation by metal cation. The properties of exopolyphosphatases of different yeast cell compartments have been described in detail in (Lichko et al. 2003; Kulaev et al. 2004).

The *PPX1* gene encoding major yeast exopolyphosphatase was cloned and sequenced, and null mutants were obtained (Wurst et al. 1995). The PPX1 belongs to the PPase C family and has no sufficient similarity to the bacterial *ppx* gene (see http://www.expasy.org). The kinetic of native and mutant forms of PPX1 were studied (Tammenkoski et al. 2007).

In the Δ*PPX1*-deficient strains (Wurst et al. 1995), 40-kDa exopolyphosphatases were not observed in the cytosol, cell envelope, and mitochondrial matrix (Lichko et al. 2003; 2006a, b, c). Although PPX1 was absent in the cytosol of the mutant, exopolyphosphatase activity in this compartment decreased only twofold. This was explained by a fivefold increase in the activity of high-molecular-mass exopolyphosphatase in this compartment, whose properties were the same as those of the high-molecular-mass exopolyphosphatase that appeared in the cytosol under phosphate overplus (Andreeva et al. 2004; Lichko et al. 2006a, b, c). Inactivation of *PPX1* did not result in any considerable changes in the content and properties of vacuolar, nuclear, and membrane-bound mitochondrial exopolyphosphatases as compared with the parent strain of *S. cerevisiae* (Lichko et al. 2003, 2006a, b, c).

The *PPN1* gene of *S. cerevisiae* was cloned and sequenced, and null mutants were obtained (Sethuraman et al. 2001). The *PPN1* gene discoverers believed it to encode endopolyphosphatase not forming P_i during PolyP hydrolysis (Sethuraman et al. 2001). Later it was shown that the high-molecular-mass exopolyphosphatase that appeared in the cytosol under phosphate overplus (Andreeva et al. 2004) and under inactivation of *PPX1* gene (Lichko et al. 2006a, b) is encoded by *PPN1* (Andreeva et al. 2006). Inactivation of the *PPN1* gene leads to elimination of exopolyphosphatases in the mitochondrial membrane, the nuclei, and the major part of activity in the vacuoles (Lichko et al. 2006a, b). In double Δ*PPX1* Δ*PPN1* mutant, only vacuoles contained a minor exopolyphosphatase encoded by the unknown gene (Lichko et al. 2006c).

The data on the properties and localization of proteins encoded by the PPX1 and PPN1 genes are summarized in the Table 3.2 and in Fig. 3.1.

As concerns the functional significance of enzymes encoded by the PPX1 and PPN1 genes, it seems to be different. The Δ*PPX1* mutants of *S. cerevisiae* were similar to parent strain in PolyP content and chain and length. The Δ*PPN1* mutants contain longer-chain PolyPs compared to the parent strains, at least in separate compartments and at certain growth stages (Lichko et al. 2006a, b).

Table 3.2 Some properties of PPX1 and PPN1 exopolyphosphatases of *Saccharomyces cerevisiae* (Andreeva et al. 1994, Lichko et al. 2003, 2006a, b, c; Andreeva et al. 2006)

Properties	PPX1	PPN1
Molecular mass	40 kDa	A tetramer of 33 kDa subunits or a high-molecular complex of 830 kDa
Specific activity U/mg protein with		
PolyP3	420	20
PolyP208	270	150
Stimulation by divalent cations		
Mg^{2+} 1 mM	39-folds	2-folds
Co^{2+} 0 mM	66-folds	6-folds
Localization	Cell envelope, cytosol, mitochondrial matrix	Vacuoles, nuclei, mitochondrial membrane, Cytosol under phosphate overplus
The peculiarities of null mutants	*PPN1* encoding polyphosphatase appears in cytosol	Defective in development of mitochondria, no growth on ethanol and lactate

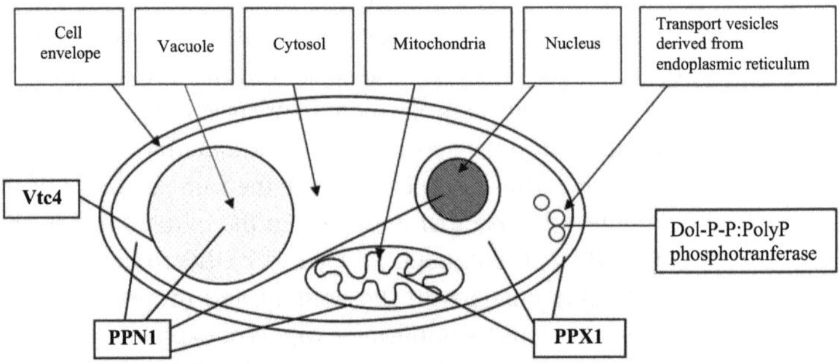

Fig. 3.1 The localization of PolyP-metabolizing enzymes in the cells of *Saccharomyces cerevisiae*

The *S. cerevisiae* strains with inactivated exopolyphosphatase-encoding genes were used for studying the effect of inactivations on the mitochondrial function (Pestov et al. 2005). As regards the *PPX1* gene, the enzyme it encodes is localized in the matrix, where PolyP is practically absent. So it is not surprising that its inactivation has no effect on PolyP metabolism. The functions and substrate of the *PPX1* enzyme in the mitochondrial matrix need further investigation. The inactivation of *PPN1* resulted in the absence of membrane-bound exopolyphosphatase and in the accumulation of long-chain PolyP in these organelles (Pestov et al. 2005). Promitochondria with defective respiratory function were present in the Δ*PPN1* strains even at the stationary growth stage. In *PPN1* mutants the formation of well-developed mitochondria was blocked. These mutants were defective in respiration functions and consumption of oxidable carbon sources such as lactate and ethanol. This fact suggests that the *PPN1* gene is essential for mitochondrial functioning in *S. cerevisiae*.

The growth of yeast strains with inactivated *PPX1* and *PPN1* genes and the strain with double mutations in these genes in a P_i-deficient medium is not disturbed (Lichko et al. 2008). All strains under study are able to maintain relatively constant P_i levels in the cytosol. In the *PPX1*-deficient strain, long-chain PolyPs are depleted first and only then are short-chain PolyPs hydrolyzed. In the double Δ*PPX1* Δ*PPN1* mutant, PolyP hydrolysis in the cytosol starts with a notable delay, and about 20 % of short-chain PolyPs are retained undergrown in P_i-deficient medium. This fact suggests that *S. cerevisiae* possesses a system providing compensation for inactivation of the *PPX1* and *PPN1* genes encoding yeast cell exopolyphosphatases.

The proteins encoded by the *PPX1* and *PPN1* genes demonstrate no homology, are differently localized in the cell, vary in substrate specificity, and seem to have different functions. As yet we have failed to find the conditions when the *PPX1* gene mutants would be different in their growth or other characteristics from the parent strain. As regards the *PPN1* gene mutants, they are unable to consume oxidized substrates, due to the disturbance of formation of mitochondria.

The databases contain the enzyme adenosine-tetraphosphate phosphohydrolase (EC 3.6.1.14), which catalyzes the reaction:

$$\text{Adenosine}\, 5'\text{-tetraphosphate} + H_2O \rightarrow \text{ATP} + \text{phosphate}$$

In the yeast, adenosine-tetraphosphate phosphohydrolase and guanosine-tetraphosphate phosphohydrolase activities are an inherent property of exopolyphosphatase PPX1. It was demonstrated both in the cytosol preparations (Kulakovskaya et al. 1997) and in the purified PPX1 enzyme (Guranowski et al. 1998). Exopolyphosphatase PPX1 from the cytosol of *S. cerevisiae* is able to hydrolyze adenosine-5′-tetraphosphate and guanosine-5′-tetraphosphate about twice more actively than PolyP_{15} with the apparent K_m value of 80–100 μM (Kulakovskaya et al. 1997).

The cloning, overexpression, purification, and characterization of exopolyphosphatase (LmPPX) from the protozoan *Leishmania major* were reported (Rodrigues et al. 2002). The gene sequence shows a similarity with *PPX1*. The product of this gene (LmPPX) has 388 amino acids and a molecular mass of 48 kDa. Heterologous expression of LmPPX in *E. coli* produced a functional enzyme that was similar to the yeast exopolyphosphatase with respect to its Mg^{2+} requirement, optimal pH, and sensitivity to cations, amino acids, and heparin (Rodrigues et al. 2002). This processive enzyme did not hydrolyze pyrophosphate, ATP, or p-nitrophenylphosphate. Immunofluorescence microscopy using affinity-purified antibodies against the recombinant enzyme indicated its acidocalcisomal and cytosolic localization (Rodrigues et al. 2002). The exopolyphosphatase TbrPPX1 of *Trypanosoma brucei* with multiple cell localization was revealed (Luginbuehl et al. 2011).

3.3.4 Endopolyphosphatase (Polyphosphate Depolymerase, EC 3.6.1.10)

Endopolyphosphatase splits long PolyP molecules into shorter ones.

$$\text{PolyP}_n + H_2O \rightarrow \text{Oligopolyphosphates}$$

The reaction was observed in yeast and fungi long ago (Malmgren 1952; Kritsky et al. 1972). The activity was identified by the low-sensitive and low-specific viscometric method. Later on, the yeast enzyme encoded by the *PPN1* gene was regarded as endopolyphosphatase (Sethuraman et al. 2001; Swiss Prot://http:/kr.expasy.org.enzyme). However, the enzyme PPN1 was shown to release P_i under PolyP hydrolysis and, hence, to be an exopolyphosphatase (Andreeva et al. 2004, 2006). The high activity of exopolyphosphatases PPX1 and PPN1 makes it difficult to identify endopolyphosphatase activity in the yeast. The activity was revealed in the cytosol preparations of the *S. cerevisiae* Δ*PPX1* Δ*PPN1* strain. The enzyme cleaves PolyP with the chain length of 208 to 15 phosphate residues to shorter chains without P_i release (Lichko et al. 2010). Partially purified endopolyphosphatase from the cytosol of the yeast *Saccharomyces cerevisiae* was inhibited by heparin and insensitive to fluoride. Mg^{2+}, Mn^{2+}, and Co^{2+} (1.5 mM) stimulated the activity, while Ca^{2+} was ineffective. The existence of endopolyphosphatase activity explains why the Δ*PPX1* Δ*PPN1* mutant uses PolyP undergrown in the P_i-deficient medium.

3.4 Polyphosphate-Metabolizing Enzymes in Mammals

Although the first evidence of the presence of PolyPs in mammalian cells was obtained long ago (Gabel and Thomas 1971), their role in the higher eukaryotes is still little studied. One of the reasons is a very small amount of PolyP in animal cells. Exopolyphosphatase and endopolyphosphatase activities were observed in mammalian tissues, including bone and brain (Kumble and Kornberg 1995, 1996; Schröder et al. 1999, 2000). The alkaline phosphatases of the intestines and bone tissues degrade PolyP to orthophosphate (Lorenz and Schröder 2001).

The role of PolyP degradation in bone tissue development is associated with the following (Omelon et al. 2009). Skeletal mineralization is associated with noncrystalline, calcium- and phosphate-containing electron-dense granules. These granules originate from mitochondria that have accumulated much calcium and lost their energy functions. The mitochondria take up P_i and condense them into PolyP and also sequester Ca^{2+}. The formed granules containing PolyP and Ca^{2+} may be transported out of the osteoclasts. Probably they can be transported into osteoblasts that build a new bone. The osteoblasts may embed the granules in the new unmineralized bone. Alkaline phosphatase releases P_i from PolyP, increasing

Table 3.3 The BLAST search of proteins similar to the known PolyP-metabolizing enzymes of *E. coli* and *S. cerevisiae* in the human genome

Gene names (as indicated in http://www.expasy.org)		Query coverage	*E* value
GPPA_ECOLI (*gppA*)	No homologous proteins found		
PPK_ECOLI (*ppk1*)	No homologous proteins found		
PPX_ECOLI (*ppx*)	Hypothetical protein FLJ38723, isoform CRA_c	39 %	0.17
PPN1_YEAST (*PPN1*)	Acid sphingomyelinase-like phosphodiesterase	41 %	e-22
PPX1_YEAST (*PPX1*)	PRUNE-like protein	91 %	4e-25

local P_i and Ca^{2+} concentrations. This results in apatite mineral formation under bone growth or posttraumatic repair and bone tissue renewal.

The BLAST search of proteins similar to the known PolyP-metabolizing enzymes of *Escherichia coli* and *S. cerevisiae* in the human genome yields the following results (Table 3.3). As expected, no proteins similar in sequence to polyphosphate kinase have been found. This fact is well known from A. Kornberg's works and gives grounds for the application of polyphosphate kinase inhibitors as novel antibiotics. The sequences similar to the gppA protein of *E. coli* have not been found either, and the data on similarity between the unknown human protein and the ppx of this bacterium are hard to interpret due to the lack of information about the activities of this protein.

The situation is quite different when searching for the sequences similar to the sequences of yeast exopolyphosphatases PPN1 and PPX1. There is certain similarity between the *PPN1* gene product and acid sphingomyelinase-like phosphodiesterase (EC 3.1.4.12) performing the reaction:

$$\text{Sphingomyelin} + \text{H(2)O} \Leftrightarrow N\text{-acylsphingosine} + \text{choline phosphate}$$

The protein is membrane bound and belongs to the nucleotide pyrophosphatase/phosphodiesterase family. Potential significance of such similarity seems to be interesting, because this enzyme is of great importance for the digestive system, and mutations in its gene have been found in cancer cells of the intestines (Duan 2006). In the colon, the enzyme may play antiproliferative and anti-inflammatory roles through generating ceramide, reducing the formation of lysophosphatidic acid and inactivating the platelet-activating factor (Duan 2006).

The similarity between the sequences of yeast exopolyphosphatase PPX1 and human protein h-prune proved to be most impressive. Human protein h-prune, a binding partner of the metastasis suppressor nm23-H1, is frequently overexpressed in metastatic cancers (Marino et al. 2011). The protein h-prune efficiently hydrolyzes short-chain PolyP including tripoly- and tetrapolyphosphates and nucleoside 5′-tetraphosphates (Tammenkoski et al. 2008). Long-chain PolyP (≥25 phosphate residues) are converted slower, whereas pyrophosphate and nucleoside triphosphates are not hydrolyzed. The reaction requires a divalent metal cofactor such as

Mg^{2+}, Co^{2+}, or Mn^{2+}, which activates both the enzyme and the substrate. Notably, the exopolyphosphatase activity of h-prune is suppressed by nm23-H1, long-chain polyphosphates and pyrophosphate, which may be potential physiological regulators. Prune is supposed to be the missing exopolyphosphatase in animals, and the metastatic effects of h-prune are probably modulated by PolyP (Tammenkoski et al. 2008).

Thus, it may be supposed that at least some of the genes encoding polyphosphate-metabolizing enzymes in the lower eukaryotes have been retained in humans as well. In the case with the prune protein, the ability for PolyP hydrolysis may be important for its function.

3.5 Conclusion

The knowledge about the enzymes of PolyP metabolism is of great importance for medicine due to the following reasons:

- PolyPs are involved in various processes of human organism: enhancement of the proliferation of fibroblasts (Shiba et al. 2003), stimulation of mammalian TOR (a kinase involved in the proliferation of mammary cancer cells) (Wang et al. 2003), regulation of Ca^{2+} uptake in mitochondria (Pavlov et al. 2010), participation in bone tissue development, and blood coagulation. Human protein h-prune, being an exopolyphosphatase, was found to be overexpressed in metastatic cancers.
- PolyPs are widely used as food additives, and therefore, it is necessary to know how these components are utilized in the digestive system. In particular, there is information that the probiotic-derived PolyP can be released from the cells of *Lactobacillus* and induce the cytoprotective activity of integrin. PolyP secretion by commensal bacteria is a factor essential for maintaining intestinal homeostasis (Segawa et al. 2011).
- In view of the great role of PolyP in cell metabolism, bone tissue development, and blood coagulation, it is necessary to control its amounts in food. Enzymatic methods using polyphosphate kinase and exopolyphosphatases are of interest for PolyP detection in biological samples (Ault-Riché et al. 1998; Vagabov et al. 2008).
- The importance of polyphosphate kinases ppk1 and ppk2 in bacterial metabolism and virulence of pathogens is of interest for the development of new inhibitors that may be prospective antibacterial compounds.

New PolyP metabolizing enzymes especially catalyzing PolyP biosynthesis need to be discovered in human organism.

Acknowledgments The authors were supported by the Russian Foundation for Basic Research (Grant 11-04-01009). We thank E. Makeeva for her help with preparation of the manuscript.

References

Ahn K, Kornberg A (1990) Polyphosphate kinase from *Escherichia coli*. J Biol Chem 265:11734–11739

Akiyama M, Crooke E, Kornberg A (1992) The polyphosphate kinase gene of *Escherichia coli*. Isolation and sequence of the ppk gene and membrane location of the protein. J Biol Chem 267:22556–22561

Akiyama M, Crooke E, Kornberg A (1993) An exopolyphosphatase of *Escherichia coli*. The enzyme and its ppx gene in a polyphosphate operon. J Biol Chem 268:633–639

Andreeva NA, Okorokov LA (1993) Purification and characterization of highly active and stable polyphosphatase from *Saccharomyces cerevisiae* cell envelope. Yeast 9:127–139

Andreeva NA, Okorokov LA, Kulaev IS (1990) Purification and certain properties of cell envelope polyphosphatase of the yeast *Saccharomyces carlsbergensis*. Biochemistry (Mosc) 55:819–826

Andreeva NA, Kulakovskaya TV, Kulaev IS (1994) Characteristics of the cytosol polyphosphatase activity of the yeast *Saccharomyces cerevisiae*. Biochemistry (Mosc) 59:1411–1417

Andreeva NA, Kulakovskaya TV, Karpov AV, Sidorov IA, Kulaev IS (1998a) Purification and properties of polyphosphatase from *Saccharomyces cerevisiae* cytosol. Yeast 14:383–390

Andreeva NA, Kulakovskaya TV, Kulaev IS (1998b) Purification and properties of exopolyphosphatase isolated from *Saccharomyces cerevisiae* vacuoles. FEBS Lett 429:194–196

Andreeva NA, Kulakovskaya TV, Kulaev IS (2000) Inorganic polyphosphates and phosphohydrolases in *Halobacterium salinarium*. Mikrobiologiia 69:499–505

Andreeva NA, Kulakovskaya TV, Kulaev IS (2004) Purification and properties of exopolyphosphatase from the cytosol of *Saccharomyces cerevisiae* not encoded by the PPX1 gene. Biochemistry (Mosc) 69:387–393

Andreeva NA, Kulakovskaya TV, Kulaev IS (2006) High molecular mass exopolyphosphatase from the cytosol of the yeast *Saccharomyces cerevisiae* is encoded by the PPN1 gene. Biochemistry (Mosc) 71:975–977

Ault-Riché D, Fraley CD, Tzeng CM, Kornberg A (1998) Novel assay reveals multiple pathways regulating stress-induced accumulations of inorganic polyphosphate in *Escherichia coli*. J Bacteriol 180:1841–1847

Ayraud S, Janvier B, Labigne A, Ecobichon C, Burucoa C, Fauchere JL (2005) Polyphosphate kinase: a new colonization factor of Helicobacter pylori. FEMS Microbiol Lett 243:45–50

Bolesch DG, Keasling JD (2000) Polyphosphate binding and chain length recognition of *Escherichia coli* exopolyphosphatase. J Biol Chem 275:33814–33819

Bonting CF, Korstee GJ, Zehnder AJ (1991) Properties of polyphosphate:AMP phosphotransferase of *Acinetobacter* strain 210A. J Bacteriol 173:6484–6488

Bonting CF, Korstee GJ, Zehnder AJ (1993) Properties of polyphosphatase of *Acinetobacter johnsonii* 210A. Antonie van Leeuwenhoek 64:75–81

Brown MR, Kornberg A (2008) The long and short of it—polyphosphate, PPK and bacterial survival. Trends Biochem Sci 33:284–290

Caen J, Wu Q (2010) Hageman factor, platelets and polyphosphates: early history and recent connection. J Thromb Haemost 8:1670–1674

Cardona ST, Chaves FP, Jerez CA (2002) The exopolyphosphatase gene from *Sulfolobus solfataricus:* characterization of the first gene found to be involved in polyphosphate metabolism in *Archaea*. Appl Environ Microbiol 68:4812–4819

Crooke E, Akiyama M, Rao NN, Kornberg A (1994) Genetically altered levels of inorganic polyphosphate in *Escherichia coli*. J Biol Chem 269:6290–6295

Dirheimer G, Ebel JP (1965) Characterisation d'une polyphosphate AMP-phosphotransferase dans *Corynebacterium xerosis*. R C Acad Sci Paris 260:3787–3790

Docampo R, Moreno SN (2001) The acidocalcisome. Mol Biochem Parasitol 114:151–159

Duan RD (2006) Alkaline sphingomyelinase: an old enzyme with novel implications. Biochim Biophys Acta 1761:281–291

Fillipovich SY, Afanasieva TP, Bachurina GP, Kritskii MS (2000) ATP and polyphosphate-dependent bacterial NAD^+-kinases. Appl Biochem Microbiol (Mosc) 36:117–121

Gabel NW, Thomas V (1971) Evidence for the occurrence and distribution of inorganic polyphosphates in vertebrate tissues. J Neurochem 18:1229–1242

Gangaiah D, Liu Z, Arcos J, Kassem II, Sanad Y, Torrelles JB, Rajashekara G (2010) Polyphosphate kinase 2: a novel determinant of stress responses and pathogenesis in *Campylobacter jejuni*. PLoS One 5(8):e12142

Gómez-García MR, Kornberg A (2004) Formation of an actin-like filament concurrent with the enzymatic synthesis of inorganic polyphosphate. Proc Natl Acad Sci USA 101:15876–15880

Guranowski A, Starzynska E, Barnes LD, Robinson AK, Liu S (1998) Adenosine 5′-tetraphosphate phosphohydrolase activity is an inherent property of soluble exopolyphosphatase from *Saccharomyces cerevisiae*. Biochem Biophys Acta 1380:232–238

Hooley P, Whitenhead MP, Brown MR (2008) Eukaryote polyphosphate kinases: is the "Kornberg" complex ubiquitous? Trends Biochem Sci 33:577–582

Hothorn M, Neumann H, Lenherr ED, Wehner M, Rybin V, Hassa PO, Uttenweiler A, Reinhardt M, Schmidt A, Seiler J, Ladurner AG, Herrmann C, Scheffzek K, Mayer A (2009) Catalytic core of a membrane-associated eukaryotic polyphosphate polymerase. Science 324:513–516

Hsieh PC, Shenoy BC, Jentoft JE, Phillips NFB (1993) Purification of polyphosphate and ATP glucose phosphotransferase from *Mycobacterium tuberculosis* H37Ra: evidence that poly (P) and ATP glucokinase activities are catalyzed by the same enzyme. Protein Expr Purif 4:76–84

Hsieh PC, Shenoy BC, Samols D, Phillips NFB (1996) Cloning, expression and characterization of polyphosphate glucokinase from *Mycobacterium tuberculosis*. J Biol Chem 271:4909–4915

Ishige K, Noguchi T (2000) Inorganic polyphosphate kinase and adenylate kinase participate in the polyphosphate:AMP phosphotransferase activity of *Escherichia coli*. Proc Natl Acad Sci USA 976:14168–14171

Ishige K, Noguchi T (2001) Polyphosphate:AMP phosphotransferase and polyphosphate:ADP phosphotransferase activities of Pseudomonas aeruginosa. Biochem Biophys Res Commun 281:821–826

Ishige K, Zhang H, Kornberg A (2002) Polyphosphate kinase (PPK2), a potent, polyphosphate-driven generator of GTP. Proc Natl Acad Sci USA 99:16684–16688

Kawai S, Mori S, Mukai T, Suzuki S, Yamada T, Hashimoto W, Murata K (2000) Inorganic polyphosphate/ATP-NAD kinase of *Micrococcus flavus* and *Mycobacterium tuberculosis* H37Rv. Biochem Biophys Res Commun 276:57–63

Kawai S, Mori S, Mukai T, Hashimoto W, Murata K (2001) Molecular characterization of *Escherichia coli* NAD kinase. Eur J Biochem 268:4359–4365

Keasling JD, Bertsh L, Kornberg A (1993) Guanosine pentaphosphate phosphohydrolase of *Escherichia coli* is a long-chain exopolyphosphatase. Proc Natl Acad Sci USA 90:7029–7033

Kornberg A, Portola V (2003) Novel antimicrobial therapies. US Patent US 0162691 A1, 28 Aug 2003

Kornberg A, Kornberg S, Simms E (1956) Methaphosphate synthesis by enzyme from *Escherichia coli*. Biochim Biophys Acta 20:215–227

Kornberg A, Rao NN, Ault-Riché D (1999) Inorganic polyphosphate: a molecule with many functions. Annu Rev Biochem 68:89–125

Kowalczyk TH, Phillips NFB (1993) Determination of endopolyphosphatase using polyphosphate glucokinase. Anal. Biochem 212:194–205

Kritsky MS, Chernysheva EK, Kulaev IS (1972) Polyphosphate depolymerase activity in cells of the fungus *Neurospora crassa* (in Russian). Biokhimiia 37:983–990

Kulaev IS (1994) Inorganic polyphosphate functions at various stages of cell evolution. J Biol Phys 20:255–273

Kulaev IS, Bobyk MA (1971) Detection of a new enzyme in *Neurospora crassa*—1,3-diphosphoglycerate:polyphosphatephosphotransferase. Biochemistry (Mosc) 36:356–359

Kulaev IS, Vagabov VM (1983) Polyphosphate metabolism in microorganisms. Adv Microbiol Physiol 24:83–171
Kulaev IS, Bobyk MA, Nikolaev NN, Sergeev NS, Uryson SO (1971) Polyphosphate synthesizing enzymes in some fungi and bacteria. Biochemistry (Moscow) 36:791–796
Kulaev IS, Vagabov VM, Kulakovskaya TV (2004) The biochemistry of inorganic polyphosphates. Wiley, Chichester
Kulakovskaya TV, Andreeva NA, Kulaev IS (1997) Adenosine-5'-tetraphosphate and guanosine-5'-tetraphosphate–new substrates of the cytosol exopolyphosphatase of *Saccharomyces cerevisiae*. Biochemistry (Moscow) 62:1180–1184
Kumble KD, Kornberg A (1995) Inorganic polyphosphate in mammalian cells and tissues. J Biol Chem 270:5818–5822
Kumble KD, Kornberg A (1996) Endopolyphosphatases for long chain polyphosphate in yeast and mammals. J Biol Chem 271:27146–27151
Kumble KD, Ahn K, Kornberg A (1996) Phosphohistidylactive sites in polyphosphate kinase of *Escherichia coli*. Proc Natl Acad Sci USA 93:14391–14395
Kuroda A (2006) A polyphosphate-Lon protease complex in the adaptation of *Escherichia coli* to amino acid starvation. Biosci Biotechnol Biochem 70:325–331
Kuroda A, Ohtake H (2000) Molecular analysis of polyphosphate accumulation in bacteria. Biochemistry (Mosc) 65:304–309
Lee SJ, Lee YS, Lee YC, Choi YL (2006) Molecular characterization of polyphosphate (PolyP) operon from Serratia marcescens. J Basic Microbiol 46:108–115
Liao H, Myung S, Zhang YH (2011) One-step purification and immobilization of thermophilic polyphosphate glucokinase from Thermobifida fusca YX: glucose-6-phosphate generation without ATP. Appl Microbiol Biotechnol. doi:10.1007/s00253-011-3458-1
Lichko LP, Kulakovskaya TV, Kulaev IS (2000) Purification and characterization of a soluble polyphosphatase from Mitochondria of *Saccharomyces cerevisiae*. Biochemistry (Mosc) 65:355–361
Lichko LP, Andreeva NA, Kulakovskaya TV, Kulaev IS (2003) Exopolyphosphatases of the yeast *Saccharomyces cerevisiae*. FEMS Yeast Res 3:233–238
Lichko LP, Kulakovskaya TV, Kulaev IS (2006a) Inorganic polyphosphates and exopolyphosphatases in different cell compartments of *Saccharomyces cerevisiae*. Biochemistry (Mosc) 71:1171–1175
Lichko LP, Kulakovskaya TV, Pestov NA, Kulaev IS (2006b) Inactivation of the pp n1 gene exerts different effects on the metabolism of inorganic polyphosphates in the cytosol and the vacuoles of the yeast Saccharomyces cerevisiae. Mikrobiologiia 75:305–311
Lichko L, Kulakovskaya T, Pestov N, Kulaev I (2006c) Inorganic polyphosphates and exopolyphosphatases in cell compartments of the yeast *Saccharomyces cerevisiae* under inactivation of PPX1 and PPN1 genes. Biosci Rep 26:45–54
Lichko LP, Kulakovskaya TV, Kulakovskaya EV, Kulaev IS (2008) Inactivation of PPX1 and PPN1 genes encoding exopolyphosphatases of *Saccharomyces cerevisiae* does not prevent utilization of polyphosphates as phosphate reserve. Biochemistry (Mosc) 73:985–999
Lichko LP, Kulakovskaya TV, Kulaev IS (2010) Properties of partially purified endopolyphosphatase of the yeast Saccharomyces cerevisiae. Biochemistry (Mosc) 75:1404–1407
Lindner SN, Niederholtmeyer H, Schmitz K, Schoberth SM, Wendisch VF (2010) Polyphosphate/ATP-dependent NAD kinase of Corynebacterium glutamicum: biochemical properties and impact of ppnK overexpression on lysine production. Appl Microbiol Biotechnol 87:583–593
Lorenz B, Schröder HC (2001) Mammalian intestinal alkaline phosphatase acts as highly active exopolyphosphatase. Biochim. Biophys. Acta 1547:254–261
Luginbuehl E, Kunz S, Wentzinger L, Freimoser F, Seebeck T (2011) The exopolyphosphatase TbrPPX1 of Trypanosoma brucei. BMC Microbiol 11:4
Magni G, Di Stefano M, Orsomando G, Raffaelli N, Ruggieri S (2009) NAD(P) biosynthesis enzymes as potential targets for selective drug design. Curr Med Chem 16:1372–1390

Magnusson LU, Farewell A, Nyström T (2005) ppGpp: a global regulator in *Escherichia coli*. Trends Microbiol 13:236–242

Malmgren H (1952) Enzymatic breakdown of polymetaphosphate. V. Purification and specificity of the enzyme. Acta Chem Scand 6:16–26

Marino N, Marchall JP, Steeg P (2011) Protein-protein interactions: a mechanism regulating the antimetastatic properties of Nm23-H1. Naunyn-Schmiedebergs Arch Pharmacol 384:351–362

McGrath JW, Kulakova AN, Kulakov LA, Quinn JP (2005) In vitro detection and characterisation of a polyphosphate synthesising activity in the yeast Candida humicola G-1. Res Microbiol 156:485–491

McMahon KD, Dojka MA, Pace NR, Jenkins D, Keasling JD (2002) Polyphosphate kinase from activated sludge performing enhanced biological phosphorus removal. Appl Environ Microbiol 68:4971–4978

Miyake T, Shiba T, Kameda A, Ihara Y, Munekata M, Ishige K, Noguchi T (1999) The gene for an exopolyphosphatase of *Pseudomonas aeruginosa*. DNA Res 6:103–108

Mori S, Yamasaki M, Maruyama Y, Momma K, Kawai S, Hashimoto W, Mikami B, Murata K (2004) Crystallographic studies of Mycobacterium tuberculosis polyphosphate/ATP-NAD kinase complexed with NAD. J Biosci Bioeng 98:391–393

Morita K, Doi K, Kubo T, Takeshita R, Kato S, Shiba T, Akagawa Y (2010) Enhanced initial bone regeneration with inorganic polyphosphate-adsorbed hydroxyapatite. Acta Biomater 6:2808–2815

Mukai T, Kawai S, Matsukawa H, Matuo Y, Murata K (2003) Characterization and molecular cloning of a novel enzyme, inorganic polyphosphate/ATP-glucomannokinase, of *Arthrobacter* sp. Strain KM. Appl Environ Microbiol 69:3849–3857

Muller O, Neumann H, Bayer MJ, Mayer A (2002) Role of the Vtc proteins in V-ATPase stability and membrane trafficking. J Cell Sci 116:1107–1115

Müller WE, Wang X, Diehl-Seifert B, Kropf K, Schloßmacher U, Lieberwirth I, Glasser G, Wiens M, Schröder HC (2011) Inorganic polymeric phosphate/polyphosphate as an inducer of alkaline phosphatase and a modulator of intracellular Ca(2+) level in osteoblasts (SaOS-2 cells) in vitro. Acta Biomater 7:2661–2671

Murata K, Kato J, Chibata I (1979) Continuous production of NADP by immobilized *Brevibacterium ammoniagenes* cells. Biotechnol Bioeng 21:887–895

Nocek B, Kochinyan S, Proudfoot M, Brown G, Evdokimova E, Osipiuk J, Edwards AM, Savchenko A, Joachimiak A, Yakunin AF (2008) Polyphosphate-dependent synthesis of ATP and ADP by the family-2 polyphosphate kinases in bacteria. Proc Natl Acad Sci USA 105:17730–17735

Ogawa N, Tzeng CM, Fraley CD, Kornberg A (2000a) Inorganic polyphosphate in *Vibrio cholerae*: genetic, biochemical, and physiologic features. J Bacteriol 182:6687–6693

Ogawa N, DeRisi J, Brown PO (2000b) New components of a system for phosphate accumulation and polyphosphate metabolism in *Saccharomyces cerevisiae* revealed by genomic expression analysis. Mol Biol Cell 11:4309–4321

Omelon SJ, Grynpas MD (2008) Relationships between polyphosphate chemistry, biochemistry and apatite biomineralization. Chem Rev 208:4694–4715

Omelon S, Georgiou J, Henneman ZJ, Wise LM, Sukhu B, Hant T, Wynnyckyj S, Holmyard D, Bielecki R, Grynpas MD (2009) Control of vertebrate skeletal mineralization by polyphosphates. PLoS One 4(5):e5634

Pavlov E, Aschar-Sobbi R, Campanella M, Turner RJ, Gómez-García MR, Abramov AY (2010) Inorganic polyphosphate and energy metabolism in mammalian cells. J Biol Chem 285:9420–9428

Pepin CA, Wood HG (1987) The mechanism of utilization of polyphosphate by polyphosphate glucokinase from *Propionibacterium shermanii*. J Biol Chem 262:5223–5226

Pestov NA, Kulakovskaya TV, Kulaev IS (2005) Effects of inactivation of the PPN1 gene on exopolyphosphatases, inorganic polyphosphates and function of mitochondria in the yeast *Saccharomyces cerevisiae*. FEMS Yeast Res 5:823–828

Petrelli R, Felczak K, Cappellacci L (2011) NMN/NaMN adenylyltransferase (NMNAT) and NAD kinase (NADK) inhibitors: chemistry and potential therapeutic applications. Curr Med Chem 18:1973–1992

Phillips NFB, Hsien PC, Kowalczyk TH (1999) Polyphosphate glucokinase. In: Schröder HC, Müller WEG (eds) Inorganic polyphosphates. biochemistry, biology, biotechnology, vol 23, Program in cell and molecular biology. Springer, Heidelberg, pp 101–127

Potrykus K, Cashel M (2008) (p)ppGpp: still magical? Annu Rev Microbiol 62:35–51

Rao NN, Kornberg A (1996) Inorganic polyphosphate support resistance and survival of stationary-phase *Escherichia coli*. J Bacteriol 178:1394–1400

Rao NN, Liu S, Kormberg A (1998) Inorganic polyphosphate in *Escherichia coli*: the phosphate regulon and the stringent response. J Bacteriol 180:2186–2193

Rao NN, Gómez-García MR, Kornberg A (2009) Inorganic polyphosphate: essential for growth and survival. Ann Rev Biochem 78:605–647

Reizer J, Reizer A, Jr MH, Saier B, Bork CS (1993) Exopolyphosphate phosphatase and guanosine pentaphosphate phosphatase belong to the sugar kinase/actin/hsp 70 superfamily. Trends Biochem Sci 18:247–248

Reusch RN (1992) Biological complexes of poly-β-hydroxybutyrate. FEMS Microbiol Rev 103:119–130

Rodrigues CO, Ruiz FA, Vieira M, Hill JE, Docampo R (2002) An acidocalcisomal exopolyphosphatase from *Leishmania major* with high affinity for short-chain polyphosphate. J Biol Chem 277:50899–50906

Schröder HC, Lorenz B, Kurz L, Müller WEG (1999) Inorganic polyphosphate in eukaryotes: enzymes, metabolism and function. In: Schröder HC, Müller WEG (eds) Inorganic polyphosphates. biochemistry, biology, biotechnology, Program in cell and molecular biology. Springer, Heidelberg, pp 45–83

Schröder HC, Kurz L, Müller WEG, Lorenz B (2000) Polyphosphate in bone. Biochemistry (Mosc) 65:296–304

Schuddemat J, de Boo R, Van Leeuwen CCM, Van den Broek PJA, Van Steveninck J (1989) Polyphosphate synthesis in yeast. Biochem Biophys Acta 100:191–198

Segawa S, Fujiya M, Konishi H, Ueno N, Kobayashi N, Shigyo T, Kohgo Y (2011) Probiotic-derived polyphosphate enhances the epithelial barrier function and maintains intestinal homeostasis through integrin-p38 MAPK pathway. PLoS One 6(8):e23278

Sethuraman A, Rao NN, Kornberg A (2001) The endopolyphosphatase gene: essential in *Saccharomyces cerevisiae*. Proc Natl Acad Sci USA 98:8542–8547

Shabalin YA, Kulaev IS (1989) Solubilization and properties of yeast dolichylpyrophosphate: polyphosphate phosphotransferase. Biokhimiya (Mosc) 54:68–75

Shabalin YA, Vagabov VM, Tsiomenko AB, Zemlianuhina OA, Kulaev IS (1977) Study of polyphosphate kinase activity in the yeast vacuoles. Biokhimia (Mosc) 42:1642–1648

Shabalin YA, Vagabov VM, Kulaev IS (1979) On the coupling mechanism of biosynthesis of high-molecular polyphosphates and mannan in *Saccharomyces carlsbergensis* yeast. Dokl Acad Nauk SSSR 249:243–246

Shabalin YA, Naumov AV, Vagabov VM, Kulaev IS (1984) The discovery of new enzyme activity dolichyldiphosphate polyphosphate phosphotransferase in yeast. Dokl Acad Nauk SSSR 278:482–485

Shabalin YA, Vagabov VM, Kulaev IS (1985) Dolichyldiphosphate mannose-an intermediate of glycoprotein biosynthesis in yeast? Dokl Acad Nauk SSSR 283:720–723

Shiba T, Tsutsumi K, Yano H, Ihara Y, Kameda A, Tanaka K, Takahashi H, Munekata M, Rao NN, Kornberg A (1997) Inorganic polyphosphate and the induction of RPOS expression. Proc Natl Acad Sci USA 94:11210–11215

Shiba T, Nishimura D, Kawazoe Y, Onodera Y, Tsutsumi K, Nakamura R, Ohshiro M (2003) Modulation of mitogenic activity of fibroblast growth factors by inorganic polyphosphate. J Biol Chem 278:26788–26792

Shum KT, Lui EL, Wong SC, Yeung P, Sam L, Wang Y, Watt RM, Tanner JA (2011) Aptamer-mediated inhibition of Mycobacterium tuberculosis polyphosphate kinase 2. Biochemistry 50:3261–3271

Smith SA, Mutch NJ, Baskar D, Rohloff P, Docampo R, Morrissey JH (2006) Polyphosphate modulates blood coagulation and fibrinolysis. Proc Natl Acad Sci USA 103:903–908

Smith SA, Choi SH, Davis-Harrison R, Huyck J, Boettcher J, Reinstra CM, Morrissey JH (2010) Polyphosphate exerts differential effects on blood clotting, depending on polymer size. Blood 116:4353–4359

Srivatsan A, Wang JD (2008) Control of bacterial transcription, translation and replication by (p)ppGpp. Curr Opin Microbiol 11:100–105

Sureka K, Sanyal S, Basu J, Kundu M (2009) Polyphosphate kinase 2: a modulator of nucleoside diphosphate kinase activity in mycobacteria. Mol Microbiol 74:1187–1197

Szymona M (1957) Utilization of inorganic polyphosphates for phosphorylation of glucose in *Mycobacterium phlei*. Bull Acad Pol Sci Ser Sci Biol 5:379–381

Szymona M, Szymona O (1961) Participation of volutin in the hexokinase reaction of *Corynebacterium diphtheriae*. Bull Acad Pol Sci Ser Sci Biol 9:371–374

Szymona O, Szymona M (1978) Multiple forms of polyphosphate-glucose phosphotransferase in various *Mycobacterium* strains. Acta Microbiol Pol 27:73–78

Tammenkoski M, Moiseev VM, Lahti M, Ugochukwu E, Brondijk TH, White SA, Lahti R (2007) Baykov AA (2007) Kinetic and mutational analyses of the major cytosolic exopolyphosphatase from *Saccharomyces cerevisiae*. J Biol Chem 282:9302–9311

Tammenkoski M, Koivula K, Cusanelli E, Zollo M, Steegborn C, Baykov AA, Lahti R (2008) Human metastasis regulator protein H-prune is a short-chain exopolyphosphatase. Biochemistry 47:9707–9713

Tanaka S, Lee SO, Hamaoka K, Kato J, Takiguchi N, Nakamura K, Ohtake H, Kuroda A (2003) Strictly polyphosphate-dependent glucokinase in a polyphosphate-accumulating bacterium, *Microlunatus phosphovorus*. J Bacteriol 185:5654–5656

Tani C, Ohtomo R, Osaki M, Kuga Y, Ezawa T (2009) ATP-dependent but proton gradient-independent polyphosphate-synthesizing activity in extraradical hyphae of an arbuscular mycorrhizal fungus. Appl Environ Microbiol 75:7044–7050

Temperton B, Gilbert JA, Quinn JP, McGrath JW (2011) Novel analysis of oceanic surface water metagenomes suggests importance of polyphosphate metabolism in oligotrophic environments. PLoS One 6:e16499

Tinsley CR, Gotschlich EC (1995) Cloning and characterization of the meningococcal polyphosphate kinase gene: production of polyphosphate synthesis mutant. Infect Immun 63:1624–1630

Tzeng CM, Kornberg A (1998) Polyphosphate kinase is highly conserved in many bacterial pathogens. Mol Microbiol 29:381–382

Vagabov VM, Trilisenko LV, Kulakovskaya TV, Kulaev IS (2008) Effect of carbon source on polyphosphate accumulation in Saccharomyces cerevisiae. FEMS Yeast Res 8:877–882

Varela C, Mauriaca C, Paradela A, Albar JP, Jerez CA, Chávez FP (2010) New structural and functional defects in polyphosphate deficient bacteria: a cellular and proteomic study. BMC Microbiol 10:7

Wang L, Fraley CD, Faridi J, Kornberg A, Roth RA (2003) Inorganic polyphosphate stimulates mammalian TOR, a kinase involved in the proliferation of mammary cancer cells. Proc Natl Acad Sci USA 100:11249–11254

Wurst H, Kornberg A (1994) A soluble exopolyphosphatase of *Saccharomyces cerevisiae*. J Biol Chem 269:10996–101001

Wurst H, Shiba T, Kornberg A (1995) The gene for a major exopolyphosphatase of *Saccharomyces cerevisiae*. J Bacteriol 177:898–906

Zhang H, Ishige K, Kornberg A (2002) A polyphosphate kinase (PPK2) widely conserved in bacteria. Proc Natl Acad Sci USA 99:16678–16683

Zhang H, Gómez-García MR, Brown MR, Kornberg A (2005) Inorganic polyphosphate in *Dictyostelium discoideum*: influence on development, sporulation, and predation. Proc Natl Acad Sci USA 102:2731–2735

Zhang H, Gómez-García MR, Shi X, Rao NN, Kornberg A (2007) Polyphosphate kinase 1, a conserved bacterial enzyme, in a eukaryote, *Dictyostelium discoideum*, with a role in cytokinesis. Proc Natl Acad Sci USA 104:16486–16491

Zhu Y, Lee SS, Xu W (2003) Crystallization and characterization of polyphosphate kinase from *Escherichia coli*. Biochem Biophys Res Commun 305:997–1001

Chapter 4
Polyoxometalates Active Against Tumors, Viruses, and Bacteria

Toshihiro Yamase

Contents

Abstract Polyoxometalates (PMs) as discrete metal-oxide cluster anions with high solubility in water and photochemically and electrochemically active property have a wide variety of structures not only in molecular size from sub-nano to sub-micrometers with a various combination of metals but also in symmetry and highly negative charge. One of the reasons for such a structural variety originates from their conformation change (due to the condensed aggregation and the structural assembly) which strongly depends on environmental parameters such as solution pH, concentration, and coexistent foreign inorganic and/or organic substances. In the course of the application of the physicochemical properties of such PMs to the medical fields, antitumoral, antiviral, and antibacterial activities have been developed for realization of a novel inorganic medicine which provides a biologically excellent activity never replaced by other approved medicines. Several PMs as a candidate for clinical uses have been licensed toward the chemotherapy of solid tumors (such as human gastric cancer and pancreatic cancer), DNA and RNA viruses (such as HSV, HIV, influenza, and SARS), and drug-resistant bacteria (such

T. Yamase (✉)
MO device, 2-14-10 Kanaiwa-higashi, Kanazawa 920-0335, Japan

Chemical Resources Laboratory, Tokyo Institute of Technology, R1-21 4259 Nagatsuta, Midori-ku, Yokohama 226-8503, Japan
e-mail: yamase.modevice@nifty.com

W.E.G. Müller et al. (eds.), *Biomedical Inorganic Polymers*, Progress in Molecular and Subcellular Biology 54, DOI 10.1007/978-3-642-41004-8_4,

as MRSA and VRSA) in recent years: $[NH_3Pr^i]_6[Mo_7O_{24}]\cdot 3H_2O$ (PM-8) and $[Me_3NH]_6[H_2Mo^V{}_{12}O_{28}(OH)_{12}(Mo^{VI}O_3)_4]\cdot 2H_2O$ (PM-17) for solid tumors; $K_7[PTi_2W_{10}O_{40}]\cdot 6H_2O$ (PM-19), $[Pr^iNH_3]_6H[PTi_2W_{10}O_{38}(O_2)_2]\cdot H_2O$ (PM-523), and $K_{11}H[(VO)_3(SbW_9O_{33})_2]\cdot 27H_2O$ (PM-1002) for viruses; and $K_6[P_2W_{18}O_{62}]\cdot 14H_2O$ (PM-27), $K_4[SiMo_{12}O_{40}]\cdot 3H_2O$ ($SiMo_{12}$), and PM-19 for MRSA and VRSA. The results are discussed from a point of view of the chemotherapeutic clarification in this review.

List of Abbreviations

5-FU	5-Fluorouracil
ABC transporter	ATP-binding cassette transporter
ACNU	1-(4-amino-2-methylpyridine-5yl)methyl-3-(2-chloroethyl)-3-nitrosourea
ACV	Acyclovir
ALP	Alkaline phosphatase
ALT	Alanine aminotransferase
AMD3100	Octahydrochloride dihydrate of 1,1′-[1,4-phenylenebis (methylene)]-bis-1,4,8,11-tetraazacyclotetradecane
AST	Aspartate aminotransferase
AZT	3′-azido-3′-deoxythymidine
BUN	Blood urea nitrogen
CC_{50}	Median cytotoxic concentration
CDDP	*cis*-Diamminedichloroplatinum(II)
CDV	Canine distemper virus
DFV	Dengue fever virus
EC_{50}	Median effective concentration
ED_{50}	50 % Effective dose
FIC	Fractional inhibitory concentration index
FIPV	Feline infectious peritonitis virus
FluV-A	Influenza virus A
FMN	Flavin mononucleotide
FUT3	α1,3-fucosyltransferase-III
Gal3ST-2	Gal:3-*O*-sulfotransferase-2
Gal6ST	Gal 6-*O*-sulfotransferase
GFP	Green fluorescent protein
Gn6ST-1	GlcNAc 6-*O*-sulfotransferase-1
H&E	Hematoxylin and eosin staining
HA	Hemagglutinin peptide
HCMV	Human cytomegalovirus
HPA-23	$[NH_4]_{17}Na[Na(SbW_7O_{24})_3(Sb_3O_7)_2]\cdot 14H_2O$
HS-058	$K_{10}[Fe_4(H_2O)_2(PW_9O_{34})_2]\cdot xH_2O$

HSV-1	Herpes simplex virus type 1
HSV-2	Herpes simplex virus type 2
i.p.	Intraperitoneal or intraperitoneally
IC_{50}	50 % Inhibitory concentration
ILS	Increase in life-span
JM1590	$K_{13}[Ce(SiW_{11}O_{39})_2]\cdot 26H_2O$
JM2766	$K_6[BGa(H_2O)W_{11}O_{39}]\cdot 15H_2O$
JM2820	$[Me_3NH]_8[(SiNb_3W_9O_{37})_2O_3]$
K_i	Inhibition constant
LC3	Light-chain3
LD_{50}	50 % Lethal dose
M1	Ribonucleoprotein sheath
M2	Matrix protein ion channel
MAGI	Multinuclear activation of the galactosidase indicator assay
MDCK	Madin–Darby canine kidney
MIC	Minimum inhibitory concentration
MRSA	Methicillin-resistant *Staphylococcus aureus*
MTT	3-(4,5-dimethylthiazol-2-yl)-2,5-diphenyltetrazolium bromide colorimetry
NA	Neuraminidase
PBP	Penicillin-binding protein as the peptidoglycan-synthetic enzyme on the membrane surface
PBS	Phosphate-buffered saline
PfluV-2	Parainfluenza virus type 2
PM-1	$K_5[BW_{12}O_{40}]\cdot 15H_2O$
PM-1001	$K_{10}Na[(VO)_3(SbW_9O_{33})_2]\cdot 26H_2O$
PM-1002	$K_{11}H[(VO)_3(SbW_9O_{33})_2]\cdot 27H_2O$
PM-104	$[NH_4]_{12}H_2[Eu_4(MoO_4)(H_2O)_{16}(Mo_7O_{24})_4]\cdot 13H_2O$
PM-1207	$K_{12}[(VO)_3(AsW_9O_{33})_2]\cdot 12H_2O$
PM-1208	$K_{12}[(VO)_3(BiW_9O_{33})_2]\cdot 29H_2O$
PM-1213	$K_{12}[(VO)_3(PW_9O_{34})_2]\cdot nH_2O$
PM-17	$[Me_3NH]_6[H_2Mo^{V}{}_{12}O_{28}(OH)_{12}(Mo^{VI}O_3)_4]\cdot 2H_2O$
PM-19	$K_7[PTi_2W_{10}O_{40}]\cdot 6H_2O$
PM-27	$K_6[P_2W_{18}O_{62}]\cdot 14H_2O$
PM-30	A-β-$Na_9[SiW_9O_{34}H]\cdot 23H_2O$
PM-32	$Na_5[IMo_6O_{24}]\cdot 34H_2O$
PM-43	$K_5[SiVW_{11}O_{40}]\cdot nH_2O$
PM-44	$K_5[PVW_{11}O_{40}]\cdot 6H_2O$
PM-46	$K_6[BVW_{11}O_{40}]\cdot nH_2O$
PM-47	$K_7[BVW_{11}O_{40}]\cdot nH_2O$
PM-504	$K_9H_5[(GeTi_3W_9O_{37})_2O_3]\cdot 16H_2O$
PM-518	$[Et_2NH_2]_7[PTi_2W_{10}O_{40}]\cdot 4H_2O$
PM-520	$[Pr^{i}{}_2NH_2]_5[PTiW_{11}O_{40}]\cdot 4H_2O$

PM-523	$[Pr^iNH_3]_6H[PTi_2W_{10}O_{38}(O_2)_2]\cdot H_2O$
PM-8	$[NH_3Pr^i]_6[Mo_7O_{24}]\cdot 3H_2O$
RNP	Ribonucleoprotein
RSV	Respiratory syncytial virus
RT-PCR	Reverse transcription polymerase chain reaction
S. aureus	*Staphylococcus aureus*
s.c.	Subcutaneous or subcutaneously
SARS-V	Severe acute respiratory syndrome coronavirus
SARS-V	Severe acute respiratory syndrome coronavirus
SDS-PAGE	Sodium dodecyl sulfate-polyacrylamide gel electrophoresis
SI	Selective index
$SiMo_{12}$	$K_4[SiMo_{12}O_{40}]\cdot 3H_2O$
SRC	Subrenal capsule in kidney assay
ST3Gal-I and -III	Gal:α2,3-sialyltransferase-1 and Gal:α2,3-sialyltransferase-1II
ST6Gal-I	Gal:α2,6-sialyltransferase-1
ST6GalNAc-I	GalNAc:α2,6-sialyltransferase-1
$TCID_{50}$	Median tissue culture infection dose
TEM	Transmission electron microscope
TGEV	Transmissible gastroenteritis virus of swine
TK^-	Thymidine kinase deficient
TUNEL	Terminal deoxynucleotidyl transferase-mediated “nick-end” labeling staining
TWI	Tumor weight inhibition
VRSA	Vancomycin-resistant *Staphylococcus aureus*
β3Gn-T2	β1,3-GlcNAc-transferase-2
β4Gal-TI	β1,4-galactosyltransferase-I

4.1 Introduction

Polyoxometalates (PMs) as discrete metal-oxide cluster anions with high solubility in water and photochemically and electrochemically active property have a wide variety of structures not only in molecular size from sub-nano to sub-micrometers with a various combination of metals but also in symmetry and highly negative charge (Pope and Müller 1994, 2001; Hill 1998; Yamase and Pope 2002; Borrás-Alamenar et al. 2003; Yamase 2003; Proust et al. 2008; Long et al. 2010). One of the reasons for such a structural variety originates from their conformation change (due to the condensed aggregation and the structural assembly) which strongly depends on environmental parameters such as solution pH, concentration, and coexistent foreign inorganic and/or organic substances. Therefore, PMs under the physiological condition let us expect to induce some modifications in a variety of biological systems such as the adsorption to receptor, the penetration of substances through the cellular membrane, and the multiple enzymes which can work independently or cooperatively. The interaction of PMs with multiple biomolecule targets in a single event or different processes simultaneously in the biological systems will provide a basis for the development of a novel inorganic medicine and greatly improve our understanding of the mechanism and timing of the interacting species, in biological and clinical studies, as well. Both in vitro and in vivo antitumor activities of polyoxomolybdates, especially $[Mo_7O_{24}]^{6-}$, have been investigated along with this line since the finding of in vivo antitumor effect against solid human (breast, lung, and colon) cancer xenografts (MX-1, OAT, and CO-4, respectively) in 1988 (Yamase et al. 1988, 1992a).

Historically, 40 years ago Raynaud and Jasmin first found the biological activity of PMs against various non-retro RNA and DNA viruses in vitro and in vivo and demonstrated the inhibition of Friend leukemia virus and Moloney murine sarcoma virus in vitro by $[SiW_{12}O_{40}]^{4-}$ as a Keggin structure PM and in vivo by $[Na(SbW_7O_{24})_3(Sb_3O_7)_2]^{18-}$ (as a French drug $[NH_4]_{17}Na[Na(SbW_7O_{24})_3(Sb_3O_7)_2]\cdot 14H_2O$ historically called HPA-23) (Jasmin et al. 1973, 1974). Also, they showed HPA-23 and other polyoxotungstates to inhibit RNA-dependent DNA polymerases of retroviruses, which let us expect the activity against human immunodeficiency virus (HIV). Although HPA-23 was tested in animal models of lentiviral infections and infected humans, HPA-23 was too toxic to be effective in clinical trials in both France and the USA (Rozenbaum et al. 1985, Moskovitz et al. 1988). In spite of the negativity of in vivo anti-HIV activity of HPA-23, the assay of the antiviral activity for PMs by other groups has been continued, and many polyoxotungstates with Keggin or Wells–Dawson related (lacunary and multiply condensed) structures, in particular, $K_{13}[Ce(SiW_{11}O_{39})_2]\cdot 26H_2O$ (JM1590), $K_6[BGa(H_2O)W_{11}O_{39}]\cdot 15H_2O$ (JM2766), $[Me_3NH]_8[(SiNb_3W_9O_{37})_2O_3]$ (JM2820), $K_7[PTi_2W_{10}O_{40}]\cdot 6H_2O$ (PM-19), $K_9H_5[(GeTi_3W_9O_{37})_2O_3]\cdot 16H_2O$ (PM-504), $[Pr^iNH_3]_6H[PTi_2W_{10}O_{38}(O_2)_2]\cdot H_2O$ (PM-523), $K_{10}[Fe_4(H_2O)_2(PW_9O_{34})_2]\cdot xH_2O$ (HS-058), $K_7[P_2W_{17}NbO_{62}]$, and $K_7[P_2W_{17}(NbO_2)O_{61}]$, were found to be in vitro active against DNA and RNA viruses (Hill et al. 1990; Inouye et al. 1990; Fukuma et al. 1991; Take et al. 1991; Weeks et al. 1992; Yamamoto et al. 1992; Ikeda et al. 1993; Shigeta et al. 1996, 1997; Barnard et al. 1997; Rhule et al. 1998; Witvrouw et al. 2000). Also, the observation of in vitro anti-HIV-1 activity of $[NH_4]_{12}H_2[Eu_4(MoO_4)(H_2O)_{16}(Mo_7O_{24})_4]\cdot 13H_2O$ (PM-104, Naruke et al. 1991) was interesting, since PM-104 is classified in polyoxomolybdates although the attachment of Eu atoms in the molecule seemed to be biologically significant in the anti-HIV-1 activity (Inouye et al. 1993). However, none of these PMs was yet advantageous enough to surpass clinically approved drugs such as 3′-azido-3-′-deoxythymidine (AZT) and ribavirin in both antiviral activity and cytotoxicity. For realization of the PM-based inorganic medicines as a novel drug, it is at least necessary to show an excellent activity of PMs against the biological target which is impossible to be replaced by the existing approved drugs, while we have been always confronted with historically conservative (or pessimistic) background for inorganic medicines in medicinal society (Yamase 2005). We have shown a potent activity of the tris(vanadyl)-18-tungsto-2-antimonate(III) anions,

$[(V^{IV}O)_3(SbW_9O_{33})_2]^{12-}$ and $[(V^VO)(V^{IV}O)_2(SbW_9O_{33})_2]^{11-}$, against a wide variety of the enveloped viruses which infect high-risk individuals such as infants born with prematurity, cardiovascular failure, pulmonary dysplasia, and HIV: their selective index (SI) values ($>10^4$) for HIV are much higher than 3–4 $\times$ 10^3 of AZT and dextran sulfate (in mol. wt. 5,000, DS5000) (Shigeta et al. 2003). In addition, most of the antiviral polyoxotungstates exhibit a great synergistic effect with β-lactam antibiotics against methicillin-resistant *Staphylococcus aureus* (MRSA) and vancomycin-resistant *Staphylococcus aureus* (VRSA) strains (Yamase et al. 1996a; Fukuda et al. 1999). Together with their potency against other bacteria such as *Streptococcus pneumoniae* and *Helicobacter pylori* (Fukuda and Yamase 1997; Inoue et al. 2005), the synergy effect of PMs lets us investigate the mechanistic details on the antibacterial activity in relevance of a clarification of the antiviral mechanism for the PMs. MRSA, VRSA, and *Streptococcus pneumoniae* as Gram-positive bacteria and *Helicobacter pylori* as a microaerophilic Gram-negative bacterium pose serious problems of hospital-acquired infections in surgical intensive care units and global infections of the latter (associated with the development of gastritis, gastric ulcer, duodenal ulcer, peptic ulcer, and gastric cancer) in approximately half of the world population. Also, PMs enhanced nerve growth factor (NGF)-induced neurite growth of PC12 cells with expression of the axonal growth associated protein 43 (GAP-43) (Oda et al. 2007). Since rat neuronal PC12 cells which differentiate upon NGF stimulation are used as the in vitro model for the characterization of neurotoxicants, this finding suggests that PMs may be one of the candidates for a novel drug against the neurodegenerative disorders such as Alzheimer's disease (Geng et al. 2011). Here we review recent developments on biological (antitumor, antiviral, and anti-MRSA and VRSA bacterial) activities of the PMs selected from a viewpoint of the clinical significance.

Figure 4.1 shows the structures of anions for some of the bioactive PMs, which will be discussed below. Structures of $[Mo^{VI}{}_7O_{24}]^{6-}$ and $[H_2Mo^V{}_{12}O_{28}(OH)_{12}(Mo^{VI}O_3)_4]^{6-}$ are depicted by atom and bond model, as shown in Fig. 4.1a, b where Mo^{VI}, Mo^V, and O atoms are indicated by yellow, red, and white color, respectively (Yamase et al. 1981; Ohashi et al. 1982; Yamase and Ishikawa 2008). The former (with C_{2v}-symmetry) is the anion for $[NH_3Pr^i]_6[Mo_7O_{24}]\cdot 3H_2O$ (PM-8), and the latter (with T_d-symmetry) for $[Me_3NH]_6[H_2Mo^V{}_{12}O_{28}(OH)_{12}(Mo^{VI}O_3)_4]\cdot 2H_2O$ (PM-17). Both compounds are antitumor compounds, and the latter as a photoreduction product for the long-term photolysis of PM-8 at pH 5–7 is more active than the former (Yamase et al. 1988). Keggin-structural $[PTi_2W_{10}O_{40}]^{7-}$ (Fig. 4.1c) and Keggin-condensed dimer structural $[(GeTi_3W_9O_{37})_2O_3]^{14-}$ (Fig. 4.1d) are anions for antiviral compounds, $K_7[PTi_2W_{10}O_{40}]\cdot 6H_2O$ (PM-19) and $K_9H_5[(GeTi_3W_9O_{37})_2O_3]\cdot 16H_2O$ (PM-504), respectively (Domaille and Knoth 1983; Yamase et al. 1993, 2007). In $[PTi_2W_{10}O_{40}]^{7-}$, two W atoms in T_d-symmetric Keggin-structural $[PW_{12}O_{40}]^{3-}$ are substituted by Ti atoms to yield a C_2-symmetric anion (Domaille and Knoth 1983; Yamase et al. 1992b), and $[(GeTi_3W_9O_{37})_2O_3]^{14-}$ can be regarded as a dimeric condensation product (with D_{3h}-symmetry) for Ti-substituted

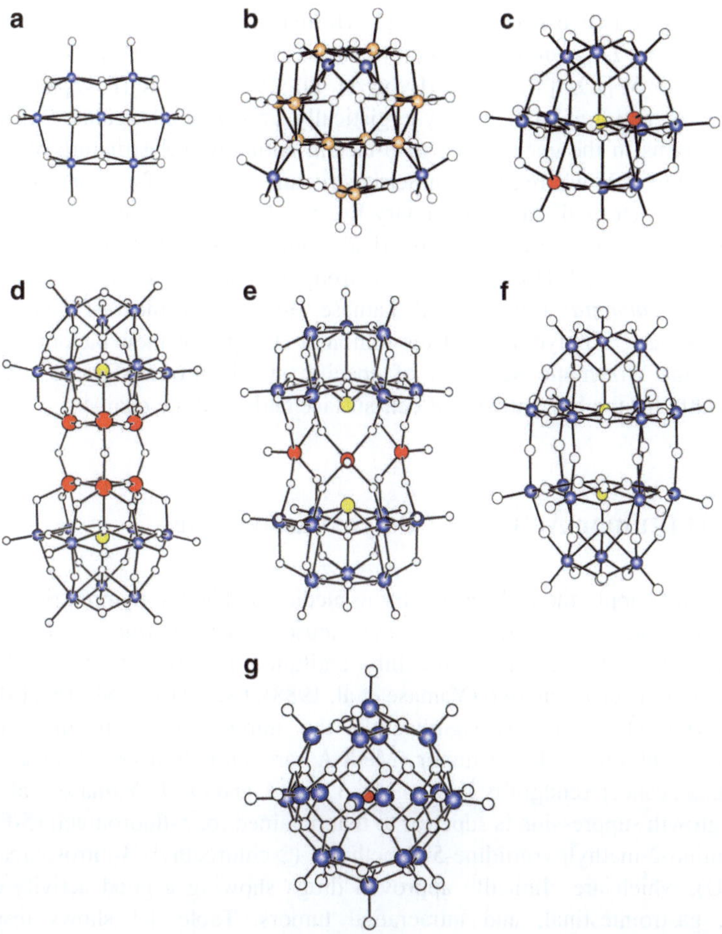

Fig. 4.1 Structures depicted by atom and bond model for anions of some of the bioactive PMs. $[Mo^{VI}{}_7O_{24}]^{6-}$ (**a**), $[H_2Mo^{V}{}_{12}O_{28}(OH)_{12}(Mo^{VI}O_3)_4]^{6-}$ (**b**), $[PTi_2W_{10}O_{40}]^{7-}$ (**c**), $[(GeTi_3W_9O_{37})_2O_3]^{14-}$ (**d**), $[(V^{IV}O)_3(SbW_9O_{33})_2]^{12-}$ (**e**), $[V^{IV}{}_{18}O_{42}(Cl)]^{13-}$ (**f**), and $[P_2W_{18}O_{62}]^{6-}$ (**g**). (**a**) and (**b**) correspond to the anion for antitumoral $[NH_3Pr^i]_6[Mo_7O_{24}]\cdot 3H_2O$ (PM-8) and $[Me_3NH]_6[H_2Mo^{V}{}_{12}O_{28}(OH)_{12}(Mo^{VI}O_3)_4]\cdot 2H_2O$ (PM-17), respectively, in which Mo^{VI}, Mo^{V}, and O atoms are indicated by *blue*, *brown*, and *white color*, respectively. (**c**), (**d**), (**e**), and (**f**) correspond to the anion for antiviral $K_7[PTi_2W_{10}O_{40}]\cdot 6H_2O$ (PM-19), $K_9H_5[(GeTi_3W_9O_{37})_2O_3]\cdot 16H_2O$ (PM-504), $K_{11}H[(VO)_3(SbW_9O_{33})_2]\cdot 27H_2O$ (PM-1002), and $K_6[P_2W_{18}O_{62}]\cdot 14H_2O$ (PM-27), respectively, in which W^{VI}, Ti^{IV} (or V^{IV}), P^{V} (Ge^{IV} or Sb^{III}), and O atoms are indicated by *blue*, *red*, *yellow*, and *white color*, respectively. (**g**) Corresponds to the anion for a strong sialyl/sulfotransferase inhibitor, $K_{10}H_3[V_{18}O_{42}(Cl)]\cdot 12H_2O$, which is also antibacterial against *Streptococcus pneumoniae*. In (**f**) V^{IV}, Cl, and O atoms are indicated by *blue*, *red*, and *white color*, respectively

A-α-$[GeTi_3W_9O_{40}]^{10-}$ within the T_d-Keggin-structural $[GeW_{12}O_{40}]^{4-}$ framework (Yamase et al. 1993). The anion of $[(V^{IV}O)_3(SbW_9O_{33})_2]^{12-}$ for $K_{11}H[(VO)_3(SbW_9O_{33})_2]\cdot 27H_2O$ (PM-1002) as a potently antiviral (especially anti-SARS) active compound is shown in Fig. 4.1e where $(V^{IV}O)_3$ triangle is sandwiched by two B-α-$[Sb^{III}W_9O_{33}]^{9-}$ (formally as a trilacunary Keggin structure) to yield a D_{3h}-symmetric anion (Yamase et al. 2001, 2002). Wells–Dawson structural $[P_2W_{18}O_{62}]^{6-}$ anion (Lyon et al. 1991) for $K_6[P_2W_{18}O_{62}]\cdot 14H_2O$ (PM-27) is antiviral and also synergistically antibacterial against MRSA and VRSA strains in the coexistence of β-lactam antibody, the structure of which is shown in Fig. 4.1f where D_{3h}-symmetric structure consists of a dimer of two α-A-$[P^VW_9O_{31}]^{3-}$ (formally as a trilacunary Keggin structure). Figure 4.1g shows D_{4d}-symmetric Cl^--encapsulated spherical anion of $[V^{IV}{}_{18}O_{42}(Cl)]^{13-}$ for $K_{10}H_3[V_{18}O_{42}(Cl)]\cdot 12H_2O$ exhibiting a strong antibacterial activity against *Streptococcus pneumoniae* (Fukuda and Yamase 1997). Also, most of such multiply reduced spherical polyoxovanadates inhibited strongly at their nanomolar levels in vitro the enzymatic activities of specific sialyl/sulfotransferases which are responsible for the biosynthesis of cell-surface carbohydrate chains.

4.2 Antitumor Activity of Polyoxomolybdates

In our trial to apply the PMs chemistry to medical fields, C_{2v}-symmetric V-shaped heptamolybdates (Fig. 4.1a) and C_i-symmetric Anderson-structural polyoxomolybdates have been recognized to exhibit antitumor activities against solid tumors at noncytotoxic doses in vivo (Yamase et al. 1988). Especially, $[NH_3Pr^i]_6[Mo_7O_{24}]\cdot 3H_2O$ (PM-8) suppresses significantly the tumor growth in mice bearing methylcholanthrene-induced tumor (Meth-A sarcoma), MM-46 adenocarcinoma, and human cancer xenografts such as MX-1, CO-4, and OAT (Yamase et al. 1992a, b). Its growth suppression is superior to that obtained for 5-fluorouracil (5-FU) and 1-(4-amino-2-methylpyrimidine-5yl)methyl-3-(2-chloroethyl)-3-nitrosourea (ACNU), which are clinically approved drugs showing a good activity against breast, gastrointestinal, and intracranial tumors. Table 4.1 shows results of antitumor activities of PM-8, $[NH_4]_6[Mo_7O_{24}]\cdot 4H_2O$, $K_6[Mo_7O_{24}]\cdot 4H_2O$, $[NH_3Pr^i]Cl$, and PM-17 as the brown-colored powder (the photoreduction product of PM-8) against Meth-A sarcoma and MM-46 adenocarcinoma. Five compounds were basically administrated intraperitoneally (*i.p.*) nine times at 1-day intervals from days 1 to 9 after the subcutaneous (*s.c.*) or *i.p.* implantation of tumor cells (1×10^5 cells of Meth-A) into 8–11 mice/group (female Balb/c and C3H/He mice) for Meth-A, respectively on day 0. Tumor weight (mg) for tumor weight inhibition (TWI) was estimated by measuring the length (l) and width (w) of each tumor with vernier caliper (mm) and using a formula of $lw^2/2$. PM-8 shows a significant inhibition of Meth-A sarcoma with high values of ILS as an increase in life-span (ILS in %) defined by $100(t - c)/c$ where t and c are mean survival times for PM-treated and untreated (control) groups, respectively. Especially, its *i.p.* administration of 50 mg/kg provides a remarkable prolongation of ILS (=111 %) for *i.p.* implants. It is easy to see the dose schedule in which ILS values for PM-8 are higher than for 5-FU and ACNU. The administration of high dose of 5-FU and ACNU

Table 4.1 Antitumor activity of PM-8 and PM-17 against Meth-A sarcoma and MM-46 adenocarcinoma

Experimental number	Route of tumor implantation	Compound	Dose, *i. p.* × 9 mg/kg^{-1}/day^{-1}	Body weight change (g) on day 14	TWI (%) on day 14	ILS (%)
Meth-A sarcoma						
1	*s.c.*	Control		(+)3.2		
		PM-8	100	(+)2.5	83*****[a]	63****
			50	(+)1.7	38****	32***
		ACNU	50	(+)1.6	57*****	38***
2	*s. c.*	Control		(+)2.5		
		PM-8	200	(+)2.2	64*****	61****
		5-FU	68[b]	(+)0.4	52***	37**
3	*s. c.*	Control		(+)2.6		
		PM-8	250	(+)0.8	44*	69****
		5-FU	20	(−)2.5	80****	19
		ACNU	10	(−)1.1	99****	47****
4	*s.c.*	Control		(+)2.8		
		$[NH_3Pr^i]Cl$	100	(+)1.7	14	19
		$[NH_4]_6[Mo_7O_{24}]\cdot 4H_2O$	100	(+)1.2	31****	33****
		$K_6[Mo_7O_{24}]\cdot 4H_2O$	100	(+)0.2	50****	58****
		PM-8	50	(+)1.7	38****	32****
		PM-17	25	(−)2.2	45****	34****
5	*i.p.*	Control		(+)5.6		
		PM-8	200	(+)4.3		44***
			100	(+)3.9		48***
			50	(+)2.0		111****
		PM-17	100[c] (*i.p.* × 3)	(−)2.2		127****
			50 (*i.p.* × 7)	(−)4.9		216****[d]
MM-46 adenocarcinoma						
6	*s.c.*	Control		(+)2.4		
		PM-8	200	(+)1.8	58****	111****
			100	(+)2.0	80****	167****
			50	(+)1.8	50****	121****

[a]Significantly different from corresponding tumor control group (*$p < 0.05$; **$p < 0.02$; ***$p < 0.01$; ****$p < 0.001$)
[b]5-FU was administrated perorally on days 1, 5, and 9
[c]Two of 11 mice per group died on days 4 and 8 after injection of the compound
[d]Two of 11 mice per group survived over 60 days after tumor implantation and were free from the tumor

leads to toxic deaths of mice, as is indicated by negative changes in body weight. PM-8 is also effective against MM-46 adenocarcinoma (with 5×10^5 cells on day 0) as added in Table 4.1. A dose effect on the inhibition of growth of both Meth-A sarcoma and MM-46 adenocarcinoma is not clear. There is no apparent toxicity of PM-8 in spite of the high dose of 250 mg/kg, as the mice maintain on average their weights throughout 14 days. This excludes the possibility that the tumor growth inhibition is due to a toxic effect on the host. To study the structure-activity relationship of PM-8, the effect of chemical variation against the Meth-A sarcoma is investigated by a use of three different ways (1) $[NH_3Pr^i]^+$ in PM-8 was replaced by $[NH_4]^+$ and K^+, (2) $[Mo_7O_{24}]^{6-}$ was replaced by Cl^-, (3) PM-17, the photoreduction product (with brown color) of PM-8, was used, the anion ($[H_2Mo^V{}_{12}O_{28}(OH)_{12}(Mo^{VI}O_3)_4]^{6-}$) of which was X-ray crystallographically characterized for $[Me_3NH]^+$ salt (Yamase and Ikawa 1977; Yamase and Ishikawa 2008). Figure 4.1a, b shows anion structures of PM-8 and PM-17, respectively. As shown in Table 4.1, $[NH_4][Mo_7O_{24}]\cdot 4H_2O$ and $K_4[Mo_7O_{24}]\cdot 4H_2O$ are effective as well as PM-8, while $[NH_3Pr^i]Cl$ (100 mg/kg) is hardly effective. The administration of PM-17 of 25 mg/kg gives a significant inhibition of the Meth-A sarcoma growth on day 14. However, the consecutive administration of PM-17 induced a negative change in the body weight. Since PM-17 is a multiply photoreduced species originated from PM-8, the toxicity reflected by the negative change in the body weight seems to be associated with the high condensation structure produced as a result of the 12-electron reduction of $[Mo_7O_{24}]^{6-}$. Because of a small size (about 8 Å) of $[Mo_7O_{24}]^{6-}$, PM-8 will enable to be transported across membranes and undergo enzyme-mediated metabolism or bind to plasma proteins or tissue. The saturation at high concentration of PM-8 through these pathways is likely to occur with a resultant nonlinear relationship between dose of PM-8 and its antitumor activity (Table 4.1).

Figure 4.2 shows the effectiveness of PM-8 and PM-17 against the progressive growth of small xenografts of human breast MX-1 neoplasms. The growth of MX-1 human breast cancers (2×2 mm) implanted *s.c.* in athymic nude mice (6-week-old female Balb/c, 4 mice/group) on day 0 proceeds and can be detected on day 17 when tumor size ranges from 350 to 491 mm^3. The tumor volume (V) is determined by measurement of two principal perpendicular diameters (in mm) as $V = (\text{length}) \times (\text{width})^2/2$. The first administration of PM-8 of 200 mg/kg is *i.p.* made on day 17. Ten administrations (once a day from days 17–27, except day 19) of PM-8 give a growth inhibition of 73 % on day 46 without any special risk to the mice as long as the mice are appropriately sterilized by filtration. This means that the size of the breast tumor on day 46 is 27 % of the tumor size (7,466 mm^3) for the control group. The averaged tumor weights measured after killing (dissection) on day 46 for PM-8-treated and untreated tumor mice were 2.54 ± 0.04 and 8.40 ± 0.33 g, respectively. PM-17 provided apparent signs of toxicity (Fig. 4.1): three *i.p.* administrations of PM-17 of 100 mg/kg on days 17, 18, and 20 resulted in a death of one mouse on day 22. Subsequent seven *i.p.* administrations of PM-17 of 25 mg/kg once a day from days 21–27 give toxic deaths of another mouse on day

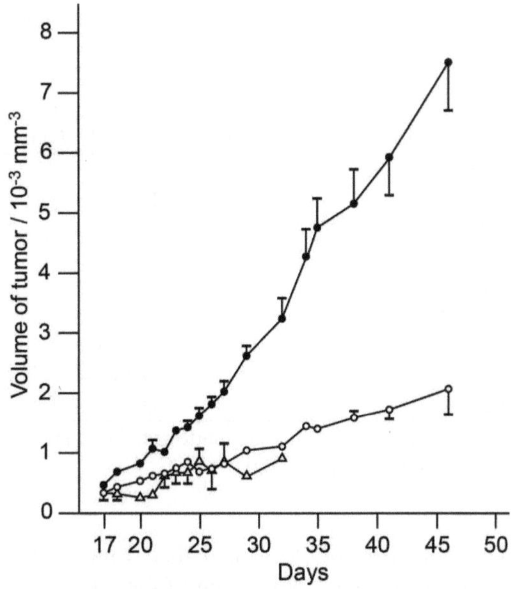

Fig. 4.2 Inhibition of tumor growth by PM-8 and PM-17 after implantation of MX-1. PM-8 (*open circles*) was *i.p.* administrated by 200 mg/kg/day on days 17–27 except 19 and 100 mg/kg/day on days 17, 18, and 20, and PM-17 (*open triangles*) was done by 25 mg/kg/day on days 21–27. Control (*filled circles*) showed no PM treatment. The change in body weights on days 17–32 for PM-17-treated mice was −5.0 g, while there was no significant change on days 17–46 for PM-8-treated mice

29 and others on day 34. However, the therapy by PM-17 exhibits a tumor growth inhibition of 75 % compared with the control group on day 29. Thus, it is possible to say that PM-17 exhibits a cancerocidal potency similar to PM-8 but is strongly toxic, indicating that an optimum of the administration schedule should be taken into consideration, as below exemplified for the nude mice loaded by AsPC-1 (human pancreatic cancer) and MKN-45 (human gastric cancer). The growth suppression by PM-8 was also observed for the nude mice bearing OAT human lung cancer xenograft and CO-4 human colon cancer. The tumor growth suppression of PMs 8 and 17 was also investigated by using the subrenal capsule in kidney (SRC) assay, in which small pieces of CO-4 tumor (approximately 1 mm^3) were implanted into subrenal capsules of ICR 5-week-old female mice under anesthesia on day 0 and PMs were *i.p.* administrated once a day on days 1–4 or 5, thereafter tumor sizes were determined on day 5 or 6. The results for by the SRC assay are shown in Table 4.2, where $Na_5[IMo_6O_{24}]\cdot 34H_2O$ (PM-32), 5-FU, ACNU, and *cis*-diamminedichloroplatinum(II) (CCDP) are also evaluated for comparison (Fujita et al. 1992; Yamase et al. 1992a, b; Yamase 1994, 1996). The antitumor potency of PM-8 against CO-4 human colon cancer is comparable to that of approved drugs and superior to PM32 (with Anderson structure).

The distribution of PM-8 to the organs in the tumor-implanted mice has been investigated by the radioactivation (to ^{101}Mo) analysis of the liver, kidney, brain, plasma, and tumor at 0.5 h after *i.p.* administration of PM-8 into the MM-46-implanted C3H/HeNCrj mice (Yamase et al. 1992a, b). The tumor mice are prepared by the implantation of 1 × 10^6 cells of MM-46/mouse on day 0 and PM-8 of 100 mg/kg was administrated *i.p.* on day 14 when the tumor growth was detectable. Amounts of the Mo atoms uptaken into organs at 0.5 h after *i.p.*

Table 4.2 Antitumor activity of PM-8 and PM-17 against CO-4 human colon cancer by using SRC assay for *i.p.* administration of compounds

Compound	Dose (mg/kg^{-1})	Administration schedule (*i.p.*)	Ratio of body weight change (day 6/day 0)	Antitumor activity (T/C, % on day 7)
Control			1.03 ± 0.01	
PM-8	200	Days 1–5	1.07 ± 0.01	62.2**[a]
	100	Days 1–5	1.03 ± 0.02	67.6***
	50	Days 1–5	1.03 ± 0.01	45.6**
PM-17	25	Days 1–5	0.70 ± 0.05	42.4**
$Na_5[IMo_6O_{24}] \cdot 34H_2O$	100	Days 1–5	1.03 ± 0.01	70.3***
5-FU	30	Days 1–4	0.99 ± 0.01	56.0****
ACNU	10	Days 1–4	0.94 ± 0.08	62.4****
CDDP	2	Days 1–4	0.97 ± 0.02	54.4****

[a]Significantly different from corresponding tumor control group (*$p < 0.05$; **$p < 0.02$; ***$p < 0.01$; ****$p < 0.001$)

administration of PM-8 and their time profile are shown in Figs. 4.3 and 4.4, respectively. The comparison of the amounts of Mo atoms between PM-8-treated and untreated mice indicates that PM-8 is preferentially distributed to the kidney and tumor but hardly to the brain or liver (Fig. 4.3). The time profile of the amount of Mo atoms in the kidney and tumor indicated a maximum within 1 h after *i.p.* administration and the excretion by urine during 6 h (Fig. 4.4). A high trapping of Mo atoms in kidney lets us presage a possible damage of the kidney. However, this unpleasant effect may be minimized by maintaining a rapid urine output after the administration through high solubility of PM-8 (1 g/ml at maximum) in water. The antitumor activity of more toxic PM-17 (Tables 4.1 and 4.2 and Fig. 4.2) compared with PM-8 gives us a clue to the mechanism of the antitumor activity of $[Mo_7O_{24}]^{6-}$, provided that PM-17 is generated as one of the metabolites of PM-8 in the biological system as well as the photolysis (Yamase and Ishikawa 2008), as denoted by Eq. (4.1).

$$[Mo_7O_{24}]^{6-} + e^- + H^+ \rightarrow \left[Mo^V Mo_6O_{23}(OH)\right]^{6-} \rightarrow\rightarrow PM - 17 \tag{4.1}$$

The biological formation of PM-17 is based on the hypothesis that the redox potential of electron-donor sites within tumor cells is negative enough to reduce $[Mo_7O_{24}]^{6-}$ to PM-17. The high toxicity of PM-17 seems to be due to the multi-electron redox reaction between PM-17 and host cells in addition to the molecular recognition arising from its T_d-symmetric condensed anion structure (Fig. 4.1b). The variety of cations in the $[Mo_7O_{24}]^{6-}$ system will modify the residence time of $[Mo_7O_{24}]^{6-}$ within the tumor cells (leading to a high amount of Mo atoms) as well as the solubility (or stability) under the physiological condition. The agarose gel electrophoretic patterns for the restriction endonuclease digestion of the pBR322 plasmid DNA treated with PM-8 showed the noncovalent binding of PM-8 with DNA (Tomita et al. 1989), which was different from the case of CDDP showing the

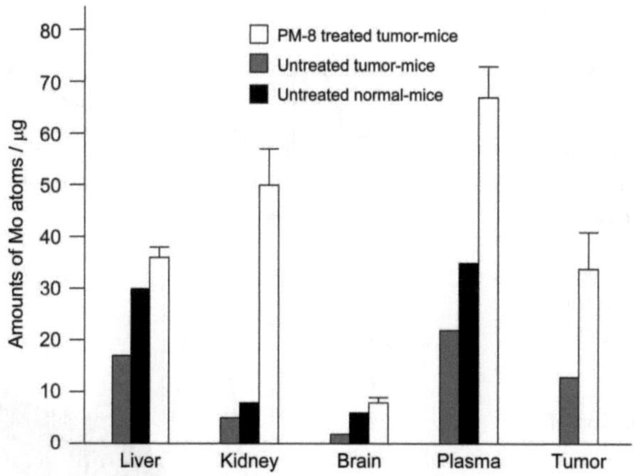

Fig. 4.3 Contents of Mo atoms in the liver, kidney, brain, plasma, and tumor for the tumor mice at 0.5 h after *i.p.* administration of PM-8, which were determined by the radioactivation (to ^{101}Mo) analysis. The contents for PM-8 untreated mice (both tumor mice and normal mice) as backgrounds are also indicated for comparison

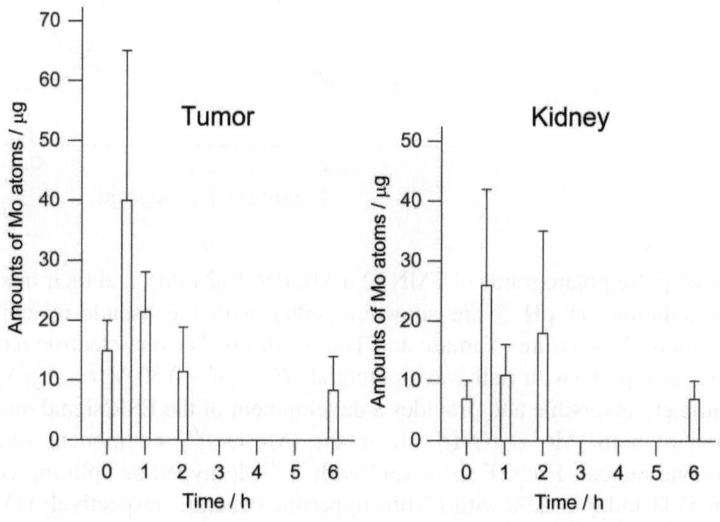

Fig. 4.4 Time profile of the Mo contents in the kidney and tumor for the tumor mice after *i.p.* administration of PM-8

covalent attachment to N-7 atoms of DNA guanine bases through the DNA interstrand cross-linking (Pinto and Lippard 1985). This might support low toxicity of PM-8. It is known that the reduction of $[Mo_7O_{24}]^{6-}$ is biologically possible through the formation of 1:1 complex with flavin mononucleotide (FMN) with the formation constant (K_f) of 8.9 × 10^3 per molar at pH 5, as shown in Fig. 4.5 where

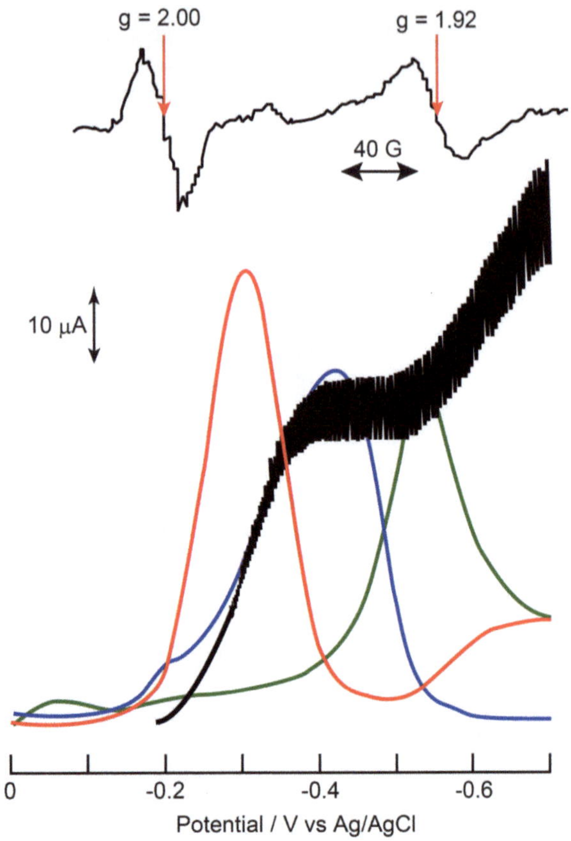

Fig. 4.5 Differential pulse polarograms of FMN (2 mM), PM-8 (2 mM), and their mixture in aqueous solutions (at pH 5) containing 0.1 M $NaClO_4$, which are depicted by *blue*, *green*, and *red curves*, respectively. DC polarogram of the mixture is shown by *black line* for comparison. *Top figure* shows ESR spectrum observed by the one-electron reduction of the mixture

differential pulse polarograms of FMN (2 mM), PM-8 (2 mM), and their mixture in aqueous solutions at pH 5 are shown together with the pseudo-reversible dc polarogram of the mixture (Yamase and Tomita 1990). The one-electron reduction of the 1:1 complex (with half-wave potential ($E_{1/2}$) of −0.30 V vs. Ag/AgCl) is approximately reversible and provides a development of the ESR signal due to the formation of both $[Mo^{V}O_5(OH)]$ site in the $[Mo_7O_{24}]^{6-}$ component and FMN semiquinone radical, FMNH$^{\bullet}$ ($g = 1.92$ with ^{95,97}Mo hyperfine splitting constant of about 57 G and $g = 2.00$ with 12-line hyperfine structure, respectively) (Yamase 1982; Yamase and Tomita 1990). The analysis of the dc polarograms indicates that the differential pulse polarographic peak of FMN at −0.42 V is shifted to more positive potential at −0.30 V in the presence of PM-8 which gives peak potential at −0.53 V, implying the catalytic reduction of PM-8 through the 1:1 complex with FMN. Since FMN is a prosthetic group in a flavoprotein and acts as an electron carrier for the electron transfer (from NADH to coenzyme Q) coupled with the ATP generation at the site 1 (Stryer 1975), it seems to be reasonable to assume that multi-electron reduction of $[Mo_7O_{24}]^{6-}$ to PM-17 starts biologically through the one-electron reduction of the 1:1 $[Mo_7O_{24}]^{6-}$-FMN complex formed on the tumor cell mitochondria with a resultant suppression of the tumor growth due to the inhibition of the ATP generation. The preferential distribution of PM-8 to the tumor (Fig. 4.3) supports the proposed mechanism. Also, the interaction between $[Mo_7O_{24}]^{6-}$ and ATP has been investigated using ^1H and ^{31}P NMR spectra and has shown that $[Mo_7O_{24}]^{6-}$ is a catalysis of the ATP hydrolysis to ADP and phosphate (Ishikawa and Yamase 2006). This implies that the therapy of PM-8 contributes to the elongation of the survival rate, since the ATP hydrolysis is the main source of energy for most biological processes.

The crystals of PM-17 isolated from the photolyte of PM-8 at pH 5–7 were X-ray crystallographically characterized, the chemical formula of which is $[Me_3NH]_6[H_2Mo^V{}_{12}O_{28}(OH)_{12}(Mo^{VI}O_3)_4]\cdot 2H_2O$. The structural insight of PM-17 in aqueous solutions has been investigated by using ^{95}Mo NMR, electronic absorption, IR, and electrospray ionization mass (ESI-MS) spectrometries (Yamase and Ishikawa 2008). In the anion of PM-17 (Fig. 4.1b), the ε-Keggin $Mo^V{}_{12}$ core (showing 12 edge-shared Mo^VO_6 octahedra in the arrangement of 6 diamagnetic $Mo^V{}_2$ pairs with a Mo^V–Mo^V bond distance of 2.6 Å) is capped by four $Mo^{VI}O_3$ units and remains intact at pH 5–9. A partial detachment of the $Mo^{VI}O_3$ unit from PM-17 occurs at high pH level with an accompanying increase in the electronic transition of the $Mo^V{}_2$ pairs. PM-17 was tested for the therapy of more aggressive cancer cells with poor outcome, in order to see the perspective for a novel inorganic drug based on PMs. Expectedly PM-17 showed the cancerocidal potency against AsPC-1 human pancreatic and MKN-45 human gastric cancer cells, with more effective potency than PM-8 (Ogata et al. 2005, 2008; Mitsui et al. 2006). In vivo cell killing by PM-17 for both AsPC-1 and MKN-45 cells is based on the apoptosis in parallel with the autophagy (Ogata et al. 2008). When the cells were suspended in RPMI-1640 culture medium (GIBCO Carlbad, CA. USA) containing PM-17, IC_{50} values of PM-17 against AsPC-1 cells and MKN-45 cells were 175 μg/ml (63 μM) and 40 μg/ml (14 μM), respectively, which are much smaller than IC_{50} values of PM-8, 1.65 mg/ml (1.12 mM) and 0.90 mg/ml (0.62 mM). Figure 4.6 shows survival rates of AsPC-1(a) and MKN-45(b) cells (1.5×10^5 cells/5ml) treated with PM-17 for 24 h. In vivo potentiality of PM-17 against the proliferation of AsPC-1 and MKN45 cells was also recognized. The results for the AsPC-1implanted mice are exemplified in Fig. 4.7, where 2×10^6 AsPC-1 cells were transplanted into the back of nude mice, and after 10 days, intratumoral injections of PM-17 (125 μg or 500 μg dissolved in 100 μl of saline) were performed for 10 days with 2-day intermission on day 6. The control mice were treated with 100 μl of saline per day under the same conditions. PM-17 inhibited tumor growth: on 41 days after implantation of the tumor, for example, it showed rates of 33.5 and 68.5 % at the doses of 125 μg and 500 μg/body/day compared with controls, respectively (Fig. 4.7a). The tumors observed at 41 day for five mice/group administrated by PM-17 500 μg/body/day are pictured (at the right side) in comparison with those of the control (at the left side), which were measured by a micrometer caliper (Fig. 4.7b). There was no loss of body weight for mice injected with PM-17 in this experiment, implying that the cytotoxicity of PM-17 is not significant as far as the present therapy of 500 μg/body/day at least was employed (Fig. 4.7c). Little

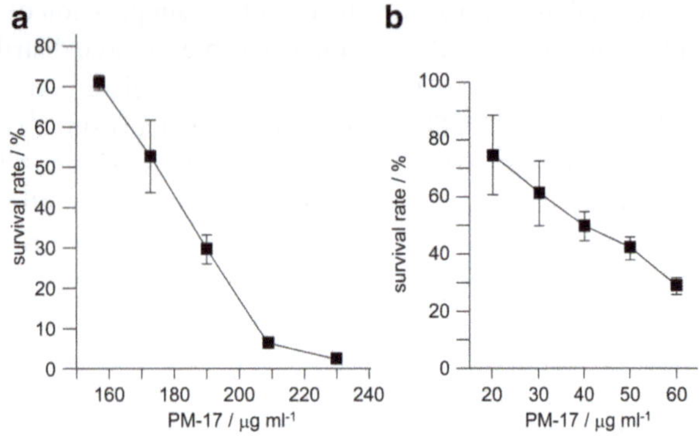

Fig. 4.6 Effect of PM-17 on in vitro survival of AsPC-1 (**a**) and MKN-45 (**b**) cells (1.5 × 10^5 cells/5 ml for each cell) for 24 h. Each point indicates the percentage of living cell in three independent experiments (bars depict s.d.)

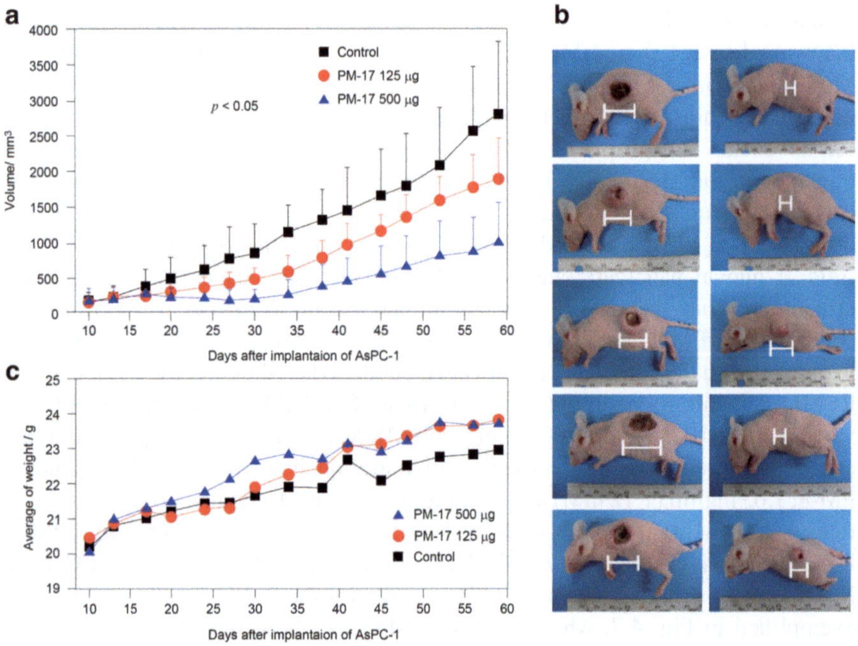

Fig. 4.7 In vivo tumor growth inhibition by PM-17 for AsPC-1 implanted mice. 2 × 10^6 AsPC-1 cells were transplanted into the back of Balb/c nude mice, and after 10 days, intratumoral injections of PM-17 (125 μg or 500 μg dissolved in 100 μl of saline) were performed for 10 days with 2-day intermission on day 6. The control mice were treated with 100 μl of saline per day under the same conditions. Changes in tumor volume (**a**) and body weight (**c**) for five mice per group are depicted with photographs (**b**) for the mice on day 41after the tumor implantation

cytotoxicity of PM-17 through the therapy was supported also by the hematological examination and blood chemistry assay for the mice treated with PM-17 (of 1.0 mg/body/day) which indicated no damage of kidney and hepatic functions as described below. Thus, PM-17 exhibits the antitumor activity which inhibits the proliferation of AsPC-1 human pancreatic cancer cells in a dose-dependent manner in vivo. The thin sections of tumors on day 10 after a single intratumoral injection of PM-17

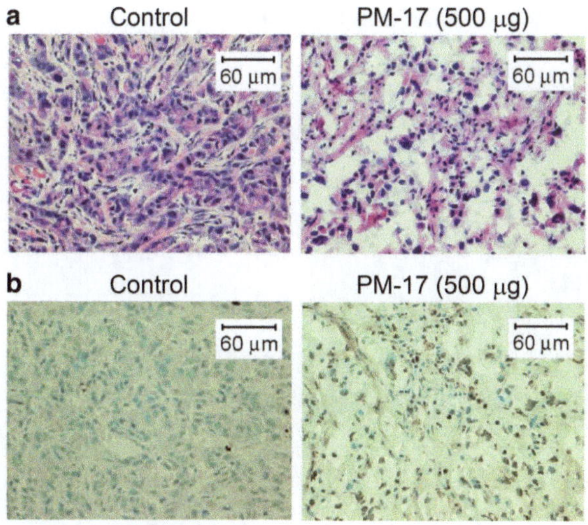

Fig. 4.8 Hematoxylin and eosin (H&E) (**a**) and terminal deoxynucleotidyl transferase-mediated "nick-end" labeling (TUNEL) (**b**) stainings for thin sections of the tumor on the 10th day after a single intratumoral injection of PM-17 (500 μg/100 μl saline) into the AsPC-1 transplanted nude mice, in comparison with the control (with the injection of 100 μl 0.9 % NaCl saline)

(500 μg/100 μl saline) or PM-8 (4 mg/100 μl saline) into the AsPC-1 transplanted nude mice showed a pathological hyalinization on both hematoxylin and eosin (H&E) staining and terminal deoxynucleotidyl transferase-mediated "nick-end" labeling (TUNEL) staining. Figure 4.8 shows the results of H&E (a) and TUNEL (b) for PM-17 in comparison with the control (with the injection of 100 μl 0.9 % NaCl saline). The genotoxicity of PM-17 and PM-8 for the AsPC-1 and MKN-45 cells during the growth indicated the formation of apoptotic small bodies for the cells due to the DNA fragmentation which was revealed by Hoechst dye 33342 (Sigma) staining of the suspension containing the cells treated with PM-17 and PM-8 (with IC_{50}) for 24 h (Ogata et al. 2005, 2008; Mitsui et al. 2006). The morphological analysis by Hoechst staining indicated a sharp DNA laddering after the PM treatment on the agarose gel electrophoresis of cell DNA extract, which allowed us to confirm the induction of apoptosis by PM-17 and PM-8.

Transmission electron microscope (TEM) was employed for the analysis of ultrastructural alterations in AsPC-1 and MKN-45 cells treated with PM-17 (with IC_{50}) for 12 h. Figure 4.9 shows the TEM images of the AsPC-1 cells treated with and without PM-17 (Ogata et al. 2008). The average dimensions of control AsPC-1 cells ranged from 13 to 15 μm in diameter, with a nucleus ranging from 7 to 10 μm. In the PM-17 untreated AsPC-1 cells (control), a number of round-shaped mitochondria make the predominant cytoplasmic feature observed along with the external microvilli (Fig. 4.9a, b). In comparison with the control, the morphological alterations of the cells treated with PM-17 were distinguished by the occurrence of both chromatin/chromosome condensation and damage of mitochondria (Clarke 1990). The former in nucleus leads to the apoptotic cell death (Fig. 4.9c), and the

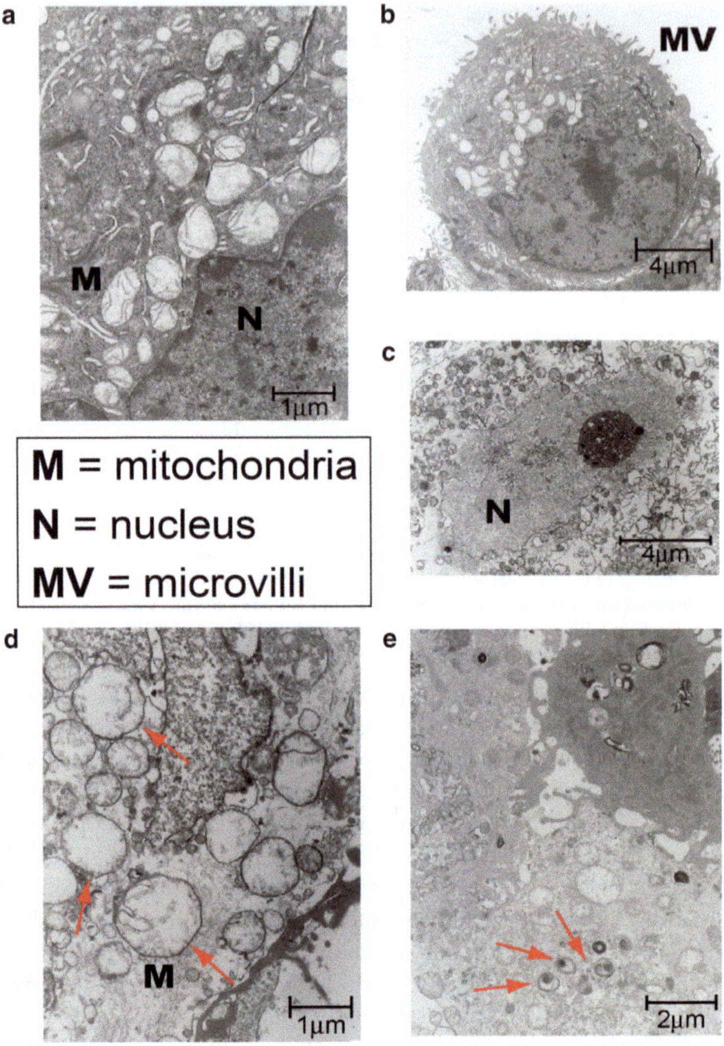

Fig. 4.9 TEM images of the AsPC-1 cells treated with (**c**–**e**) and without (**a** and **b**) PM-17

latter in the cytoplasm indicates the formation of smaller round organelles which results from damage throughout swelling of mitochondria (Fig. 4.9d). Some of PM-17-treated AsPC-1 cells contained a number of autophagosomes/autolysosomes in the cytoplasm (Fig. 4.9e), which are a characteristic feature of autophagy (Ogata et al. 2008). No significant decrease in the number of microvilli on the surfaces of the treated cells suggests little alteration in membrane permeability on the exposure of the cells to PM-17. To visualize the autophagy, the green fluorescent protein (GFP)-tagged light chain3 (LC3) was expressed in the AsPC-1 cells: during the autophagic process LC3 is concentrated in autophagosomes, and thereby the punctate fluorescence emitted by GFP-LC3 is a good indicator of

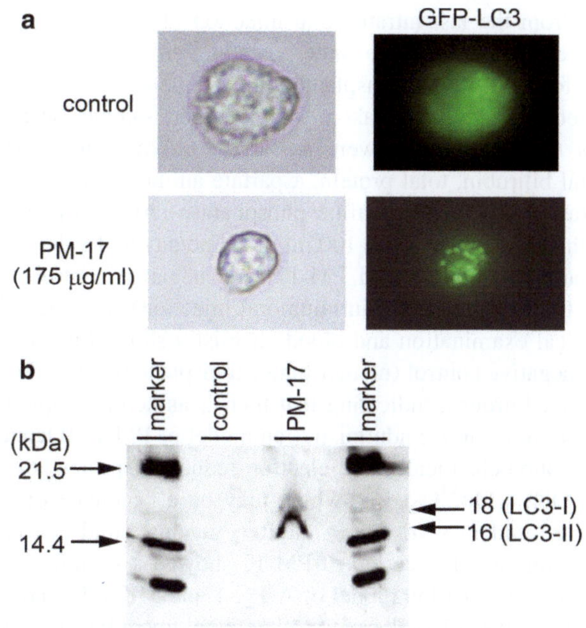

Fig. 4.10 In vitro expression of autophagy by the green fluorescent protein (GFP)-tagged light-chain 3 (LC3) in the PM-17-treated AsPC-1 cells. 2×10^5 AsPC-1 cells were cultured with the BRMI-1640 growth medium for 24 h in 24-wells plate, transfected with GFP-LC3 expression plasmid (0.8 μg/well) with lipofectamine 2000 transfection reagent (2 μg/well). After 24 h the cells were treated with and without PM-17 (175 μg/ml saline) for 24 h, and the distribution (**a**) of fluorescence of GFP-LC3 under fluorescence microscope and a Western blotting picture (**b**) were investigated

autophagy (Takeuchi et al. 2005). The result is shown in Fig. 4.10 (Ogata et al. 2008). While the diffuse cytoplasmic localization of GFP-LC3 was observed for the control, the PM-17 (with IC_{50} and for 24 h)-treated AsPC-1 cells transfected by GFP-LC3 expression plasmid showed the punctate fluorescence (Fig. 4.10a). The expression of LC3 (I and II) proteins for the PM-17-treated AsPC-1 cells indicated the autophagy in contrast to the control showing no expression of LC3 on a Western blotting picture (Fig. 4.10b). Similar result was also obtained for the PM-17-treated MKN-45 cells (Ogata et al. 2008). Since LC3-II, as a truncated form of LC3, is necessary to bind with the seclusion membrane at the primary stage of the autophagy (Takeuchi et al. 2005), the above results reveal that PMs induce the activation of not only apoptosis but also autophagy as cell death pathways. The determination factor of the magnitude of both apoptosis and autophagy in the PM-17-induced cell death remains still unclear, although the activation of caspase-3 and caspase-12 as a regulatory factor could be identified by using a Caspase-Glo™ 3/7 assay kit (Lamparska-Przybysz et al. 2005; Golstein and Kroemer 2007).

After 24 h from the last intratumoral injection of PM-17 (500 μg or 1.0 mg/100 μl saline per day) through the same procedure employed in the tumor growth inhibition test for the AsPC-1 transplanted mice (Fig. 4.7), the mice were anesthetized and blood samples were taken. Whole blood samples were used for the hematological test. Blood sera were assessed for blood urea nitrogen (BUN), creatinine, total bilirubin, total protein, aspartate aminotransferase (AST), alanine aminotransferase (ALT), and alkaline phosphatase (ALP). Table 4.3 shows the results for both control mice (with 100 ml of saline/day under the same conditions used for the animals treated with PM-17) and negative control mice (received neither tumor transplantation nor intratumoral injection) are added. The results of the hematological examination and blood chemistry showed no significant difference between negative control (neither tumor transplantation nor PM-17 injection) and PM-17-treated groups, indicating that there was neither impaired kidney nor hepatic function in the above administration model of PM-17 (Ogata et al. 2008).

Thus, the photochemically 12-electron-reduced species of $[Mo_7O_{24}]^{6-}$, $[H_2Mo^{V}{}_{12}O_{28}(OH)_{12}(Mo^{VI}O_3)_4]^{6-}$, which may be a candidate of metabolites of PM-8, is in vitro and in vivo more inhibitory against solid cancer growth than $[Mo_7O_{24}]^{6-}$. Intratumoral injection of PM-17 showed a significant efficacy in an experimental mice inoculation model of AsPC-1 and MKN-45, indicating a therapeutic usefulness of PM-17 as alternative chemotherapy of PM-8. Together with the fact that the pancreatic cancer is an aggressive form of cancer and has one of the poorest outcomes of all cancers, the development of the drug-delivery system for PM-17 is required for the realization of PM-17 as a novel anticancer drug.

4.3 Antiviral Activity of Polyoxotungstates

In the course of in vitro antiviral assay of a variety of PMs, especially polyoxotungstates, PM-19 (the anion of which is shown in Fig. 4.1c) exhibited a potent inhibition both in vitro and in vivo against a broad spectrum of DNA viruses including herpes simplex virus (HSV) type 1 (HSV-1) and HSV type 2 (HSV-2), thymidine kinase-deficient (TK^-) HSV mutant, and human cytomegalovirus (HCMV) (Fukuma et al. 1991; Dan et al. 2003). PM-19 inhibited the viral penetration of the host cells and the second round of infection without its direct interaction with HSV virions (Dan et al. 1998, 2003). PM-19-pretreated cells provided approximately 10-fold enhancement of the anti-HSV potency of PM-19 compared with the cells treated with PM-19 only after infection (Dan et al. 2002). Together with the fact that PM-19 inhibited the interaction between HSV envelope protein (gD, glycosylated ectodomain) and cell-surface membrane proteins (HVEM) (Dan and Yamase 2006), the pretreatment of the cell with PM-19 suggests a strong binding of PM-19 with HVEM. Dose response of PM-19 to its in vivo activity for the mice (12–24/group) infected *i.p.* with 2.8×10^2 pfu of HSV-2 is shown in Table 4.4, where PM-19 at different dose was *i.p.* administrated to the HSV-2-infected mice once daily for 3 days by starting immediately after virus

Table 4.3 Toxicity test for PM-17 in mice by serum chemistry

Group	Total protein (g/dl^{-1})	AST (U/l^{-1})	ALT (U/l^{-1})	ALP (U/l^{-1})	Total bilirubin (mg/dl^{-1})	BUN (mg/dl^{-1})	Creatine (mg/dl^{-1})
Control	5.33 ± 0.68	653 ± 627	144 ± 133	288 ± 28	0.02 ± 0.01	23.0 ± 3.6	0.12 ± 0.02
PM-17							
500 μg	4.94 ± 0.21	236 ± 162	117 ± 117	263 ± 21	0.02 ± 0.02	19.4 ± 2.6	0.10 ± 0.02
1,000 μg	4.97 ± 0.23	301 ± 93	96 ± 66	250 ± 33	0.01 ± 0.02	23.9 ± 1.0	0.13 ± 0.02
Negative							
Control	4.76 ± 0.08	337 ± 114	271 ± 156	357 ± 40	0.00 ± 0.00	29.6 ± 0.4	0.13 ± 0.02

Data indicate mean ± s.d. per group

Table 4.4 PM-19 dose response for the mice infected with HSV-2

Compound	Dose ($mg/kg^{-1}/kg^{-1}$)	Number of survivors/total	Survival days
Experiment 1			
No treatment		0/24 (0)[a]	6.83 ± 0.46 (0)[a]
PM-19	0.03 (on days 0–2)	0/12 (0)	6.83 ± 0.42 (0)
	0.1 (on days 0–2)	2/12 (17)	8.00 ± 1.03 (17)
	0.3 (on days 0–2)	7/12 (58)**[b]	11.50 ± 0.94 (68)**[b]
	1.0 (on days 0–2)	9/12 (75)**	12.08 ± 1.02 (77)**
ACV	25.0 (on days 0–5)	2/12 (17)	7.64 ± 1.22 (12)
	50.0 (on days 0–5)	2/12 (17)	9.08 ± 0.84 (33)*
Experiment 2			
No treatment		2/24 (8)	7.13 ± 0.47 (0)
PM-19	1.0 (on days 0–2)	9/12 (75)**	13.08 ± 0.48 (84)**
	2.5 (on days 0–2)	8/12 (67)**	11.50 ± 1.07 (61)**
	5.0 (on days 0–2)	7/12 (58)**	11.40 ± 1.02 (60)**
	10.0 (on days 0–2)	9/12 (75)**	12.50 ± 0.87 (75)**
	25.0 (on days 0–2)	11/12 (92)**	14.00 ± 0.00 (97)**
	50.0 (on days 0–2)	9/12 (75)**	12.33 ± 0.92 (73)**
ACV	25.0 (on days 0–2)	0/12 (0)	7.92 ± 0.78 (11)
	50.0 (on days 0–2)	0/12 (0)	7.33 ± 0.38 (3)
	100.0 (on days 0–2)	1/12 (8)	3.85 ± 1.54 (−50)

Mice were treated *i.p.* with indicated doses of the test compounds once daily from day 0 (immediately after infection) until day 2 or 5 after *i.p.* infection with HSV-2 (2.8×10^2 pfu)
[a]Number in parenthesis indicates the percentage of mice that were still alive on day 14 after infection
[b]Significantly different from corresponding tumor control group ($*p < 0.05$; $**p < 0.01$)

challenge. As shown in Table 4.4, a significant protection was achieved with doses of PM-19 at the range of 0.3–50 mg/kg. Acyclovir (ACV) as an approved drug (viral DNA-polymerase inhibition by ACV-triphosphate formed in the cell) against HSV (Elion et al. 1977), tested in parallel with PM-19, showed only weak anti-HSV activity even if administrated at doses of 100 mg/kg/day for 6 days. 50 % effective dose (ED_{50}) values on the survival rates (on day 14 after infection) were 0.25 mg/kg/day for PM-19 and >100 mg/kg/day for ACV. 50 % lethal dose (LD_{50}) values of PM-19 and ACV on *i.p.* administration were 384 and 739 mg/kg/day, respectively. The mechanism of in vivo anti-HSV activity of PM-19 is attributed to an interference with the virus uptake by the cells and/or a clearance of HSV and also partially to the activation of macrophages (Ikeda et al. 1994). The latter effect was recognized by measuring the phagocytic activity of peritoneal macrophages, which was increased by *i.p.* administration of PM-19. The marked in vivo efficacy of PM-19 due to such multiple mechanisms involving a direct inhibitory effect on virus replication (inhibition of virus adsorption) as well as host-mediated antiviral response (activation of macrophage activity) was also observed for the HSV-2-infected immunosuppressed mice (Ikeda et al. 1993).

CC_{50} and EC_{50} (median cytotoxic and effective concentrations, respectively) values for seven PMs including four V/W-mixed PMs (PM-43, PM-47, PM-1001,

Table 4.5 Cytotoxicity of PMs for cell culture

		CC_{50} (μM)[a]			
PM	Structural type	MDCK	Vero	HEp-2	MT-4
PM-43	Si/W Keggin	>200	>200	>200	45.6 ± 0.43
PM-47	V/W Keggin	>200	>200	>200	47.4 ± 0.48
PM-518	Ti/W Keggin	>200	>200	>200	>100
PM-520	Ti/W Keggin	>200	>200	>200	>100
PM-523	Ti/W Keggin	>400	>400	>400	>100
PM-1001	V/W Sandwich	229.2 ± 36	362.7 ± 38.1	233.6 ± 21.5	41.9 ± 2.9
PM-1002	V/W Sandwich	>200	>200	>200	45.9 ± 0.3

[a]Average values for four independent experiments. CC_{50} values were determined by MTT assay method

Table 4.6 EC_{50} (in μM unit) of PMs for several RNA viruses in vitro

	EC_{50} (μM)[a]					
Compound	DFV (Vero)	FluV (MDCK)	RSV (HEp-2)	PfluV-2 (HMV-2)	CDV(Vero)	HIV-1 (MT-4)
PM-43	10.7 ± 6.7	8.4 ± 6.5	1.6[b]	>100	7.5 ± 0.95	0.3 ± 0.12
PM-47	10.5 ± 6.9	11.5 ± 0.6	29.0[b]	67.1[b]	6.0 ± 0.6	0.03 ± 0.01
PM-518	36.8 ± 3.0	62.3 ± 26.5	26.5[b]	53.2 ± 39.2	>50	2.0 ± 0.8
PM-520	11.7 ± 7.1	45.2 ± 25.8	0.74 ± 0.58	23.2 ± 2.4	7.4 ± 0.4	2.0 ± 0.5
PM-523	>61.5	5.6 ± 2.0	1.3 ± 0.46	2.5 ± 1.0	7.3 ± 1.1	0.3 ± 0.07
PM-1001	0.45[b]	1.75 ± 1.6	<0.16	1.1 ± 0.9	5.7 ± 0.5	0.14 ± 0.17
PM-1002	1.95 ± 1.4	4.6 ± 1.7	0.75 ± 0.05	0.75 ± 0.05	2.8 ± 1.0	0.03 ± 0.01
Ribavirin	>100	5.0 ± 2.75	3.9 ± 3.1	14.0 ± 4.8	73.6 ± 34.5	ND[c]

[a]EC_{50} values for the virus-infected cells (indicated in parenthesis) were determined by the MTT method, but for CDV by the plaque reduction method
[b]Average of two experiments
[c]Not determined

and PM-1002) and three Ti/W-mixed PMs (PM-518, PM-520, and PM-523) against RNA viruses are listed in Tables 4.5 and 4.6, respectively (Shigeta et al. 2003). In Table 4.5, CC_{50} values of PMs for Madin-Darby canine kidney (MDCK), Vero, HEp-2, and MT-4 cells were determined by using the 3-(4,5-dimethylthiazol-2-yl)-2,5-diphenyltetrazolium bromide (MTT) colorimetry for the cells cultured in the presence of PMs (with a variety of concentrations) at 35 °C during 5 days (Pauwels et al. 1988; Watanabe et al. 1994; Inouye et al. 1992, 1995; Mori and Shigeta 1995). All V/W- and Ti/W-mixed Keggin or Wells–Dawson PMs (Fig. 4.1c–f) show little toxicity for MDCK, Vero, and HEp-2 cells in CC_{50} values more than 200 μM, while they are slightly toxic for MT-4 cell: Ti/W-mixed PMs show $IC_{50} > 100$ μM and V/W-mixed PMs show $CC_{50} = 41.9$–47.4 μM. The listed PMs were inhibitory to the replication of HIV-1(IIIb) in EC_{50} values at the range of 0.03–2.0 μM and were selectively inhibitory against other RNA viruses such as dengue fever virus (DFV), influenza virus A (FluV-A), respiratory syncytial virus (RSV), parainfluenza virus type 2 (PfluV-2), and canine distemper virus (CDV) with more inhibitive values than ribavirin as an inhibitor of the viral RNA synthesis due to its inhibition of inosine monophosphate dehydrogenase activity in cells (Gilbert and Knight 1986): PM-43, PM-518, and PM-523 were not inhibitory against PfluV-2, CDV, and DFV, respectively, as indicated by EC_{50} values more than 50 μM. PM-1001 and PM-1002 with the same anion structure (Fig. 4.1e), in which the former anion is the one-electron oxidation species of the latter anion consisting of $(V^{IV}O)_3$ triangle sandwiched by two α-B $[SbW_9O_{33}]^{9-}$ ligands, were significantly potent inhibitors of HIV-1. These PMs were also inhibitory against DFV (which is causative of dengue fever recently migrated from the tropical zone to the mild zone) and RSV (which causes bronchitis and/or pneumonia through a respiratory infection among infants) in EC_{50} values of 0.45–1.95 μM and 0.16–0.75 μM, respectively. DFV and RSV have been known as clinically important pathogens. A difference in the antiviral activity of PMs among a variety of the RNA viruses may be associated with the dependence on the composition of the target amino acid sequence of the viral envelope glycoproteins, as indicated for the different inhibition of dextran sulfate compounds (with a variety of molecular weight) (Hosoya et al. 1991).

$K_{13}[Ce(SiW_{11}O_{39})_2]\cdot 26H_2O$ (JM1590) and $K_6[BGa(H_2O)W_{11}O_{39}]\cdot 15H_2O$ (JM2766), which inhibited HIV-1 and simian immunodeficiency viruses at concentrations as low as 0.008–0.8 μM, seem to be the most potent anti-HIV compounds (Yamamoto et al. 1992). They also inhibited the formation of giant cells (syncytium) in co-cultures of HIV-infected HUT-78 cells and uninfected Molt-4 cells. The anti-HIV activity for most of PMs could be based on inhibition of virus-to-cell binding (due to a high affinity for gp120 as the viral glycoprotein involved in virus adsorption to the cells), since there was a good correlation between the inhibitory effects on both HIV-1-induced cytopathicity and syncytium formation and also a close correlation between their inhibitory effects on both the syncytium formation and the interaction with the viral envelope glycoprotein gp-120. Selective indices, SI ($=CC_{50}/EC_{50}$), of the V/W-mixed Keggin PMs (especially PM-1001 and PM-1002) against HIV-1 in MT-4 cell line were higher than those of AZT (as the inhibitor for the reverse transcriptase) (Mitsuya et al. 1985) and dextran sulfate in mol. wt. 5,000 (DS5000) (as the inhibitor for the syncytium formed by the cell-to-cell infection) (Mitsuya et al. 1988, Baba et al. 1988, Callahan et al. 1991). Table 4.7 shows the result of the multinuclear activation of the galactosidase indicator (MAGI) assay for HeLa CD4/LTR-β-Gal cell line (Kimpton and Emerman 1992), which indicates SI values (>10,000 and >5,500 for PM-1001 and PM-1002, respectively) are higher than for DS5000 (>3,000) and AZT (>2,700), strongly supports their high antiviral activity against HIV-1. The fact that the cocultivation of Molt-4 cells and HIV-1-infected Molt-4/III_b cells with PM-1001 (in 10 μM) and PM-1002 (in 2 μM) provides a perfect inhibition of the syncytium formation (the multinuclear giant cell formation) assists the plausible binding of PMs with HIV-1 gp120, in contrast to the binding of AMD3100, a bicyclam derivative of octahydrochloride dihydrate of 1,1′-[1,4-phenylenebis(methylene)]-bis-1,4,8,11-tetraazacyclotetradecane, to X4 receptor (CD4 and/or CXCR4) (Donzella et al. 1998). This lets us predict clinically synergistic anti-HIV activity by a use of a combination of PMs with AMD3100.

Table 4.7 Anti-HIV activity of PM-1001 and PM-1002

Cell[a]	Compound	EC_{50} (μM)[b]	EC_{90} (μM)[b]	CC_{50} (μM)[b]	SI[c]
MT-4	PM-1001	0.14 ± 0.07	ND[d]	41.9 ± 2.9	304
	PM-1002	0.03 ± 0.01	ND	45.9 ± 0.3	1,530
	DS5000	0.65 ± 0.29	ND	>20	>30.8
	AMD3100	ND	ND	ND	ND
	AZT	0.032	ND	29.4 ± 8.4	919
HeLa	PM-1001	0.00965 ± 0.0056	0.11 ± 0.06	>100	>10,417
CD4/LTR-β-Gal	PM-1002	0.018 ± 0.007	0.11 ± 0.05	>100	>5,556
	DS5000	0.006 ± 0.0018	0.13 ± 0.055	>20	>3,333
	AMD3100	0.0003 ± 0.0001	0.0036 ± 0.0012	>100	>333,000
	AZT	0.037 ± 0.023	0.42 ± 0.33	>100	2,703

[a]The results were determined using the MTT colorimetric method for MT-4 cells and the MAGI assay for MT-4 cells HeLa CD4/LTR-β-Gal cells
[b]Average values for three independent experiments
[c]Selectivity index = (CC_{50}/EC_{50})
[d]Not determined

In vivo anti-HIV activity was carried out by using recombinant vaccinia viruses (rVV), vPE16 expressing HIV-1 *env* gene (vPE16-env), vP1206 expressing HIV-IIIb *gag*/*pol* gene (vP1206-gag/pol), and wild WR strain (WR, control rVV). After Balb/c mice (8-week-old, female, five mice/group) were infected with 5.0×10^6 pfu of vPE16-env and vP1206-gag/pol on day 0, 12.5 or 125 mg/kg/day of each of PM-523, PM-1002, and $K_{12}[(VO)_3(BiW_9O_{33})_2]\cdot 27H_2O$ (PM-1208, Botar et al. 2001) was *i.p.* once daily administrated for 3 days from day 2. On day 5 after infection, ovaries of mice were removed, homogenized, and the virus activity was evaluated by the β-galactosidase activity in the supernatants. In vivo results of the cytotoxicity (a) and the antiviral activity (b) of three PMs are shown in Fig. 4.11 (Matsui K, Yamase T, Kobayashi K, Saha S, Xin Q, Takeshita F, Okuda K Unpublished results in paper preparation). As shown in Fig. 4.11a, the PM dose for the challenge was determined along with the result of in vivo cytotoxicity of PMs: each of PM-523, PM-1002, and PM-1208 provided a100 % survival rate on the 4th day after its *i.p.* administration of less than 125 mg/kg/day. All PM-523, PM-1002, and PM-1208 suppressed the viral activity of both vPE16-env and vP1206-gag/pol in vivo with almost the same activity for administrations of 12.5 and 125 mg/kg/day/mouse. Especially, PM-523 and PM-1208 exhibited a potent antiviral activity against vPE16-env (with $p < 0.01$) and vP1206-gag/pol (with $p < 0.05$), respectively (Fig. 4.11b). Since no HIV antigen-specific immune responses were detected in the mice treated with PMs after virus infection, it was inferred that the PMs interact with HIV antigen to lead to the suppression of viral replication along with above the in vitro results.

HIV-1 inhibition based on the protease inhibition has been proposed for Wells–Dawson structural PMs containing one peroxoniobium or oxoniobium unit substituted for one tungsten unit in the parent framework, $K_7[P_2W_{17}(NbO_2)O_{61}]$ and $K_7[P_2W_{17}NbO_{62}]$: the Nb-substitution in the parent $[P_2W_{18}O_{62}]^{6-}$ (Fig. 4.1f) provided the high stability in the physiological pH level (Judd et al. 2001). These

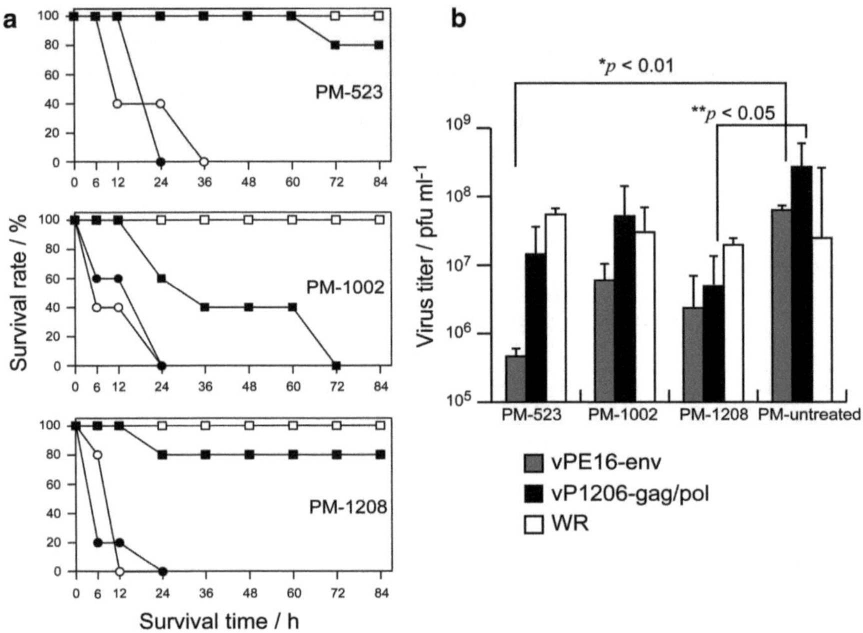

Fig. 4.11 In vivo cytotoxicity (**a**) and antiviral activity (**b**) for three PMs, 523, 1,002, and 1,208. Survival time indicates the time (in h) after *i.p.* administration of PM of 125 (*open square*), 250 (*filled square*), 500 (*open circle*), and 1,000 (*filled circle*) mg/kg/mouse. The activity of the infected virus (vPE 16, vP 1,206, or WR) in mice was evaluated on day 5, by the β-galactosidase activity for ovaries of mice which were *i.p.* administrated by 12.5 and 125 mg/kg/day/mouse of each PM (PM -523, PM -1002, or PM -1208) for 3 days from days 2 to 4 after the virus infection on day 0

compounds exhibited $EC_{50} = 0.17$–0.83 μM and $CC_{50} = 50$ to >100 μM (for peripheral blood mononuclear cells) and showed the noncompetitive inhibition of purified HIV-1 protease (with inhibition constants of 1.1 ± 0.5 and 4.1 ± 1.8 nM in 0.1 and 1.0 M NaCl, respectively). Unfortunately, the estimated EC_{50} values are inferior to those (0.3-0.5 nM) of the peptidemimetic PI containing a bis-tetrahydrofuranyl urethane and 4-methoxybenzenesulfonamide (TMC 126) as a clinically developed protease inhibitor (Yoshimura et al. 2002; DeClerq 2005).

It is well known that the initial steps of influenza virus replication start with a very unique interaction between the virus particle and the cellular membrane (Shigeta 1999; Blumenthal et al. 2003) Influenza virus acquires its infectivity after cleavage of the hemagglutinin (HA) peptide in its envelope into HA1 and HA2 by the action of the host cell protease. Under conditions of low endosome pH, the cleaved HA molecule changes its stereoscopic conformation, and the N-terminus of the HA2 cleaved site comes in contact with the cellular membrane which fuses with the virus envelope. Simultaneously, influx of protons through the matrix protein (M2) ion channels occurs and results in uncoating of the inner coat of ribonucleoprotein (M1) sheath. The liberated ribonucleoprotein (RNP) then moves to the nucleus. Influenza viruses are classified to three types (A, B, and C) based on the differences in their RNP antigenicity. Influenza A virus is further subdivided into subtypes based on differences in the antigenicity of both components of HA and neuraminidase (NA) for the virus. The current nomenclature system for human influenza virus takes the geographical location of the first isolation, strain number, year of isolation, and antigenic description of HA and NA, for example, A/Ishikawa/7/82/(H3N2) or B/Singapore/222/79 (Shigeta 1999). Most of PMs are inhibitory only at the virus adsorption stage, namely, the administration of PMs from 1.5 h after the virus infection shows little efficacy. Interestingly, PM-523 (as Keggin structure), a Ti/W-mixed PM, did not inhibit the adsorption of the influenza A virus (FluV-A) onto the cell membrane (0–1.5 h after infection) but inhibits the fusion between the FluV-A envelope and the cellular membrane (1.5–120 h after infection) (Shigeta et al. 1996). Such a different mechanism of anti-influenza activity for PM-523 led to both in vitro and in vivo synergistic efficacies in combination with ribavirin (as a viral RNA synthesis inhibitor) or zanamivir (NA inhibitor). The synergistic therapeutic effect in mice infected with FluV-A (H1N1/PR8) using PM-523 and ribavirin was examined, as shown in Fig. 4.12 (Shigeta et al. 1997, 2006). All of the control mice (inbred 8-week-old female Balb/c) were infected with 10 lethal dose of FluV-A, which corresponds to 1.6×10^4 median tissue culture infection dose ($TCID_{50}$) titers in MDCK cell line. From 8 h after infection, the mice (10 mice/group) were treated with aerosols of PM-523 and ribavirin either singly or in combination every 12 h for 4 days. Compound solutions were prepared in phosphate-buffered saline (PBS) solutions (at pH 7.2), and the infected mice were exposed to these compounds by using a continuous aerosol generator for 2 h at a rate of distribution of 120 μl of solution/h to each chamber. Each infected 20-g mouse in 10-mice/group was placed in a chamber separately. The average diameter of the aerosol particle was set to be 2.1 μm. In this experiment, the dose of the compound given to the mice was expressed in terms of the concentration of compound in the stock solution used for the aerosol exposure. The compound-untreated mice died on day 9 after infection. Similarly, all the 2.4-mM PM-523 and 40-mM ribavirin-treated mice died on days 9 and 8, respectively. 50 and 60 % of the mice treated with 4.8-mM PM-523 and 40-mM ribavirin, respectively, survived on day 9 after infection. EC_{50} values of PM-523 and ribavirin on day 9 after infection could be evaluated to be 4.8 and 70 mM, respectively. When a 2.4-mM PM-523/40-mM ribavirin combination was treated for the infected mice, 80 % of the mice survived on day 9 after infection, and 60 % of the mice survived still at the end of the experiment (on day 14 after infection), as shown in Fig. 4.12. 4.8-mM PM-523 and 80-mM ribavirin-treated mice showed 30 and 20 % survivals on day 14 after infection, respectively. Thus, the survival rate for the mice treated with the combination of PM-523 and ribavirin at a ratio of 1:16 was significantly high compared with the one with either 4.8-mM PM-523 or 80-mM ribavirin. The lungs of both the infected and untreated mice and the mice treated with either 2.4-mM PM-523 or 40-mM ribavirin became swollen and reddened due to congestion. On the other hand, the lungs of the infected mice treated with the combination of 2.4 mM PM-523 and 40-mM ribavirin remained

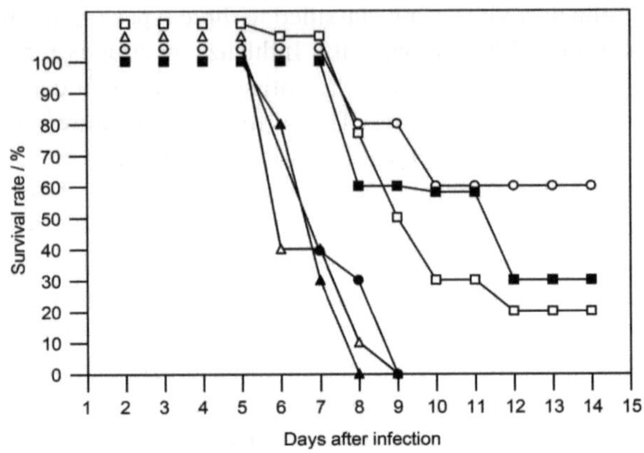

Fig. 4.12 Synergistic therapeutic effect in mice infected with FluV-A (H1N1/PR8) using PM-523 and ribavirin. All of the control mice (inbred 8-week-old female Balb/c) were infected with ten lethal doses of FluV-A, which corresponds to 1.6×10^4 median tissue culture infection dose ($TCID_{50}$) titers in MDCK cell line. From 8 h after infection, the mice (ten mice/group) were treated with aerosols of PM-523 and ribavirin either singly or in combination every 12 h for 4 days. Compound solutions were prepared in phosphate-buffered saline (PBS) solutions (at pH 7.2), and the infected mice were exposed to the compounds by using a continuous aerosol generator for 2 h at a rate of distribution of 120 μl of solution/h to each chamber. Each infected 20-g mouse was placed in a chamber separately; no compound (*filled circle*)-treated, 40 mM ribavirin (*filled triangle*)-treated, 80 mM ribavirin (*filled square*)-treated, 2.4 mM PM-523 (*open triangle*)-treated, 4.8 mM PM-523 (*open square*)-treated, and 40 mM ribavirin/2.4 mM PM-523 combination (*open circle*)-treated mice. Levels of significance of $p < 0.001$ for 2.4-mMPM-523/40-mM ribavirin combination, $p < 0.01$ for 4.8-mM PM-523, and $p < 0.05$ for 80-mM ribavirin-treated mice on day 14 after infection

normally grayish white and were not congested. The viral titer in the lungs for the infected mice treated with and without PM-523 and/or ribavirin was measured in every 24 h after infection, and the lungs for the measurement were removed from three mice for each group prepared by the same procedure of both infection and administration as the therapeutic effect experiment. The change of the viral titer in the lung is shown in Fig. 4.13 where the results for the group of the uninfected and untreated mice are added into the results for four groups of the infected mice with and without compounds for comparison (Shigeta et al. 1997). While there was no detection of virus for the uninfected mice, the viral titers in the infected mice lungs increased to $10^{5.4}$ $TCID_{50}$/g at 48 h after infection and decreased to $10^{3.5}$ and $10^{3.2}$ $TCID_{50}$/g at 72 and 96 h, respectively. The viral titers in the lungs for the infected mice treated with the combination of 2.4-mM PM-523 and 40-mM ribavirin were consistently lower than for the infected mice treated with either 2.4-mM PM-523 or 40-mM ribavirin. To identify the target of the antifusion activity for PM-523, PM-523-resistant FluV-A strains were selected after several passages of FluV-A H3N2/Ishikawa in MDCK cells in the presence of an effective dose of PM-523. The mutations conferring resistance to PM-523 indicated that isoleucine 202 and lysine 189, which were located at the interface edges of the trimer molecules of HA1, were

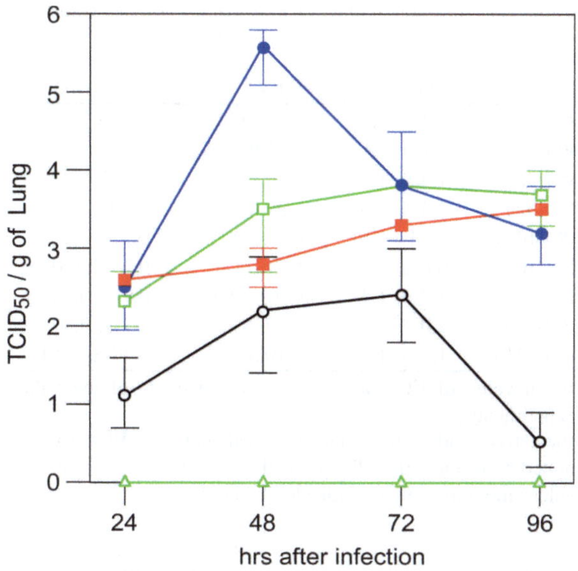

Fig. 4.13 Change of viral titers in the lungs of the infected mice with and without treatment of PM-523 and/or ribavirin after infection (1.6×10^4 $TCID_{50}$ of FluV/A/PR8/ mouse). The values for the uninfected and untreated mice are plotted for comparison. Uninfected and untreated (*open triangle*), infected and untreated (*filled circle*), 40-mM ribavirin-treated (*filled square*), 2.4-mMPM-523-treated (*open square*), and 40 mM ribavirin/ 2.4 mM PM-523 combination-treated (*open circle*) mice

substituted with threonine and asparagine, respectively. This strongly implies that PM-523 binds to the interface edges of HA trimers and inhibits the opening of HA1 trimers to lead to the inhibition of the fusion of viral envelope to cellular membrane by HA2 hydrophobic amino acids at the edge of the cleavage site (Shigeta 2001; Shigeta et al. 2006). The fusion inhibition by PM-523 is unique and the synergistic anti-influenza virus activity of PM-523 with ribavirin or zanamivir is promising as a novel therapeutic drug for the influenza virus infection, as exemplified in Figs. 4.12 and 4.13. PM-523 and its congeners Ti/W-mixed PMs (e.g., PM-19 and PM-504) and V/W-mixed PMs have a broad antiviral activity against enveloped RNA viruses (Table 4.6) and may be developed as broad-spectrum drugs for acute respiratory infection (ARI) caused by orthomyxoviruses (influenza viruses A, B, and C) and paramyxoviruses (human parainfluenza viruses 1, 3, 2, and 4; measles virus; and RSV).

Since most of the antiviral PMs inhibit the adsorption of the enveloped RNA viruses toward the host cells, these PMs are expected also to inhibit severe acute respiratory syndrome coronavirus (SARS-V) (Shigeta and Yamase 2005). SARS is severe and highly contagious, which first emerged in China in 2002 and spread within a few months from its origin to all over the world. In vitro antiviral activities of Ti/W- and V/W-mixed PMs were assayed for SARS-V together with other coronaviruses of transmissible gastroenteritis virus of swine (TGEV) and feline infectious peritonitis virus (FIPV) (Yamase 2005; Shigeta et al. 2006). As shown in Table 4.8, all of the PMs exhibited anti-TGEV activities. The V/W-mixed PMs displayed the anti-SARS-V activity, while they showed no apparent relationship between the structure and the anti-FIPV activity. Overall, all (PM-1002, PM-1207, PM-1208, and PM-1213) of tris(vanadyl)-18-tungstate anion,

Table 4.8 Antiviral activity of PMs against coronaviruses in vitro

	EC_{50} (μM)[a]			SI[b]		
PM	TGEV	FIPV	SARS-V	TGEV	FIPV	SARS-V
PM-504	1.4 ± 0.5	0.4 ± 0.3	3.5 ± 2.4	>12.8	**>46.7**	>14.3
PM-518	2.1 ± 1.1	>32.9	>50	>15.6	<1.0	<1.0
PM-520	1.4 ± 1.3	0.76 ± 0.5	4.9 ± 1.4	>20.4	**>38.5**	>10.3
PM-523	2.5 ± 0.6	>32.0	>50	**>32.0**	<1.0	<1.0
PM-1001	0.14 ± 0.1	0.2 ± 0.1	0.7 ± 0.4	**83.6**	**84.4**	**>70.4**
PM-1002	0.6 ± 0.4	ND	0.25 ± 0.07	18.6	–	**>400**
PM-1207	1.9 ± 1.4	0.09 ± 0.04	0.9 ± 0.24	>9.3	**35.2**	**>54.3**
PM-1208	2.0 ± 1.8	>3.6	2.7 ± 0.6	>8.1	<1.0	>18.4
PM-1213	1.7 ± 1.0	>10.5	0.47 ± 0.14	>18.2	<1.0	**106.4**

[a]Each value of EC_{50} and CC_{50} is the average of the values obtained by three independent experiments

[b]Selectivity index (SI) values were calculated by EC_{50}/CC_{50}, in which CC_{50} values were determined by using CPK cells (for TGEV), fewf-4 cells (for FIPV), and Vero cells (for SARS-V). SI values more than 30 are indicated by bold

$[(V^{IV}O)_3(XW_9O_{33})_2]^{12-}$ (X = As^{III}, Sb^{III}, Bi^{III}) and $[(V^{IV}O)_3(P^VW_9O_{34})_2]^{12-}$ (which consists of $(V^{IV}O)_3$ ring sandwiched by two A-α-$[P^VW_9O_{34}]^{9-}$ ligands), in addition to PM-1001 as the one-electron oxidation species ($[(V^{IV}O)_2(V^VO)(XW_9O_{33})_2]^{11-}$) of PM-1002, exhibit potent and selective activities (Table 4.8). Since $[\{Mn(H_2O)\}_3(SbW_9O_{33})_2]^{12-}$, $[Co_3(AsW_9O_{33})_2]^{12-}$, $[Cu_3(AsW_9O_{33})_2]^{12-}$, and $[Co_4(H_2O)_2(PW_9O_{34})_2]^{10-}$, as similar derivative anions, did not exhibit any efficacy against the coronaviruses (data not shown), the structural significance of tris(vanadyl) ring moiety is pointed out for observation of in vitro antiviral activity against the coronaviruses.

The first step of the infection of host cells by influenza virus is the binding (adsorption) of the influenza virus HA to sialic acid residues of the carbohydrate side chains of cellular proteins projecting from the plasma membrane that the virus exploits as receptors. The second step is the conformation change of HA as a membrane fusogen at the acidic pH (pH 5.0 at 37 °C for optimal fusogenic activity) which induces the fusion of the viral envelope membrane with the membrane of the endosome. This expels the viral RNA into the cytoplasm, where it can begin to replicate (Shigeta 1999; Blumenthal et al. 2003a). Together with the specific binding of PM-19 with HVEM as a TNF receptor superfamily (which was suggested by the investigation of anti-HSV activity of PM-19) (Dan and Yamase 2006), therefore, the antiviral mechanism of the PMs, which inhibit the adsorption of the enveloped RNA (or DNA) viruses toward the host cells, is likely to involve the functional quenching of the sialic acid residues of the carbohydrate side chains of cellular proteins, if we considered that binding between multiple HA ligands of influenza virus and sialic acid surface receptors of an erythrocyte for the cell during viral infection occurs with extremely strong affinity of 10^{15} per molar, while the association constant for a single sialic acid-HA interaction is only 10^3 per molar (Mammén et al. 1998). Sialyl/sulfotransferases are responsible for the biosynthesis of these carbohydrate chains: sialyltransferases transfer sialic acid residues from cytidine-5-monophosphato-sialic acid, CMP-Neu5Ac, to specific acceptor sugar chains, whereas sulfotransferases transfer sulfate groups from adenosine 3′-phosphate 5′-phosphosulfate (PAPS) to specific acceptor sugar chains. Most of Ti/W- and V/W-mixed PMs and multiply reduced spherical polyoxovanadates such as $K_5[H_6KV_{13}O_{31}(MePO_3)_3]\cdot16.5H_2O$ (Yamase et al. 2000), $Na_{12}H_2[H_2V_{18}O_{44}(N_3)]\cdot30H_2O$ (Yamase et al. 1997), $K_{10}[H_2V_{18}O_{42}(H_2O)]\cdot16H_2O$, and $K_{10}H_3[V_{18}O_{42}(Cl)]\cdot12H_2O$ (Fig. 4.1g, Yamase et al. 1996b) inhibited in vitro enzymatic activities of specific sialyl/sulfotransferases such as α2,3-sialyltransferases (ST3Gal-I and ST3Gal-III), α2,6-sialyltransferases (ST6GalNAc-I and ST6Gal-I), Gal 3-*O*-sulfotransferase-2 (Gal3ST-2), GlcNAc 6-*O*-sulfotransferase-1 (Gn6ST-1), Gal 6-*O*-sulfotransferase (Gal6ST), heparan sulfate 6-*O*-sulfotransferase, β1,4-galactosyltransferase-I (β4Gal-TI), β1,3-GlcNAc-transferase-2 (β3Gn-T2), and α1,3-fucosyltransferase-III (FUT3), at sub-nano and nanomolar concentrations with the noncompetitive inhibitory mode for both donor and acceptor substrates (Seko et al. 2009): for example, PM-504 inhibited ST3Gal-1 activity with inhibition constant (K_i) ~ 0.5 nM (0.47 and 0.53 nM for core 1 as an acceptor and CMP-Neu5Ac as a donor, respectively) which were much lower than K_i = 29 nM for the previously reported lowest inhibitors of glycosides analogues (Schwörer and Schmidt 2002). The inhibition mode of PM-504 on ST3Gal-1 is shown in Fig. 4.14, where plots of the reciprocal velocity versus the reciprocal substrate concentration with and without PM-504. The steady-state kinetics results indicate the noncompetitive inhibition of PM-504 against ST3Gal-1, suggesting the interaction of PM-504 with a variety of the binding sites of ST3Gal-1. 50 % inhibitory concentration (IC_{50}) values of a variety of PMs for two sialyl/sulfotransferases, ST3Gal-1 and Gal3ST-2, are listed in Table 4.9 (Seko et al. 2009). IC_{50} values of PM-19, PM-43, PM-518, PM-523, and PM-1002 for ST3Gal-1 were extremely small, 0.4, 0.2, 0.3, 0.6, and 0.4 nM, respectively. These values were 2–3 orders of magnitude smaller than for Gal3ST-2 (40, 50, 20, 100, and 60 nM, respectively). On the other hand, the spherical polyoxovanadates inhibited both enzymatic activities in the same order of magnitude of nanomolar concentrations: IC_{50} values of $K_5[H_6KV_{13}O_{31}(MePO_3)_3]\cdot16.5H_2O$, $Na_{12}H_2[H_2V_{18}O_{44}(N_3)]\cdot30H_2O$, $K_{10}[H_2V_{18}O_{42}(H_2O)]\cdot16H_2O$, and $K_{10}H_3[V_{18}O_{42}(Cl)]\cdot12H_2O$ were 7, 9, 7, and 7 nM, respectively, for ST3Gal-1 and were all 3 nM for ST3Gal-1. It should be noted that the inhibition of the above PMs for in vitro enzymatic activities of sialyl/sulfotransferases is at least 100- to 1,000-fold stronger than any other known inhibitors (Schwörer and Schmidt 2002; Rath et al. 2004). The result for the inhibitory effect of PM-504 on the mutated ST3Gal-1 enzymes implied that the inhibition of PMs against the adsorption of the enveloped RNA viruses toward the host cells is due to the strong interaction of the PMs not only with ^{335}Arg residue in the C-terminal region of ST3Gal-1 but also with other basic amino acids residue (Seko et al. 2009). Thus, some types of PMs have the ability to inhibit specific sialyl/sulfotransferases, indicating that the PMs directly inhibit the activities of enzymes involved in carbohydrate metabolism to lead to the inhibition of the binding of the HA ligands

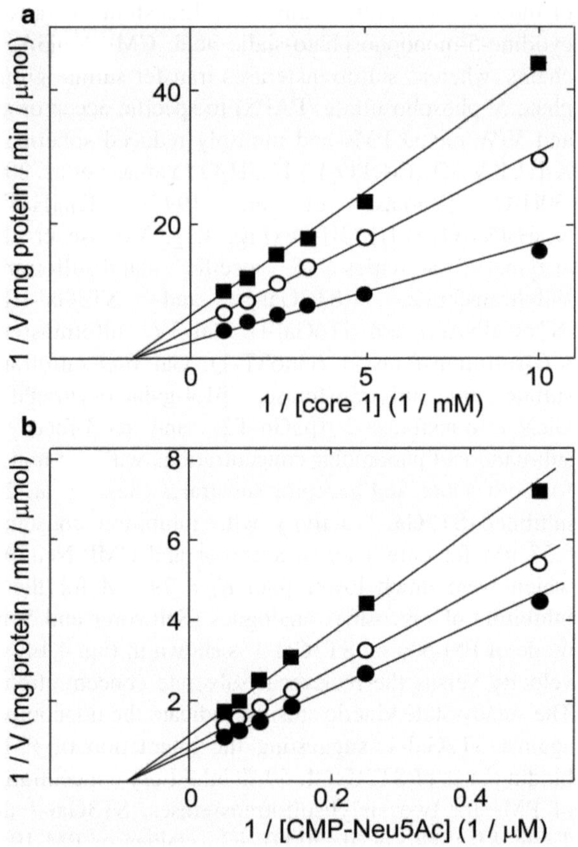

Fig. 4.14 Inhibition mode of PM-504 on the enzymatic activity of ST3Gal-1 at various concentrations of core 1 (**a**) and CMP-Neu5Ac (**b**) in the absence and presence of PM-504. (**a**): 0 (*filled circle*), 0.15 nM (*open circle*), and 0.3 nM (*filled square*) of PM-504 in the presence of 1.6-μM CMP-Neu5Ac. (**b**): 0 (*filled circle*), 0.1 nM (*open circle*), and 0.2 nM (*filled square*) of PM-504 in the presence of 1-mM core 1

of influenza virus with sialic acid surface receptors. As shown in Table 4.9, no activity of PM-8 and PM-17 to the sialyl/sulfotransferases enzymes corresponds to missing of the antiviral activity for these polyoxomolybdates.

4.4 Antibacterial Activity Against Gram-Positive Methicillin-Resistant *Staphylococcus aureus* and Vancomycin-Resistant *Staphylococcus aureus*

Most polyoxotungstates show a synergistic effect with β-lactam antibiotics against methicillin-resistant *Staphylococcus* (*S.*) *aureus* (MRSA) as Gram-positive bacterium which was first isolated in 1961 after the introduction of methicillin into clinical use against *S. aureus* strains (Jevons 1961) and have induced clinically serious problems due to their strong resistance to many antibiotics. The inhibition by methicillin against *S. aureus* strains is based on the binding of β-lactam

Table 4.9 IC_{50} (in nM unit) of PMs for ST3Gal-1 and Gal3ST-2

PM	IC_{50} (nM)	
	ST3Gal-1	Gal3ST-2
Molybdates		
PM-8	>500	>500
$Na_2[Mo_8O_{26}(L\text{-}Lys)_2]\cdot 8H_2O$	>500	>500
PM-17	>500	>500
Tungstates		
PM-19	0.4	40
PM-43	0.2	50
PM-504	0.3	20
PM-518	0.3	20
PM-523	0.6	100
PM-1002	0.4	60
Vanadates		
$K_5[H_6KV_{13}O_{31}(MePO_3)_3]\cdot 16.5H_2O$	7	3
$K_{10}H_3[V_{18}O_{42}(Cl)]\cdot 12H_2O$	7	5
$K_{10}H_3[V_{18}O_{42}(H_2O)]\cdot 16H_2O$	9	3
$K_{12}H_2[H_2V_{18}O_{44}(N_3)]\cdot 30H_2O$	7	3

antibiotics to penicillin-binding proteins (PBPA, PBP2, PBP3, and PBP4) as the peptidoglycan-synthetic enzyme on the membrane surface. Different from such a methicillin-susceptible *Staphylococcus aureus* (MSSA), MRSA possesses PBP2′ (sometimes called PBP2a or MecA) as an additional PBP on the membrane surface, which is coded by *mecA* gene and exhibits low affinity to the β-lactams. Concerning the existence of PBP2′ on the surface, MRSA strain can be classified into two, constitutive (PBP2′ exists irrespective of the β-lactams) and inducible (PBP2′ is induced on the surface in the presence of the β-lactams) (Ubukata et al. 1985; Matsuhashi et al. 1986). Vancomycin as a glycopeptide has become the antibiotic of last resort for the hard-to-treat infection of MRSA infection. However, eventfully bacteria developed resistance to vancomycin, too, in some cases by changing an amide to an ester in a bacterial glycopeptide cell wall precursor. Since Mu50, as a vancomycin-resistant *S. aureus* (VRSA) strain, was first isolated in 1997 (Hanaki et al. 1998a, b), other VRSA strains have emerged which produce structurally unusually complicated cell wall precursor with few interaction with vancomycin (Boneca and Chiosis 2003). Thus, novel chemotherapeutics against MRSA and VRSA strains are strongly required.

The effect of PMs against both MRSA strains of SR3605 (constitutive) and ATCC43300 (inducible) has been investigated (Yamase et al. 1996a, b, c), as shown in Table 4.10 where both the minimum inhibitory concentration (MIC in μM unit) and the fractional inhibitory concentration (FIC) index for PMs were measured by the agar dilution checkerboard method for the evaluation of the effect in combination with oxacillin (Hallander et al. 1982). As shown in Table 4.10, all of the polyoxotungstates tested in this study exhibited a synergistic effect with oxacillin, as a FIC index less than 1/2 is defined as a synergistic effect. In particular, the synergistic potential of Keggin-structural polyoxotungstates and their lacunary

Table 4.10 Antibacterial activity of polyoxotungstates alone and in combination with oxacillin against SR3605 (constitutive MRSA) and ATCC43300 (inducible MRSA)[a]

PM	MIC (μM)[a] SR3605	FIC index	MIC (μM) ATCC43300	FIC index
$Na_2WO_4{\cdot}2H_2O$	102,400	2.0	102,400	0.15
Keggin				
PM-1	800	0.156	800	0.094
$Na_4[SiW_{12}O_{40}]{\cdot}nH_2O$	3,200	0.094	3,200	0.019
PM-47	12,800	0.063	3,200	0.063
Monovacant Keggin				
α-$K_7[PW_{11}O_{39}]{\cdot}nH_2O$	12,800	0.063	6,400	0.035
Trivacant Keggin				
B-α-$Na_{10}[PW_9O_{34}]{\cdot}nH_2O$	12,800	0.047	12,800	0.031
A-β-$Na_9[HSiW_9O_{34}]{\cdot}23H_2O$	12,800	0.094	6,400	0.018
A-α-$Na_{10}[SiW_9O_{34}]{\cdot}18H_2O$	12,800	0.156	12,800	0.010
B-α-$Na_9[SbW_9O_{33}]{\cdot}19.5H_2O$	3,200	0.156	1,600	0.156
Decatungstates				
$Na_9[EuW_{10}O_{36}]{\cdot}32H_2O$	3,200	0.188	1,600	0.313
$K_9[GdW_{10}O_{36}]{\cdot}20H_2O$	3,200	0.313	1,600	0.281
Anderson				
$K_{5.5}H_{1.5}[SbW_6O_{24}]{\cdot}6H_2O$	12,800	0.156	12,800	0.063
$K_6Na_2[MnW_6O_{24}]{\cdot}12H_2O$	6,400	0.281	3,200	0.281
Others				
$K_{18}[KSb_9W_{21}O_{86}]{\cdot}nH_2O$	200	0.313	400	0.281
$Na_{27}[NaAs_4W_{40}O_{140}]{\cdot}nH_2O$	400	0.281	400	0.156

[a]The unit (mg/ml) of all the MIC values on Tables 4.1, 4.2, and 4.3 in Yamase et al. (1996a) and Fukuda et al. (1999) should be corrected to μM

species, $[XW_{11}O_{39}]^{n-}$ and $[XW_9O_{34}]^{n-}$, were high: FIC indices of these compounds against SR3605 and ATCC43300 were 0.010–0.156. No synergistic activity was observed for $Na_2WO_4{\cdot}2H_2O$. Dodeca- and nanotungstosilicates such as $Na_4[SiW_{12}O_{40}]{\cdot}nH_2O$, A-$\beta$-$Na_9[SiW_9O_{34}H]{\cdot}23H_2O$ (PM-30), and A-α-$Na_{10}[SiW_9O_{34}]{\cdot}18H_2O$ were highly synergistic (with FIC = 0.019, 0.018, and 0.010, respectively) against ATCC43300, suggesting a significance of the Si atom as a hetero atom in polyoxotungstates against this strain. Most of the PMs of molybdenum (with an exception of $[SiMo_{12}O_{40}]^{4-}$ as discussed below) and vanadium showed little effect against SR3605, as shown in Table 4.11 (Yamase et al. 1996a, b, c). A Keggin-structural polyoxomolybdate, $Na_3[PMo_{12}O_{40}]{\cdot}nH_2O$, provided FIC = 0.516, although this compound exhibited lower toxicity compared with the same structural polyoxotungstates such as $K_5[BW_{12}O_{40}]{\cdot}15H_2O$ (PM-1), $K_5[PVW_{11}O_{40}]{\cdot}6H_2O$ (PM-44), and $K_6[BVW_{11}O_{40}]{\cdot}nH_2O$ (PM-46) (Table 4.10). Similarly, most of the polyoxovanadates seemed to exhibit no significant synergism, although $K_7[MnV_{13}O_{38}]{\cdot}18H_2O$ with FIC = 0.28 was synergistic. Since neither $MnCl_2$ nor $CoCl_2$ showed any synergistic effect (data not shown), the synergistic activity of polyoxotungstates indicates that the location of W atoms in the PM lattice is an important factor in its combination with oxacillin (Yamase

Table 4.11 Antibacterial activity of polyoxomolybdates or polyoxovanadates alone and in combination with oxacillin against SR3605 (constitutive MRSA)

PM	MIC (μM)	FIC index
Molybdates		
$Na_3[PMo_{12}O_{40}]\cdot nH_2O$	25,600	0.516
$Na_6[Mo_7O_{24}]\cdot 4H_2O$	1,600	0.531
$[NH_4]_6[Mo_7O_{24}]\cdot 4H_2O$	1,600	0.531
$[NH_3Pr^i]_4[Mo_8O_{26}]\cdot nH_2O$	1,600	0.531
$K_5[IMo_6O_{24}]\cdot nH_2O$	800	1.008
Vanadates		
Na_3VO_4	51,200	2.0
$[NH_3Bu^t]_4[V_8O_{12}]$	51,200	1.0
$K_7[MnV_{13}O_{38}]\cdot nH_2O$	800	0.28
$K_5H_2[V_{15}O_{36}(CO_3)]\cdot 15.5H_2O$	1,600	1.016

The unit (mg/ml) of all the MIC values on Tables 4.1, 4.2, and 4.3 in Yamase et al. (1996a) and Fukuda et al. (1999) should be corrected to μM

et al. 1996a, b, c; Fukuda et al. 1999). The sodium dodecyl sulfate-polyacrylamide gel electrophoresis (SDS-PAGE) of the membrane proteins separated from MRSA revealed that the synergism of the polyoxotungstates is due to the inhibition of PBP2′ expression on the membrane surface by the PMs (Yamase et al. 1996a, b, c; Fukuda et al. 1999).

Mechanistic details on the synergistic activity of PMs against MRSA and VRSA strains have been investigated by using $K_6[P_2W_{18}O_{62}]\cdot 14H_2O$ (PM-27) as a Wells–Dawson-structural polyoxotungstate (Fig. 4.1f), $K_4[SiMo_{12}O_{40}]\cdot 3H_2O$ ($SiMo_{12}$) as a Keggin-structural polyoxomolybdate, and $K_7[PTi_2W_{10}O_{40}]\cdot 6H_2O$ (PM-19) as a Ti/W-mixed Keggin-structural polyoxotungstate (Inoue et al. 2005). The three PMs were tested against five strains of two constitutive MRSA (SR3605 and MRS394-1), an inducible MRSA (ATCC43300), and two constitutive VRSA (Mu3 and Mu50) strains. The observable enhancement of the antibacterial activity of β-lactam antibiotic, oxacillin, with a help of three kinds of PMs, PM-27, $SiMo_{12}$, and PM-19, is summarized in Table 4.12, where MIC values of the PMs determined by the microdilution method are also added. The synergism between the PMs and oxacillin with the antibacterial activity was investigated by the oxacillin-incorporating disk method using Müller-Hinton (MH) agar medium where the PMs were included at tested concentrations. An observable growth-inhibitory zone around the oxacillin disk indicates the susceptibility of the tested strains to β-lactam. Figure 4.15 shows typical images of the growth-inhibitory zone around the disk for the Mu50 strain treated with PM-19. No appearance of the growth-inhibitory zone indicates the "homo-resistance of the strain" (Fig. 4.15a), the observation of the growth-inhibitory zone with colonies inside "hetero-resistance of the strain" (Fig. 4.15b), and the appearance of growth-inhibitory zone without colonies indicates "oxacillin-susceptible strain" (Fig. 4.15c). All of the tested strains but ATCC43300 were homo-resistant to β-lactam in the absence of the PMs: ATCC43300 was hetero-resistant. At the concentration of less than MIC value, PM-27 and PM-19 showed the growth-inhibitory zone against all of the

Table 4.12 Susceptibility of three MRSA and two VRSA strains to oxacillin in combination with PM at various concentrations

PM and its MIC (μM) and concentrations (μM)		MRSA			VRSA	
		ATCC43300	SR3605	MRS394-1	Mu3	Mu50
PM-27[a]	MIC	100	400	200	400	200
	50	S[b]	S	S	Hetero	Hetero
	100	S	S	S	Hetero	S
	500	n.c.[b]	n.c.	n.c.	n.c.	n.c.
	1,000	n.c.	n.c.	n.c.	n.c.	n.c.
$SiMo_{12}$	MIC	400	800	800	800	800
	50	S	Homo	S	Homo	Homo
	100	S	Homo	S	Homo	S
	500	S	S	S	S	S
	1,000	n.c.	n.c.	n.c.	n.c.	n.c.
PM-19	MIC	12,800	25,600	25,600	>25,600	25,600
	50	S	Homo	Hetero	Homo	Homo
	100	S	Homo	S	Homo	Hetero
	500	S	S	S	Hetero	S
	1,000	S	S	S	Hetero	S

[a]PM-27 = $K_6[P_2W_{18}O_{62}]\cdot 14H_2O$ (Wells–Dawson), $SiMo_{12}$ = $K_4[SiMo_{12}O_{40}]\cdot 3H_2O$ (Keggin), and PM-19 = $K_7[PTi_2W_{10}O_{40}]\cdot 6H_2O$ (Keggin)
[b]The notation of "homo," "hetero," and "S" represent homo-resistance, hetero-resistance, and susceptibility to oxacillin, respectively. The notation of "n.c." indicates the result that no colony was seen on the agar medium, because the concentration of PM exceeded MIC

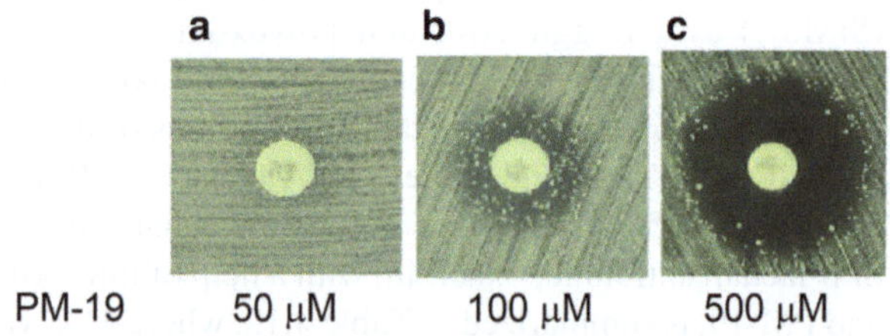

Fig. 4.15 Growth-inhibitory zone around the disk for the Mu50 strain treated with PM-19 at concentrations of 50 (a), 100 (b), and 500 (c) μM

strains except for Mu3, indicating the transformation of MRSA and VRSA-Mu50 from β-lactam resistance into β-lactam susceptibility (Table 4.12). Also, PM-27 and PM-19 caused Mu3 to be changed from homo- to hetero-resistance with a decrease of resistance to β-lactam (phenotype change). Interestingly, $SiMo_{12}$ enabled to enhance the antibacterial activity of β-lactam against all the strains, since the growth-inhibitory zone without colonies was observed at the concentration of less than MIC. This is in a strong contrast to the case of the same Keggin-structural $Na_3[PMo_{12}O_{40}]\cdot nH_2O$, which did not enhance the antibacterial activity of β-lactam against MRSA (Table 4.11). The difference in the synergism between PM-27 and $Na_3[PMo_{12}O_{40}]\cdot nH_2O$ is associated with the structural stability of the anion at physiological pH level, which was verified by the cyclic voltammetric measurements shown below.

Does the inhibition of PBP2′ expression by PMs result from missing of *mecA* gene? When the MRSA strains were cultured for 24 h at 37 °C in MH broth containing the PMs at MIC, and thereafter an aliquot of the broth was applied onto MH agar medium in the absence of the PMs according to the standard oxacillin-disk method and was incubated again for 24 h at 37 °C, however, there was no observation of the growth-inhibitory zone around the oxacillin disk on the MH agar medium. This result excludes the possibility of the missing of the *mecA* gene for the PMs-treated MRSA strains, since the PM-treated strain would exhibit the oxacillin susceptibility if the *mecA* gene was missed by PMs. As a next step, the possibility of the inhibition of the transcription from *mecA* gene to mRNA by PMs with and without oxacillin was investigated by using the reverse transcription polymerase chain reaction (RT-PCR) analysis of the MRSA and VRSA strains which were cultured until the initial stage of logarithmic growth phase (OD_{660} = 0.1–0.2) both in the presence of PM (PM-27, $SiMo_{12}$, or PM-19) alone (at less than MIC) and in the coexistence of PM and oxacillin (at 1/4 MIC for each). Figures 4.16 and 4.17 show effects of PM alone (a) and the coexistence of PM and oxacillin (b) on the electrophoresis results of the RT-PCR products for MRSA SR3605 and VRSA Mu50, respectively: *mecA* (258 bp)-, *pbpA* (411 bp)-, *pbp2* (618 bp)-, *pbp3* (814 bp)-, and *pbp4* (1,122 bp)-induced mRNAs, together with 16SrRNA (174 bp) as a control, in which the PCR primer for each of *mecA-pbp* genes was designed and the fragments of genes were amplified by the procedures of 40 cycles of the amplification consisting of 15-s denaturation (at 94 °C), 30-s primer annealing (at 55 °C), 60-s primer extension (at 68 °C), and 300-s final extension (Ryffel et al. 1990; Zhao et al. 2001). The inhibitions of the transcription to *mecA*-, *pbpA*-, and *pbp2-4*-induced mRNAs and 16S rRNA were quantitatively evaluated from the band intensities analyzed with image-analysis software. As shown in Figs. 4.16a and 4.17a, each of PM-27, $SiMo_{12}$, and PM-19 showed a decrease in the band intensities of the *mecA*-induced mRNA at concentrations less than MIC (1/2 MIC PM-27 displayed the inhibition of 30 % for SR3605 and 10 % for Mu50, although the depression of the band intensities by PM-27 alone at its concentrations less than 1/4 MIC), and the extent of the decrease increased with an increase in its concentration in comparison with that of 16S rRNA, indicating that the transcription process from *mecA* to mRNA was inhibited by the PMs. Also the RT-PCR results showed that PMs depressed also the expression for *pbpA*- and *pbp2-4*-induced mRNAs, and the extent of the depression depends on both PM and *pbp* genes (Inoue et al. 2005). As exemplified by the fact that the inhibition of the expression of the *pbp3*-induced mRNA was observed for all of the above PMs at their 1/4 MIC concentrations. Such inhibition of mRNAs by the PMs demonstrates nicely the morphological change (swelling and aggregation) of the strain cells during the cultivation in the presence of the PMs as observed on the microscope (Fukuda et al. 1999). It is notable that the inhibition of the transcription from *mecA* and *pbp* genes to mRNA by the PMs was enhanced by β-lactams. Figures 4.16b and 4.17b show the electrophoresis results of the RT-PCR products for MRSA SR3605 and VRSA Mu50, respectively, in the coexistence of oxacillin of 1/4 MIC and PMs of 1/32–1 MIC (Hino K, Inoue M, Oda M, Nakamura Y, Yamase T Unpublished

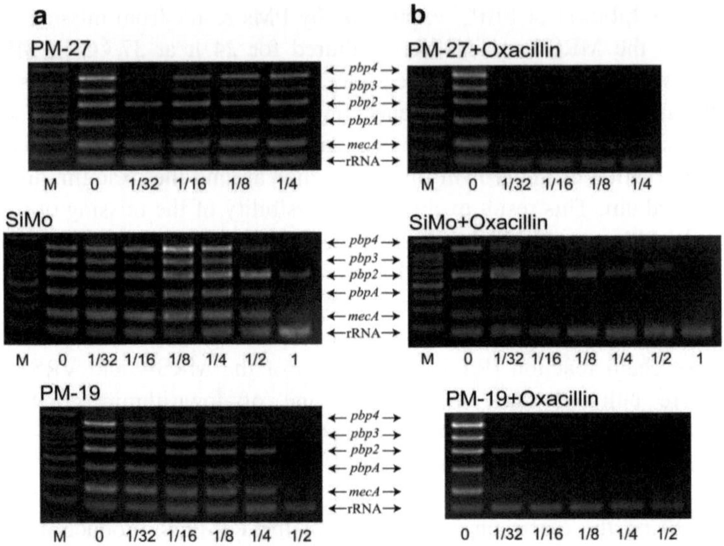

Fig. 4.16 Effects of PM alone (**a**) and the coexistence of PM and oxacillin (**b**) on the electrophoresis results of the RT-PCR products for MRSA SR3605. SR3605 strain cells were incubated in Muller–Hinton broth in the presence (**a**) of PM (PM-27, SiMo12, or PM-19) at various concentrations indicated by MIC units or in the coexistence (**b**) of PM (at various concentrations less than MIC) and oxacillin (at 1/4 MIC). Lane M indicates 100 bp DNA ladder as molecular weight marker

results in paper preparation). The coexistence of 1/4 MIC oxacillin in MRSA SR3601 enhanced the inhibition of the transcription to *mecA*-induced mRNA by PMs, approximately 16-fold (from 1/2 to 1/32 MIC) for PM-27 and 8-fold (from 1/4 to 1/32 MIC) for $SiMo_{12}$ and PM-19. Similarly, 1/4 MIC oxacillin in VRSA Mu50 enhanced 8-fold (1/2 to 1/16 MIC) for PM-27 and 4-fold (1/4 to 1/16 MIC) for $SiMo_{12}$ and PM-19. The enhancement of the PM inhibition of the transcription of *mecA* for the SR3605 strain by a help of 1/4 MIC oxacillin was approximately 2-fold high compared with the case for the Mu50 strain.

The treatment (for 15 min) of both the cells of MRSA SR3605 and VRSA Mu50 with PM-27 (or $SiMo_{12}$) induced the blue coloration of the cells, in a strong contrast with the case (no coloration) of the (dead) cells treated with ethanol for 1 day. Figure 4.18 shows the blue coloration by PM-27 for live cells of both SR3605 and Mu50. The protoplast of such blue-stained cells, obtained by the treatment with lysostaphin and mutanolysin as cell wall lysis enzymes in hypertonic solution for 12 h, remained still blue. The fact that the biological reduction of PM-27 or $SiMo_{12}$ occurred not only for the MRSA cells but also for the VRSA cells with much thicker cell walls indicates that the PMs penetrating through the cell wall are uptaken into the electron transfer system of the cellular membrane probably into the respiration system involving NADH/ubiquinone/cytochrome-c, as supported by the dominant distribution of W atoms at the periphery in the elemental spectrum analysis of the

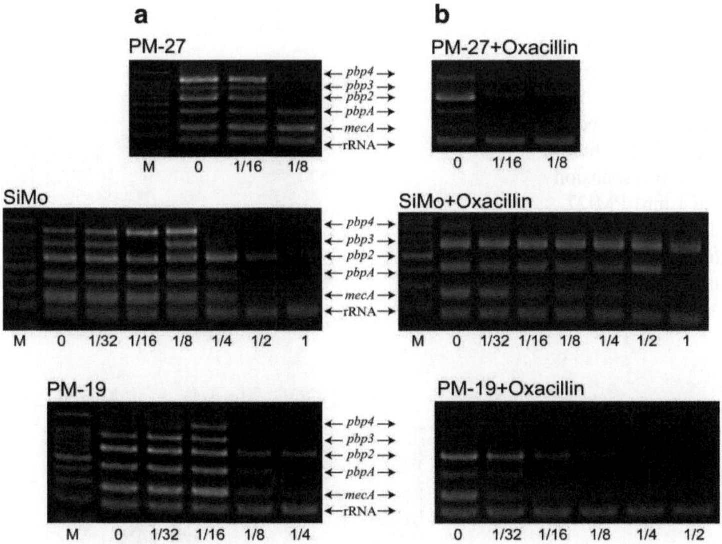

Fig. 4.17 Effects of PM alone (**a**) and the coexistence of PM and oxacillin (**b**) on the electrophoresis results of the RT-PCR products for VRSA Mu50. Mu50 strain cells were incubated in MH broth in the presence (**a**) of PM (PM-27, SiMo12, or PM-19) at various concentrations indicated by MIC units or in the coexistence (**b**) of PM (at various concentrations less than MIC) and oxacillin (at 1/4 MIC). Lane M indicates 100 bp DNA ladder as molecular weight marker

PM-27-treated MRSA cells (Fukuda et al. 1999). Redox potentials of NADH, ubiquinone, and cytochrome-c (−0.96, −0.54, and −0.39 V vs. Ag/AgCl, respectively) are negative enough to reduce PM-27 and $SiMo_{12}$, since the cyclic voltammograms of PM-27 and $SiMo_{12}$ at pH 7.4 showed their first half potentials ($E_{1/2}$) for the reversible one-electron reduction at +0.058 V and +0.237 V vs. Ag/AgCl, respectively (Inoue et al. 2005). The biological reduction of $SiMo_{12}$ was noted also in the photosystem II of the photosynthesis, where the electron transfer from photoreduced pheophytin to primary plastoquinone acceptor occurred through $SiMo_{12}$ (Giaquinta and Dilley 1975; Zilinskas and Govindjee 1975). Thus, both PM-27 and $SiMo_{12}$ are biologically easily reduced to the blue color showing the broad absorption band around 700 nm which is assigned to the intervalence charge transfer of M^{V}–O–M^{VI} ↔ M^{VI}–O–M^{V} (M = W and Mo) (Fukuda et al. 1999). Figure 4.19 shows the TEM images of the MRS396-1 cells treated with and without PM-19 and the elemental spectra of local points of the cells. The observation of the W/Ti atomic ratio of 5.3 (which is close to the one (5) of PM-19) at the periphery of the cell treated with PM-19 strongly implies that the PMs enter the cells and are localized at the cell periphery to keep their structure of the anion to be intact. This is in strong contrast to the case of the PM-19 untreated cells, which indicates no significant intensity of peaks due to W and Ti atoms. That the anion structure of PM is kept intact inside the cell was also pointed out for

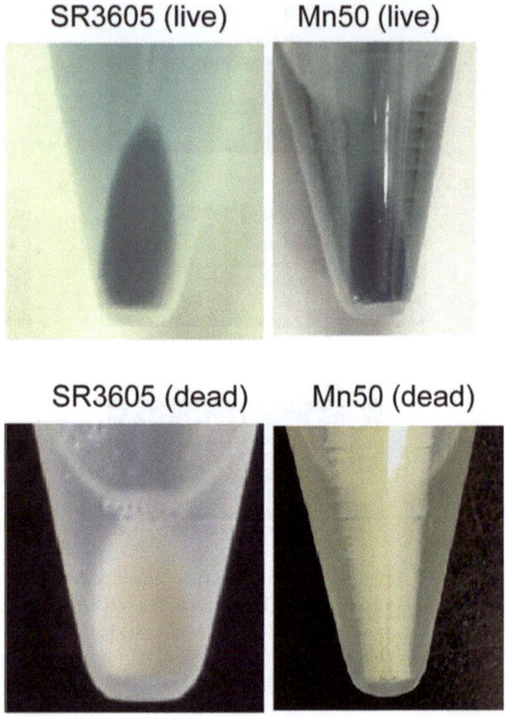

Fig. 4.18 *Blue coloration* of live cells of both SR3605 and Mu50 treated with PM-27. Photographs indicate the centrifuged cells of SR3605 or Mu50 strain in saline suspension containing 1 mM PM-27. Dead cells were prepared by the treatment of the cells with ethanol for 1 day

$K_{10}[Co_4(H_2O)_2(PW_9O_{34})_2]\cdot nH_2O$ uptaken into HIV cells (Cholewa et al. 1994). The cyclic voltammograms of PM-19 at pH 7.4 showed the first half potentials ($E_{1/2}$) for pseudo-reversible one-electron reduction at −1.02 V vs. Ag/AgCl, indicating that PM-19 was biologically redox inactive. The cyclic voltammogram of PMo_{12} showed the structural decomposition at physiological pH level of 7.4, whereas at pH 1.0 it provided the reversible two-electron reduction at +0.186 V vs. Ag/AgCl (Inoue et al. 2005). The decomposition of PMo_{12} at pH 7.4 rationalizes the lack of its synergistic activity. The bacterial ATP within the cytoplasm membrane is produced during the cell growth by the electron transfer from NADH to dissolved oxygen through cytochrome-*c*, which is triggered by the proton gradient between inside and outside of the cell (Inoue et al. 2005). Therefore, the fact that PM-27 and $SiMo_{12}$ had two orders of magnitude lower MIC than PM-19 (Table 4.12) is associated with the blocking of the electron transfer system for the respiration leading to the inhibition of the ATP production. Namely, the biologically redox-active PM-27 and $SiMo_{12}$ uptaken in the cytoplasm membrane inhibits the electron transfer system of MRSA and VRSA to lead to lowering of MIC values.

The cell wall of MRSA is constructed from the mesh-like structural peptidoglycan as polymer chains which comprise alternate sugar units of N-acetylglucosamine and *N*-acetylmuramic acid. L-Ala residue of the tetrapeptide (L-Ala-D-Glu-L-Lys-D-Ala) of the peptidoglycan is bound to the *N*-acetylmuramic acid, and the terminal D-Ala residue of the tetrapeptide is linked to L-Lys residue of the tetrapeptide of the

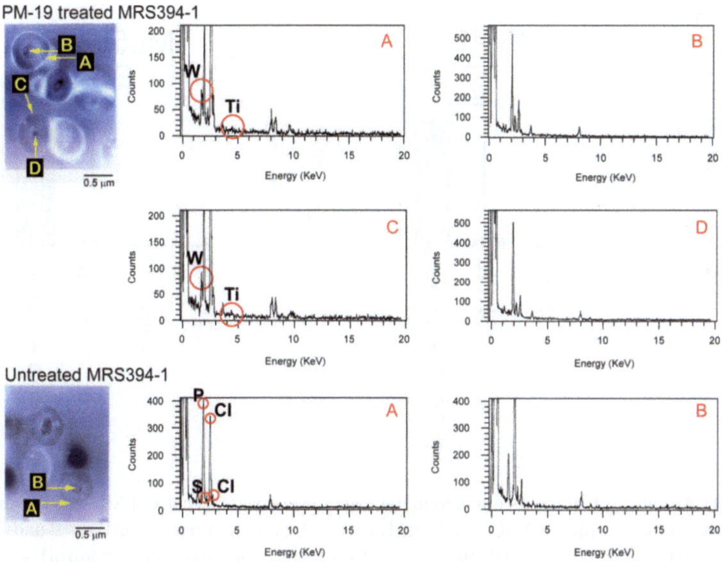

Fig. 4.19 TEM images of the MRS396-1 cells treated with and without PM-19 and elemental spectra of local points of the cells (A–D for PM-treated and A and B for untreated ones). Inoculum (10^8 cfu, colony-forming unit/ml) of MRS394-1 was inoculated in MH broth containing 40 mM PM-27 for 5 h at 34 °C. The cells harvested by centrifugation (with 5,000 × *g* for 15 min at 4 °C) and washed twice with PBS were fixed with 2.5 % glutaraldehyde in PBS and followed by dehydration with a graded series of ethanol, embedding into a resin, and sectioning by conventional method. The spectrum observed on JEOL JEM-2100 F was examined by the distribution of W and Ti atoms in the cells with a help of energy dispersive X-ray analysis (JEOL JED-2300 T)

neighboring polymer chain through pentaglycine peptide. The PMs with 1–2 nm size are small enough to pass through the void of the peptidoglycan layer and reach the cytoplasm membrane. Thus, it is reasonable to say that the inhibition of the transcription from the *mecA* and *pbp* genes to mRNA is exerted by the PMs uptaken into the cytoplasm membrane, the amounts of which are strongly associated with the synergistic activity. The enhancement of the PM-induced inhibition of the transcription from *mecA* and *pbp* genes to mRNA by β-lactams (Figs. 4.16b and 4.17b) gives us a clue to the mechanism of the synergistic activity of PMs. Amounts of W atoms uptaken into the cells in the presence of PMs and/or oxacillin were measured by an inductively coupled plasma atomic emission spectroscopy (ICP) analysis of the MRSA SR3605 (or VRSA Mu50) cells treated with 1/4 MIC PM-27, $SiMo_{12}$, or PM-19 with and without 1/4 MIC oxacillin. Inoculum of 1×10^7 cfu (colony-forming unit)/ml bacteria and 1/4 MIC additive in MHB medium were cultured at 37 °C until the stage of a logarithmic growth phase ($OD_{660} \approx 0.2$), harvested by the centrifugation, washed three times in ultra pure water, and then freeze-dried. The weighted dry samples were dissolved in 50 % nitric acid, boiled, and filtrated. The resulting filtrates were used for the ICP analysis. Both MRSA SR3605 and VRSA Mu50 provided the highest amounts of W (or Mo) atoms for the culture with PM and

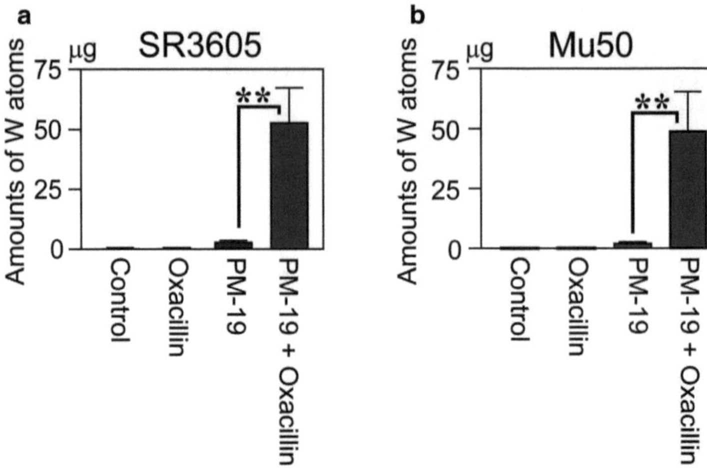

Fig. 4.20 Amounts of W atoms uptaken in the cells for the bacteria cells of SR3605 (**a**) and Mu50 (**b**) strains, cultured with 1/4 MIC oxacillin, 1/4 PM-19, or a combination of PM-19 and oxacillin (1/4 MIC for each). $^{**}p < 0.01$ for $n = 3$. Compound-untreated cells (control) showed no observable amount of W atoms in the cells

oxacillin compared with the case with PM alone. The most highlighted result was obtained for PM-19, as shown in Fig. 4.20 where the cultivation of the bacteria cells in the coexistence of oxacillin and PM-19 (at 1/4 MIC for each) enhanced the uptake of W atoms (of PM-19) into both MRSA SR3605 (a) and VRSA Mu50 (b) cells with $p < 0.01$ for $n = 3$ (Hino K, Inoue M, Oda M, Nakamura Y, Yamase T Unpublished results in paper preparation). Interestingly, when 1×10^7 cfu/ml Mu50 cells with 1/4 MIC oxacillin alone were at first cultured at 37 °C in MHB until the initial stage of a logarithmic growth phase and then cultured again with 1/4 MIC PM for 24, the amounts of the uptaken PM were extremely small compared with other three culture modes with 1/4 MIC PM alone until the initial stage of a logarithmic growth phase, at first 1/4 MIC PM until the logarithmic growth phase and subsequent with 1/4 MIC oxacillin for 24 h, and a combination of 1/4 MIC PM and 1/4 MIC oxacillin. There was no significant difference in the amounts of the uptaken PM between modes of the latter two. Figure 4.21 exemplified amounts of W atoms (of PM-27) uptaken into the Mu50 cells for a variety of cultivation modes of the Mu50 cells (Hino K, Inoue M, Oda M, Nakamura Y, Yamase T Unpublished results in paper preparation). As shown in Fig. 4.21, the amount of uptaken W atoms for the mode with at first 1/4 MIC PM until the logarithmic growth phase and subsequent with 1/4 MIC oxacillin for 24 h is higher than with 1/4 MIC PM alone until the initial stage of a logarithmic growth phase, implying that the increase of the amount of the PMs uptaken into the cells by oxacillin strongly associated with the enhancement of the PM inhibition of the transcription from the *mecA* and *pbp*s genes to mRNA by oxacillin (Figs. 4.16b and 4.17b). It provides a mechanistic key that the enhancement of the PM uptake into the membrane by β-lactam antibiotics which will be captured by PBP proteins on the membrane surface was transient and diminished within 24 h until the initial stage of

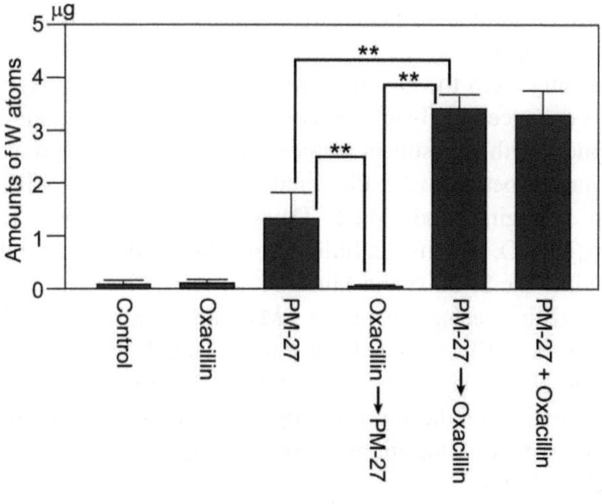

Fig. 4.21 Amounts of W atoms (of PM-27) uptaken into the Mu50 cells for six cultivation modes of the Mu50 cells. Six cultivation modes are (1) cultivation without substrate (control), (2) 1/4 MIC oxacillin alone until the stage of a logarithmic growth phase ($OD_{660} \approx 0.2$), (3) 1/4 MIC PM-27 alone until the stage of a logarithmic growth phase ($OD_{660} \approx 0.2$), (4) at first 1/4 MIC oxacillin until the logarithmic growth phase and subsequent 1/4 MIC PM-27 for 24 h, (5) at first 1/4 MIC PM-27 until the logarithmic growth phase and subsequent 1/4 MIC oxacillin for 24 h, and (6) a combination of PM-19 and oxacillin (1/4 MIC for each) until the stage of a logarithmic growth phase ($OD_{660} \approx 0.2$). $^{**}p < 0.01$ for $n = 3$

the logarithmic growth ($OD_{660} \approx 0.2$) of bacteria. One should remark that an ATP-binding cassette (ABC) transporter as a family of membrane transporter systems is encoded by the transcription of *abcA* gene and contributes to the import or export of a wide range of substances of proteins, peptides, polysaccharides, vitamins, and drugs through a use of ATP (Higgins 1992; Maupin-Furlow et al. 1995; Grunden and Shanmugam 1996). It is well known that the methicillin (or other β-lactam antibiotics such as oxacillin and cefoxitin) stimulates the transcription of *abcA* gene: *abcA* expression was induced by increasing concentrations of methicillin, and its induction was obtained only after 3 h of exposure to methicillin (since the transcription of *abcA* reached its maximum later in the exponential growth phase) (Schrader-Fischer and Berger-Bächi 2001). Therefore, the difference in PMs-uptake patterns among the above different culture modes (Fig. 4.21) seems to be associated with the maximum observed for the transcription of *abcA* induced by oxacillin. The oxacillin-induced increase of the PMs uptaken in the cytoplasm membrane makes a base of the synergistic activity of the PMs, since the PMs within the cytoplasm membrane indirectly inhibit the *mecA* and *pbp*s transcription to lead to the inhibition of PBP2′, PBPA, and PBP2-4 expressions on the membrane surface (Figs. 4.16 and 4.17). The ICP analysis of the MRSA SR3605 (or VRSA Mu50) cells treated with 1/4 MIC PMs also revealed the preferential uptake of Na^+ (with ratios more than ten, compared with the case of the absence of PMs) into the cells as counter cations of PM anions, although Mg^{2+}, K^+, and Ca^{2+} as other cations were detected (Hino K,

Inoue M, Oda M, Nakamura Y, Yamase T Unpublished results in paper preparation). Together with the location of the PMs within the cytoplasmic membrane, thus, the electric field arising from the electrochemical PMs/Na^+ double layer over the cell inside, which is enhanced by β-lactams, contributes to the inhibition of the *mecA* and *pbps* transcription with a resultant depression of the transcription process. This reflects the synergy between the PMs and β-lactams against MRSA and VRSA strains, when Keggin- and Wells–Dawson-structural PMs are employed. $(NH_4)_6[Mo_7O_{24}]\cdot 4H_2O$, which was little synergistic (Table 4.11), exhibited no significant uptake into the SR3605 and Mu50 cells irrespective of oxacillin, supporting the above synergistic mechanism of the PMs. However, other mechanisms of the synergy are not necessarily excluded, since the RT-PCR result for a double Keggin-structural PM, $K_9H_5[(GeTi_3W_9O_{37})_2O_3]\cdot 16H_2O$ (PM-504) with a larger ionic size suggests the inhibition of the posttranscription process of either translation from mRNA to PBP2′ or posttranslations of the folding of the translated polypeptide chain (Inoue et al. 2005).

4.5 Conclusions

The antitumor activity of polyoxomolybdates, in particular, PM-8 (containing $[Mo_7O_{24}]^{6-}$) and PM-17 (containing $[H_2Mo^V_{12}O_{28}(OH)_{12}(Mo^{VI}O_3)_4]^{6-}$) against solid tumors such as human (breast, lung, and colon) cancer xenografts (MX-1, OAT, and CO-4, respectively) has been extended to both AsPC-1 human pancreatic cancer and MKN-45 human gastric cancer. $[H_2Mo^V_{12}O_{28}(OH)_{12}(Mo^{VI}O_3)_4]^{6-}$ as 12-electron-reduced species of $[Mo_7O_{24}]^{6-}$ was in vitro and in vivo more inhibitory against solid cancer growth than $[Mo_7O_{24}]^{6-}$, due to the activation of both apoptosis and autophagy though the activation of caspase-3 and caspase-12. Intratumoral injection of PM-17 showed a significant efficacy in an experimental mice inoculation model of AsPC-1 and MKN-45, which indicates a therapeutic usefulness of PM-17 as an alternative chemotherapy of PM-8. Together with the fact that the pancreatic cancer is an aggressive form of cancer and has one of the poorest outcomes of all cancers, the development of the drug-delivery system for PM-17 is required for the realization of PM-17 as a novel anticancer drug. It is assumed that the multi-electron reduction of $[Mo_7O_{24}]^{6-}$ to $[H_2Mo^V_{12}O_{28}(OH)_{12}(Mo^{VI}O_3)_4]^{6-}$ is biologically possible (namely, PM-17 is one of metabolites of PM-8) through the one-electron reduction of the 1:1 $[Mo_7O_{24}]^{6-}$-FMN complex formed on the tumor cell mitochondria with a resultant suppression of the tumor growth due to the inhibition of the ATP generation. Many polyoxotungstates with Keggin or Wells–Dawson related (lacunary and multiply condensed) structures, in particular, $K_{13}[Ce(SiW_{11}O_{39})_2]\cdot 26H_2O$ (JM1590), $K_6[BGa(H_2O)W_{11}O_{39}]\cdot 15H_2O$ (JM2766), $[Me_3NH]_8[(SiNb_3W_9O_{37})_2O_3]$ (JM2820), $K_7[PTi_2W_{10}O_{40}]\cdot 6H_2O$ (PM-19), $K_9H_5[(GeTi_3W_9O_{37})_2O_3]\cdot 16H_2O$ (PM-504), $[Pr^iNH_3]_6H[PTi_2W_{10}O_{38}(O_2)_2]\cdot H_2O$ (PM-523), $K_{10}[Fe_4(H_2O)_2(PW_9O_{34})_2]\cdot xH_2O$ (HS-058), $K_7[P_2W_{17}NbO_{62}]$, $K_7[P_2W_{17}(NbO_2)O_{61}]$, and $K_{11}H[(VO)_3(SbW_9O_{33})_2]\cdot 27H_2O$ (PM-1002), were in vitro active against DNA and RNA viruses. PM-19 showed a potent inhibition both in vitro and in vivo against a broad spectrum of DNA viruses including herpes simplex virus (HSV-1 and HSV-2), TK-HSV, and HCMV. In vivo anti-HSV activity of PM-19 was attributed to an interference with the virus uptake by the cells and/or a clearance of HSV and also partially to the activation of macrophages. Ti/W- and V/W-mixed PMs have a broad antiviral activity against enveloped RNA viruses such as HIV, FluV-A, and SARS-V, the infections of which in high-risk individuals cause severe manifestations with often resultant fatalities. A significant in vivo potentiality against HIV was recognized for PM-523 and PM-1002. While most of PMs are inhibitory only at the virus adsorption stage (e.g., due to the binding of PMs with HIV-1 gp120), PM-523 did not inhibit the adsorption of the influenza A virus (FluV-A) onto the cell membrane (0–1.5 h after infection) but inhibited the fusion between the FluV-A envelope and the cellular membrane (1.5–120 h after infection). Such a different mechanism of anti-influenza activity for PM-523 led to both in vitro and in vivo synergistic efficacies in combination with ribavirin (as a viral RNA synthesis inhibitor) or zanamivir (NA inhibitor). The antifusion activity of PM-523 was investigated by using PM-523-resistant FluV-A strains and was attributed to the binding of PM-523 to the interface edges of HA trimers, which inhibits the opening of HA1 trimers of FluV-A. It is found that some of above PMs inhibited specific sialyl/sulfotransferases with in vitro enzymatic activities of at least 100- to 1,000-fold stronger than any other known inhibitors, implying that the PMs directly inhibit the activities of enzymes involved in carbohydrate metabolism to lead to the inhibition of the binding of the HA ligands of influenza virus with sialic acid surface receptors.

The synergistic effect of PMs with β-lactam antibiotics is the basis of the development of a novel chemotherapy for MRSA/VRSA-infected disease. Keggin-structural polyoxotungstates and their lacunary species, $[XW_{11}O_{39}]^{n-}$ and $[XW_9O_{34}]^{n-}$, exhibited a synergistic effect with oxacillin, with FIC indices of 0.010–0.156. Mechanistic details on the synergistic activity of PMs against five strains of two constitutive MRSA (SR3605 and MRS394-1), an inducible MRSA (ATCC43300), and two constitutive VRSA (Mu3 and Mu50) strains were investigated by using $K_6[P_2W_{18}O_{62}]\cdot 14H_2O$ (PM-27), $K_4[SiMo_{12}O_{40}]\cdot 3H_2O$ ($SiMo_{12}$), and PM-19. The ABC transporter, which participates in the transport of PMs into the cytoplasm membrane with an accompanying preferential flow of Na^+ as counter cations, is activated by β-lactams captured on the membrane surface, and thereby the uptaken PMs within the membrane make an electrochemical double layer with Na^+ cations inside of the cytoplasm. The transcription of *mecA* and *pbp*s genes to mRNA will be depressed within such PM/Na^+ electrical double layer to lead to depression of the formation of the PBP enzymes on the membrane surface. Together with the PBP binding with β-lactams on the membrane surface, the β-lactams-induced enhancement of the PM-induced suppression of the PBPs expression provides a base of the synergistic activity in combination of PMs and β-lactams. In vivo experiments are required for both the development of a novel chemotherapy for MRSA/VRSA-infected disease and the certification of the above mechanism.

Acknowledgments I would like to thank Professors Masazumi Eriguchi and Hironobu Yanagië (University of Tokyo), Professor Kenji Okuda and Dr. Kiyohiko Matsui (Yokohama City University), Professors Katsuko Yamashita and Akira Seko (Tokyo Institute of Technology), Professor Shiro Shigeta (Fukushima Medical University), and Dr. Yutaka Tajima (Juntendo University) for their valuable discussions and help in my research to date. Also, I acknowledge Grants-in-Aid for Scientific Research, No. 14204067 from the Ministry of Education, Science, Sports, and Culture and No. 99P01201 from RFTF/JSPS, and "Creation of bio-devices and bio-systems with chemical and biological molecules for medical use," CREST, Japan Science and Technology Agency (JST).

References

Baba M, Snoeck R, Pauwels R, DeClerq E (1988) Sulfated polysaccharides are potent and selective inhibitors of various enveloped viruses including herpes simplex viruses, cytomegalovirus, vesicular stomatitis viruses, and human immunodeficiency virus. Antimicrob Agents Chemother 32:1742–1745

Barnard DL, Hill CL, Gage T, Matheson JE, Huffmann JH, Sidwell RW, Otto MI, Schinazi RF (1997) Potent inhibition of respiratory syncytial virus by polyoxometalates of several structure classes. Antiviral Res 34:27–37

Blumenthal R, Clgue MJ, Durell SR, Epand RM (2003) Membrane fusion. Chem Rev 103:53–69

Boneca IG, Chiosis G (2003) Vancomycin resistance: occurrence, mechanism and strategies to combat it. Expert Opin Ther Targets 7:311–328

Borrás-Alamenar JJ, Coronado E, Müller A, Pope MT (eds) (2003) Polyoxometalate molecular science, NATO science series. Kluwer, Dordrecht

Botar B, Yamase T, Ishikawa E (2001) Synthesis and crystal structure of a novel vanadium-containing tungstobismutate(III) $K_{12}[(VO)_3(BiW_9O_{33})_2]\cdot 30H_2O$. Inorg Chem Commun 4:551–554

Callahan LN, Phelan M, Mallinson M, Norcross MA (1991) Dextran sulfate blocks antibody binding with gp120-CD4 interactions J. Virol 65:1543–1550

Cholewa M, Legge GJF, Weigold H, Holan G, Birch CJ (1994) The use of a scanning proton microprobe to observe anti-HIV drugs within cells. Life Sci 54:1607–1612

Clarke PGH (1990) Developmental cell death: morphological diversity and multiple mechanisms. Anat Embryol 181:195–213

Dan K, Miyashita K, Seto Y, Yamase T (1998) Quantitation of herpes simplex viral DNA in Vero cells for evaluation of an antiviral agent using the polymerase chain reaction. J Virol Methods 76:73–79

Dan K, Miyashita K, Seto Y, Fujita H, Yamase T (2002) The memory effect of heteropolyoxotungstate (PM-19) pretreatment on infection by herpes simplex virus at the penetration stage. Pharmacol Res 46:357–362

Dan K, Miyashita K, Seto Y, Fujita H, Yamase T (2003) Mechanism of the protective effect of heteropolyoxotungstate (PM-19) against herpes simplex virus type 2. Pharmacology 67:83–89

Dan K, Yamase T (2006) Prevention of the interaction between HVEM, herpes virus entry mediator, and gD, HSV envelope protein, by a Keggin polyoxotungstate, PM-19. Biomed Pharmacother 60:169–173

DeClerq E (2005) New approaches toward anti-HIV chemotherapy. J Med Chem 48:1313–1297

Domaille PJ, Knoth WH (1983) $Ti_2W_{10}PO_{40}{}^{7-}$ and $(CpFe(CO)_2Sn)_2\ W_{10}PO_{38}{}^{5-}$. Preparation properties and structure determination by tungsten-183 NMR. Inorg Chem 22:818–822

Donzella GA, Schols D, Lin SW, Esté JA, Nagashima A, Maddon PJ, Allaway GP, Sakmar TP, Henson G, DeClerq E, Moore JP (1998) AMD3100, a small molecule inhibitor of HIV-1 entry via the CXCR4 co-receptor. Nat Med 4:72–77

Elion GB, Furman PA, Fyfe JA, Maranda P, Beauchamp L, Schaeffer HJ (1977) Selectivity of action of an antiherpetic agent, 9-(2-hydroxyethoxymethyl)guanine. Proc Natl Acad Sci USA 74:5716–5720

Fujita H, Fujita T, Sakurai T, Yamase T, Seto Y (1992) Antitumor activity of new antitumor substance, polyoxomolybdate, against several human cancers in athymic nude mice. Tohoku J Exp Med 168:421–426

Fukuda N, Yamase T (1997) *In vitro* antibacterial activity of vanadate and vanadyl compounds against *Streptococcus pneumoniae*. Biol Pharm Bull 20:927–930

Fukuda N, Yamase T, Tajima Y (1999) Inhibitory effect of polyoxotungstates on the production of penicillin-binding proteins and β-lactamase against methicillin-resistant Staphylococcus aureus. Biol Pharm Bull 22:463–470, In this ref. all the MIC values in Tables 1-3 should be corrected to μM unit instead of μg/ml unit

Fukuma M, Seto Y, Yamase T (1991) *In vitro* antiviral activity of polyoxotungstate (PM-19) and other polyoxometalates against herpes simplex viruses. Antiviral Res 16:327–339

Geng J, Li M, Ren J, Wang E, Qu X (2011) Polyoxometalates as inhibitors of the aggregation of amyloid β peptides associated with Alzheimer's disease. Angew Chem Int Ed 50:4184–4188

Giaquinta RT, Dilley RA (1975) A partial reaction in photosystem II: reduction of silicomolybdate prior to the site of dichlorophenyldimethylurea inhibition. Biochim Biophys Acta 387:288–305

Gilbert BE, Knight V (1986) Biochemistry and clinical application of ribavirin. Antimicrob Agents Chemother 30:201–205

Golstein P, Kroemer G (2007) Cell death by necrosis: towards a molecular definition. Trends Biochem Sci 32:37–43

Grunden AM, Shanmugam KT (1996) Molybdate transport and regulation in bacteria. Arch Microbiol 168:345–354

Hallander HO, Dornbusch K, Gezelius L, Jacobson K, Karlsson I (1982) Synergism between aminoglycosides and cephalosporins with antipseudomonal activity: interaction index and killing curve method. Antimicrob Agents Chemother 22:743–752

Hanaki H, Kuwahara-arai K, Boyle-Vavra S, Daum RS, Labischinski H, Hiramatsu K (1998a) Activated cell-wall synthesis is associated with vancomycin resistance in methicillin-resistant *Staphylococcus aureus* strains Mu3 and Mu50. J Antimicrob Chemother 42:199–209

Hanaki H, Labischinski H, Inaba Y, Kondo N, Murakami H, Hiramatsu K (1998b) Increase in glutamine-non-amidated muropeptides in the peptidoglycan of vancomycin-resistant *Staphylococcus aureus* strain Mu50. J Antimicrob Chemother 42:315–320

Higgins CF (1992) ABC transporters: from microorganism to man. Annu Rev Cell Biol 8:67–113

Hill CL (ed) (1998) Polyoxometalates. Chem Rev 98:1–387

Hill CL, Weeks MS, Shinazi RF (1990) Anti-HIV activity, toxicity, and stability studies of representative structural families of polyoxometalates. J Med Chem 33:2767–2772

Hosoya M, Balzarini J, Shigeta S, DeClerq E (1991) Differential inhibitory effect of sulfated polysaccharides and polymers on the replication of various myxoviruses and retroviruses, depending on the composition of the target amino acid sequences of the viral envelope glycoproteins. Antimicrob Agents Chemother 35:2515–2520

Ikeda S, Nishiya S, Yamamoto A, Yamase T, Nishimura C, DeClercq E (1993) Activity of the Keggin polyoxotungstate PM-19 against herpes simplex virus type 2 infection in immunosuppressed mice; role of peritoneal macrophage activation. J Med Virol 41:191–195

Ikeda S, Nishiya S, Yamamoto A, Yamase T, Nishimura C, DeClercq E (1994) Antiviral activity of a keggin polyoxotungstate PM-19 against herpes simplex virus in mice. Antivir Chem Chemother 5:47–50

Inoue M, Segawa K, Matsunaga S, Matsumoto N, Oda M, Yamase T (2005) Antimicrobial activity of highly negative charged polyoxotungstates, $K_{27}[KAs_4W_{44}O_{140}]$ and $K_{18}[KSb_9W_{21}O_{86}]$, and Keggin-structural polyoxotungstates against *Helicobacter pylori*. J Inorg Biochem 99:1023–1031

Inoue M, Suzuki T, Fujita Y, Oda M, Matsumoto N, Yamase T (2005) Enhancement of antibacterial activity of β-lactam antibiotics by $[P_2W_{18}O_{62}]^{6-}$, $[SiMo_{12}O_{40}]^{4-}$, and

$[PTi_2W_{10}O_{40}]^{7-}$ against methicillin-resistant and vancomycin-resistant *Staphylococcus aureus*. J Inorg Biochem 100:1225–1233

Inoue M, Suzuki T, Fujita Y, Oda M, Matsumoto N, Iijima J, Yamase T (2006b) Synergistic effect of polyoxometalates in combination with oxacillin against methicillin-resistant and vancomycin-resistant *Staphylococcus aureus*: a high initial inoculum of 1 x 10^8 cfu/ml for *in vivo* test. Biomed Pharmacother 60:220–226

Inouye Y, Take Y, Tokutake Y, Yoshida T, Yamamoto A, Yamase T (1990) Inhibition of replication of human immunodeficiency virus by a heteropolyoxotungstate (PM-19). Chem Pharm Bull 38:285–287

Inouye Y, Tokutake Y, Kunihara J, Yoshida T, Yamase T, Nakata A, Nakamura S (1992) Suppressive effect of polyoxometalates on the cytopathogenicity of human immunodeficiency virus type 1 (HIV-1) *in vitro* and their inhibitory activity against HIV-1 reverse transcriptase. Chem Pharm Bull 40:805–807

Inouye Y, Tokutake Y, Yoshida T, Seto Y, Fujita H, Dan K, Yamamoto A, Nishiya S, Yamase T, Nakamura S (1993) In vitro antiviral activity of polyoxomolybdates. Mechanism of inhibitory effect of PM-104 $(NH_4)_{12}H_2(Eu_4(MoO_4)(H_2O)_{16}(Mo_7O_{24})_4)\cdot 13H_2O$ on human immunodeficiency virus type 1. Antiviral Res 20:317–331

Inouye Y, Fujimoto Y, Sugiyama M, Yoshida T, Yamase T (1995) Structure-activity correlationship and strain specificity of polyoxometalates in anti-human immunodeficiency virus activity. Biol Pharm Bull 18:996–1000

Ishikawa E, Yamase T (2006) ^{31}P NMR and isothermal titration calorimetry studies on polyoxomolybdates-catalyzed hydrolysis of ATP. J Inorg Biochem 100:344–350

Jasmin C, Raybaud N, Chermann JC, Haapala D, Sinoussi F, Boy Loustau C, Bonissol C, Kona P, Raynaud M (1973) *In vivo* effects of silicotungstate on some RNA viruses. Biomedicine 18:319–327

Jasmin C, Chermann JC, Hervé G, Teze A, Souchay P, Boy Loustau C, Raybaud N, Sinoussi F, Raynaud M (1974) *In vivo* inhibition of murine leukemia and sarcoma viruses by the heteropolyanion 5-tungsto-2-antimonate. J Natl Cancer Inst 53:463–474

Jevons MP (1961) "Celbenin"-resistant Staphylococci. Br Med J 1:124–125

Judd DA, Nettles JH, Nevins N, Snyder JP, Liotta DC, Tang J, Ermolieff J, Schinazi RF, Hill CL (2001) Polyoxometalate HIV-1 protease inhibitors. A new mode of protease inhibition. J Am Chem Soc 123:886–897

Kimpton J, Emerman M (1992) Detection of replication-component and pseudotyped human immunodeficiency virus with a sensitive cell line on the basis of activation of an integrated β-galactosidase gene. J Virol 66:2232–2239

Lamparska-Przybysz M, Gajkowska B, Motyl T (2005) Cathepsins and BID are involved in the molecular switch between apoptosis and autophagy in breast cancer MCF-7 cells exposed to camptothecin. J Physiol Pharmacol 56:159–179

Long DL, Thusnashima R, Cronin L (2010) Polyoxometalates: building blocks for functional nanoscale systems. Angew Chem Int Ed 49:1736–1758

Lyon DK, Miller WK, Novet T, Domaille PJ, Evitt E, Johnson DC, Finke RG (1991) Highly oxidation resistant inorganic-porphyrin analog polyoxometalate oxidation catalysts. I. The synthesis and characterization of aqueous-soluble potassium salts of α_2-$P_2W_{17}O_{61}(M^{n+}\cdot OH_2)^{(n-10)}$ and organic solvent soluble tetra-n-butylammonium salts of α_2-$P_2W_{17}O_{61}(M^{n+}\cdot OH_2)^{(n-11)}$ ($M = Mn^{3+}$, Fe^{3+}, Co^{2+}, Ni^{2+}, Cu^{2+}). J Am Chem Soc 113:7209–7221

Mammén M, Choi SK, Whitesides GM (1998) Polyvalent interactions in biological systems: implications for design and use of multivalent ligands and inhibitors. Angew Chem Int Ed 37:2754–2794

Matsuhashi M, Dong Song M, Ishino F, Washi M, Doi M, Inoue M, Ubukata K, Yamashita N, Konno M (1986) Molecular cloning of the gene of a penicillin-binding protein supported to cause high resistance to β-lactam antibiotics in *Staphylococci aureus*. J Bacteriol 167:975–980

Maupin-Furlow JA, Rosentel JA, Lee J, Deppenmeier U, Gunsalus RP, Shanmugam KT (1995) Genetic analysis of the *modABCD* (molybdate transport) operon of *Escherichia coli*. J Bacteriol 177:4851–4856

Mitsui S, Ogata A, Yanagie H, Kasano H, Hisa T, Yamase T, Eriguchi M (2006) Antitumor activity of polyoxomolybdate, $[NH_3Pr^i]_6[Mo_7O_{24}]\cdot 3H_2O$ against human gastric cancer model. Biomed Pharmacother 60:353–358

Mitsuya H, Weinhold KJ, Furman FA, St Clair MH, Lehrman SN, Gallo RC, Bolognesi D, Barry DW, Broder S (1985) 3'-Azido-3'deoxythymidine (BW A509U): an antiviral agent that inhibits the infectivity and cytopathic effect of human T-lymphotropic virus type III/lymph-adenopathy-associated virus in vitro. Proc Natl Acad Sci USA 82:7096–7100

Mitsuya H, Looney DJ, Kuno S, Ueno R, Wong-Staal F, Broder S (1988) Dextran sulfate suppression of viruses in the HIV family: inhibition of virion binding to $CD4^+$ cells. Science 240:646–649

Mori S, Shigeta S (1995) A colorimetric LDH assay for the titration of infectivity and the evaluation of anti-viral activity against ortho- and paramyxoviruses. Tohoku J Exp Med 177:315–325

Moskovitz B, HPA-23 Cooperative Study Group (1988) Clinical trial of tolerance of HPA-23 in patients with acquired immune deficiency syndrome. Antimicrob Agents Chemother 32:1300–1303

Naruke H, Ozeki T, Yamase T (1991) Structure of photoluminescent polyoxomolybdoeuropate $(NH_4)_{12}H_2[Eu_4(MoO_4)(H_2O)_{10}(Mo_7O_{24})_4]\cdot 13H_2O$. Acta Crystallogr C47:693–696

Oda M, Inoue M, Hino K, Nakamura Y, Yamase T (2007) Enhancement effect of polyoxometalates on NGF-induced neurite-outgrowth for PC-12 cells. Biol Pharm Bull 30:787–790

Ogata A, Mitsui S, Yanagie H, Kasano H, Hisa T, Yamase T, Eriguchi M (2005) A novel anti-tumor agent, Polyoxomolybdate induces apoptotic cell death in AsPC-1 human pancreatic cancer cells. Biomed Pharmacother 59:240–244

Ogata A, Yanagie H, Ishikawa E, Morishima Y, Mitsui S, Yamashita A, Hasumi K, Takamoto S, Yamase T, Eriguchi M (2008) Anti-tumoral effect of polyoxomolybdates: induction of apoptotic cell death and autophagy in *in vitro* and *in vivo* models. Br J Cancer 98:399–409

Ohashi Y, Yanagi K, Sasada Y, Yamase T (1982) Crystal structure and photochemistry of isopolymolybdates. I. The crystal structures of hexakis(propylammonium) hept Hexakis (propylammonium)heptamolybdate(VI) trihydrate and hexakis(isopropylammonium) heptamolybdate(VI) trihydrate. Bull Chem Soc Jpn 55:1254–1260

Pauwels R, Balzarini J, Baba M, Snoeck R, Schols D, Herdewijin P, Desmyter J, DeClercq E (1988) Rapid and automated tetrazolium-based colorimetric assay for the detection of anti-HIV compounds. J Virol Methods 20:309–321

Pinto AL, Lippard SJ (1985) Binding of the antitumor drug *cis*-diamminedichloroplatinum (II) (cisplatin) to DNA. Biochim Biophys Acta 780:167–180

Pope MT, Müller A (eds) (1994) Topics in molecular organization and engineering. Polyoxometalates: from platonic solids to anti-retroviral activity. Kluwer, Dordrecht

Pope MT, Müller A (eds) (2001) Polyoxometallates: from topology to industrial applications. Kluwer, Dordrecht

Proust A, Thouvenot R, Gouzerh P (2008) Functionalization of polyoxometalates: towards advanced applications in catalysis and material science. Chem Commun: 1837–1852

Rath VL, Verdugo D, Hemmerich S (2004) Sulfotransferase structural biology and inhibitor discovery. Drug Discov Today 9:1003–1011

Rhule JT, Hill CL, Judd DA (1998) Polyoxometalates in medicine. Chem Rev 98:327–357

Rozenbaum W, Dormont D, Spire B, Vilmer E, Gentilini M, Griscelli C, Montagnier L, Barr-Sinoucci F, Chermann JC (1985) Antimoniotungstate (HPA 23) treatment of three patients with AIDS and one with prodrome. Lancet 23:450–451

Ryffel C, Tesch W, Birch-Machin I, Reynolds PE, Barberis-Maino L, Kayser FH, Berger-Bächi B (1990) Sequence comparison of *mecA* genes isolated from methicillin-resistant *Staphylococcus aureus* and *Staphylococcus epidermidis*. Gene 94:137–138

Schrader-Fischer G, Berger-Bächi B (2001) The AbcA transporter of *Staphylococcus aureus* affects cell autolysis. Antimicrob Agents Chemother 45:407–412

Schwörer R, Schmidt RR (2002) Efficient sialyltransferase inhibitors based on glycosides of N-acetylglucosamine. J Am Chem Soc 124:1632–1637

Seko A, Yamase T, Yamashita K (2009) Polyoxometalates as effective inhibitors for sialyl- and sulfotransferases. J Inorg Biochem 101:1061–1066

Shigeta S (1999) Recent progress in anti-influenza chemotherapy. Drugs R D 2:153–164

Shigeta S (2001) Targets of anti-influenza chemotherapy other than neuraminidase and proton pump. Antivir Chem Chemother 12(Suppl):179–188

Shigeta S, Yamase T (2005) Current status of anti-SARS agents. Antivir Chem Chemother 16:23–32

Shigeta S, Mori S, Watanabe J, Yamase T, Schinazi RF (1996) *In vitro* antimyxovirus activity and mechanism of anti-influenza virus activity of polyoxometalates PM-504 and PM-523. Antivir Chem Chemother 7:346–352

Shigeta S, Mori S, Watanabe J, Soeda S, Takahashi K, Yamase T (1997) Synergistic anti-influenza virus A (H1N1) activities of PM-523 (polyoxometalate) and ribavirin in vitro and in vivo. Antimicrob Agents Chemother 41:1423–1427

Shigeta S, Mori S, Kodama E, Kodama J, Takahashi K, Yamase T (2003) Broad spectrum anti-RNA virus activities of titanium or vanadium substituted polyoxotungstates. Antiviral Res 58:265–271

Shigeta S, Mori S, Yamase T, Yamamoto N, Yamamoto N (2006) Anti-RNA virus activity of polyoxometalates. Biomed Pharmacother 60:211–219

Stryer L (1975) Biochemistry. Freeman, San Francisco, CA

Take Y, Tokutake Y, Inouye Y, Yoshida T, Yamamoto A, Yamase T, Nakamura S (1991) Inhibition of proliferation of human immunodeficiency virus type 1 by novel heteropolyoxotungstates *in vitro*. Antiviral Res 15:113–124

Takeuchi H, Kondo Y, Fujiwara K, Kanzawa T, Aoki H, Mills GB, Kondo S (2005) Synergistic augmentation of rapamycin-induced autophagy in malignant glioma cells by phosphatidylinositol 3-kinase/protein kinase B inhibitors. Cancer Res 65:3336–3346

Tomita K, Yamase T, Shishido K (1989) Medical chemistry of polyoxometalates. Part 2. Enzymatic study of binding of heptamolybdate to DNA. Inorg Chim Acta 157:L167–L169

Ubukata K, Yamashita N, Konno M (1985) Occurrence of a β-lactam-inducible penicillin-binding protein in methicillin-resistant Staphylococci. Antimicrob Agents Chemother 27:851–857

Watanabe W, Konno K, Ijichi K, Inoue H, Yokota T, Shigeta S (1994) MTT colorimetric assay system for the screening of anti-orthomyxo- and anti-paramyxoviral agents. J Virol Methods 48:257–265

Weeks MS, Hill CL, Schinazi RF (1992) Synthesis, characterization, and anti-human immunodeficiency virus activity of water-soluble salts of polyoxotungstate anions with covalently attached organic groups. J Med Chem 35:1216–1221

Witvrouw M, Weigold H, Pannecouque C, Schols D, DeClercq E, Holan G (2000) Potent anti-HIV (type 1 and type 2) activity of polyoxometalates: structure-activity relationship and mechanism of action. J Med Chem 43:778–783

Yamamoto N, Schols D, DeClercq E, Debyser Z, Pauwels R, Barlzarini J, Nakashima H, Baba M, Hosoya M, Snoeck R, Neyts J, Andrei G, Murrer BA, Theobalt B, Bossard G, Henson G, Abrams M, Picker D (1992) Mechanism of anti-human immunodeficiency virus action of polyoxometalates, a class of broad-spectrum antiviral agents. Mol Pharmacol 42:1109–1117

Yamase T (1982) Photochemical studies of the alkylammonium molybdates. Part 6. Photoreducible octahedron site of $[Mo_7O_{24}]^{6-}$ as determined by electron spin resonance. JCS Dalton Trans: 1987–1991

Yamase T (1994) Polyoxometalates for molecular devices: antitumor activity and luminescence. In: Pope MT, Müller A (eds) Topics in molecular organization and engineering. Polyoxometallates: from platonic solids to anti-retroviral activity. Kluwer, Dordrecht

Yamase T (1996) Antitumoral and antiviral polyoxometalates (inorganic discrete polymers of metal oxide). In: Salamone JC (ed) Polymeric materials encyclopedia: synthesis, properties, and applications. CRC, Boca Raton, FL

Yamase T (2003) Photoredox chemistry of polyoxometalates as a photocatalyst. Catal Surv Asia 7:203–217

Yamase T (2005) Anti-tumor, antiviral, and antibacterial activities of polyoxometalates for realizing an inorganic drug. J Mater Chem 15:4773–4782

Yamase T, Ikawa T (1977) Photochemical studies of the alkylammonium molybdates. III. Preparation and properties. Bull Chem Soc Jpn 50:746–749

Yamase T, Ishikawa E (2008) Photoreductive self-assembly of $[Mo^{VI}{}_{7}O_{24}]^{6-}$ to anti-tumoral $[H_2Mo^{V}{}_{12}O_{28}(OH)_{12}(Mo^{VI}O_3)_4]^{10-}$ in aqueous media. Bull Chem Soc Jpn 81:983–991

Yamase T, Pope MT (eds) (2002) Polyoxometalate chemistry for nanocomposite design. Kluwer, Dordrecht

Yamase T, Tomita K (1990) Medical chemistry of polyoxometalates. Part 3. Electrochemical study of a 1:1 polyoxomolybdate-flavin mononucleotide complex in aqueous solutions. Inorg Chim Acta 169:147–150

Yamase T, Sasaki R, Ikawa T (1981) Photochemical studies of the alkylammonium molybdates. Part 5. Photolysis in weak acid solutions. JCS Dalton Trans: 628–634

Yamase T, Fujita H, Fukushima K (1988) Medical chemistry of polyoxometalates. Part 1. Potent antitumor activity of polyoxomolybdates in animal transplantable tumors and human cancer xenograft. Inorg Chim Acta 151:L15–L18

Yamase T, Tomita K, Seto Y, Fujita H (1992a) Antitumor and antiviral activities of certain polyoxometalates. In: Ottenbrite RM, Chiellini E (eds) Polymers in medicine: biomedical and pharmaceutical applications. Technomic Publishing Company Inc., Lancaster, PA, pp 187–212

Yamase T, Ozeki T, Motomura S (1992b) ^{183}W NMR and x-ray crystallographic studies on the peroxo complexes of the Ti-substituted α-Keggin typed tungstophosphates. Bull Chem Soc Jpn 65:1453–1459

Yamase T, Ozeki T, Sakamoto H, Nishiya S, Yamamoro A (1993) Structure of hexatitanooctadecatungstodigermanate. Bull Chem Soc Jpn 66:103–108

Yamase T, Fukuda N, Tajima Y (1996a) Synergistic effect of polyoxotungstates in combination with *β*-lactam antibiotics on antibacterial activity against methicillin-resistant Staphylococcus aureus. Biol Pharm Bull 19:459–465, In this ref. all the MIC values in Tables 1-3 should be corrected to μM unit instead of μg/ml unit

Yamase T, Ohtaka K, Suzuki M (1996b) Structural characterization of spherical octadecavanadates encapsulating Cl^- and H_2O. JCS Dalton Trans: 283–289

Yamase T, Suzuki M, Ohtaka K (1997) Structures of photochemically prepared mixed-valence polyoxovanadate clusters: oblong $[V_{18}O_{44}(N_3)]^{14-}$, superkeggin $[V_{18}O_{42}(PO_4)]^{11-}$ and doughnut-shaped $[V_{12}B_{32}O_{84}Na_4]^{15-}$ anions. JCS Dalton Trans: 2463–2472

Yamase T, Makino H, Naruke H, San José Wéry AM (2000) A spherical potassium-capped vanadium methylphosphonate as another *ε*-Keggin fragment, $[H_6KV_{12}O_{27}(VO_4)(PO_3CH_3)_3]^{5-}$. Chem Lett: 1350–1351

Yamase T, Botar B, Ishikawa E, Fukaya K (2001) Chemical structure and intramolecular spin-exchange interaction of [(VO)3(SbW9O33)2]12-. Chem Lett: 56–57

Yamase T, Botar B, Ishikawa E, Fukaya K, Shigeta S (2002) Magnetic exchange coupling and potent antiviral activity of $[(VO)_3(SbW_9O_{33})_2]^{12-}$. In: Yamase T, Pope MT (eds) Polyoxometalate chemistry for nanocomposite design. Kluwer, Dordrecht

Yamase T, Cao X, Yazaki S (2007) Structure of double keggin-Ti/W-mixed polyanion, $[A\text{-}\beta\text{-}GeTi_3W_9O_{37}]_2O_3]^{14-}$ and multielectron-transfer-based photocatalytic H_2-generation. J Mol Catal A 262:119–127

Yoshimura K, Kato R, Kavlick MF, Nguyen A, Maroun V, Maeda K, Hussain KA, Ghosh AK, Gulnik SV, Erickson JW, Mitsuya H (2002) A potent human immunodeficiency virus type 1 protease inhibitor, UIC-94003 (TMC-126), and selection of a novel (A28S) mutation in protease active site. J. Virol 76:1349–1358

Zhao W, Hu ZQ, Okubo S, Hara Y, Shimamura T (2001) Mechanism of synergy between epigallocatechin gallate and β-lactam against methicillin-resistant *Staphylococcus aureus*. Antimicrob Agents Chemother 45:1737–1742

Zilinskas BA, Govindjee (1975) Silicomolybdate and silicotungstate mediated dichlorophenyl dimethylurea-insensitive photosystem II reaction: electron flow, chlorophyll a fluorescence and delayed light emission changes. Biochim Biophys Acta 387: 306–319

Chapter 5
Modeling Biosilicification at Subcellular Scales

Narjes Javaheri, Carolina M. Cronemberger, and Jaap A. Kaandorp

Contents

Abstract Biosilicification occurs in many organisms. Sponges and diatoms are major examples of them. In this chapter, we introduce a modeling approach that describes several biological mechanisms controlling silicification. Modeling biosilicification is a typical multiscale problem where processes at very different temporal and spatial scales need to be coupled: processes at the molecular level, physiological processes at the subcellular and cellular level, etc. In biosilicification morphology plays a fundamental role, and a spatiotemporal model is required. In the case of sponges, a particle simulation based on diffusion-limited aggregation is presented here. This model can describe fractal properties of silica aggregates in first steps of deposition on an organic template. In the case of diatoms, a reaction–diffusion model is introduced which can describe the concentrations of chemical components and has the possibility to include polymerization chain of reactions.

N. Javaheri • C.M. Cronemberger • J.A. Kaandorp (✉)
Section Computational Science, Faculty of Science, University of Amsterdam, Science Park 904, 1098 XH Amsterdam, The Netherlands
e-mail: J.A.Kaandorp@uva.nl

W.E.G. Müller et al. (eds.), *Biomedical Inorganic Polymers*, Progress in Molecular and Subcellular Biology 54, DOI 10.1007/978-3-642-41004-8_5,

5.1 Introduction

5.1.1 Challenges in Modeling Biomineralization

Biomineralization is the process by which specific minerals are deposited by living organisms. Modeling biomineralization consists of many subprocesses which only can be captured by coupling models defined at different spatiotemporal scales and requires the use of different computational methods. It is where hard matter (minerals) and soft matter (organic materials) are in the same system of modeling, and they interact in the presence of a solvent (Mann 2009). In most biominerals, the organic part not only controls the start of the process (nucleation) but also controls the rest, up to directing the shape and size of the crystal or amorphous material.

Another key point about biomineralization which makes its modeling as a challenge is that it contains processes which have significant behavior in several spatial and temporal scales (Harding et al. 2008). It means that different types of computational methods are needed in order to simulate the formation of the whole biomineral. For example, in the first steps of deposition or formation of precursor particles and in very small length scales, particles have to be considered at an atomic level. This demands atomic calculations such as ab initio and molecular dynamics, which have high computational costs. But if one is interested in modeling the larger scales, it is neither reasonable nor feasible to use atomic and molecular simulations to reconstruct the whole biominerals. In other words, at some point we need to consider another type of computational method like mesoscale and macroscale models. For example, one type of a mesoscale models is based on aggregate formation in which the initial units are smaller aggregates or precursor particles of order of few nanometers or more. Therefore, in this type, we are not considering calculations at the level of one by one molecular interactions. Moreover, we may need to build a model as a particle-based one or a continuous one based on distributions and concentrations of different chemical species.

During biomineralization various mechanisms are responsible for producing complex architectures of the emergent skeleton. In this chapter we will briefly describe two spatiotemporal models of silicification in sponges and in diatoms which includes some of the most important involved mechanisms.

5.2 Biosilicification in Sponges

5.2.1 About Sponges

Sponges (Porifera) represent the most basal metazoan phylum. The major skeletal elements in sponges are spicules formed from inorganic material. Demosponges, a large group of sponges, deposit amorphous hydrated silica in their spicules.

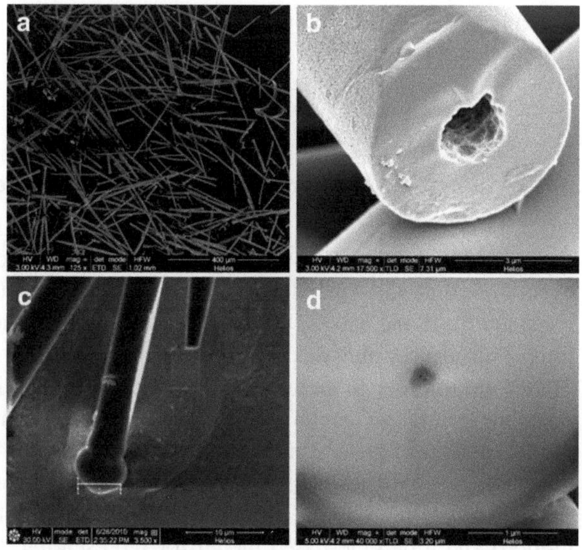

Fig. 5.1 SEM images of spicules of the sponge *Suberites domuncula*. (**a**) An image of a population of spicules. (**b**) A closer view of a broken spicule; central canal which is empty now is seen clearly. (**c**) Targeting an area of the spicule for high-energy ion beam cutting. (**d**) The new cross section of spicule after ion abrasion

The model organism used here is *Suberites domuncula*, a demosponge. The electron microscope images of the skeletal elements (called *spicules*) of this sponge are shown in Fig. 5.1a. This kind of spicule is of monaxonal styles type with a bobble shape at one end and a sharp part at the other.

Figure 5.1b shows an SEM image of the inside of a broken spicule. It shows that there is a central canal inside a spicule which had been filled with organic matter (in Fig. 5.1b it has been removed due to the sample preparation process). In order to have an image from a surface which is not affected by the preparation process, the spicule has been cut by a high-energy ion beam in a dual-beam SEM device. In Fig.5.1d the central canal is filled; however, due to the difference in the material properties, the border with surrounding silica is still visible. This axial filament is the initial template for deposition of the mineral inside the cell.

There have been recently many studies regarding the contribution and function of organic materials in biominerals. It has been shown that the growth of biosilica is a highly hierarchical process. Spicules of *S. domuncula* start their growth inside the special cells, called sclerocytes, and in subcellular level, inside the special vesicles, called silicasomes (Schröder et al. 2007). Sclerocytes produce influential proteins for silicification: silicatein, the enzyme that initiates the catalytic bio-condensation (Shimizu et al. 1998; Krasko et al. 2000), and silintaphin, a silicatein-interacting protein that it is believed to assist the formation of fibrillar structure of axial filament (Wiens et al. 2009, 2011). These proteins together make the initial organic template for biosilica deposition.

Besides sclerocytes which have a key role in formation of skeleton, other cell types, archaeocytes, chromocytes and lophocytes, have been identified to have an effect on controlling the later steps of spicule growth, both through producing proteins and affecting the aging process of biosilica (Schröder et al. 2000, 2006; Harrison and Davis 1982).

5.2.2 *Particle Simulation of the Beginning of Spicule Growth*

In systems biology, usually processes are only modeled using ordinary differential equations, and the spatial component is neglected (cells, etc. are usually not considered). In the case of silicification, emerging minerals have a complex structure. This causes the demand for including spatial information in the model. Therefore, in order to understand the process of biosilicification and to be able to compare simulation results with the actual growth forms, we need a spatiotemporal model capable of generating biological morphologies. To generate such simulated morphologies, one needs to combine two computational methods: partial differential equations for solving the diffusion equation (or its equivalent, Brownian motion) and a particle-based model for representing the morphological structure. In this section, we present a three-dimensional particle simulation, based on diffusion-limited aggregation approach, to model the beginning of the spicule growth in sponge *S. domuncula*.

5.2.2.1 Diffusion

Diffusion describes the spread of particles by random motion from regions of higher concentration to regions of lower concentration. The time dependence of the statistical distribution in space is given by the diffusion equation.

The equation is usually written as

$$\frac{\partial \phi\left(\vec{r},t\right)}{\partial t} = \nabla \cdot \left[D\left(\phi,\vec{r}\right)\nabla\phi\left(\vec{r},t\right)\right],$$

where $\phi\left(\vec{r},t\right)$ is the density of the diffusing material at location $\vec{r}$ and time t and $D\left(\phi,\vec{r}\right)$ is the diffusion coefficient and ∇ represents the vector differential operator acting on the space coordinates. If the diffusion coefficient depends on the density, then the equation is nonlinear; otherwise it is linear. If D is constant, then the equation reduces to the following linear equation:

$$\frac{\partial \phi\left(\vec{r},t\right)}{\partial t} = D\nabla^2\phi\left(\vec{r},t\right).$$

By discretizing time and space in numerical calculations, one obtains the well-known random walks.

5.2.2.2 Diffusion-Limited Aggregation Model

DLA (diffusion-limited aggregation) is a computational model for aggregate formation based on random walks of primary particles. It is applicable in many systems where diffusion is the main method of transportation. The theory was first proposed in 1981 (Witten and Sander 1981). It has had many applications, especially for growth models. There are numerous examples from physics, chemistry, and technology where aggregates and crystals like dielectric breakdown (Niemeyer et al. 1984; Arrayás et al. 2002), formation of electrodeposits (López-Tomàs et al. 1993), viscous fingering (Bensimon 1986; Park and Durian 1994), or crystal growth (Brener et al. 1996) have been studied using a combination of experiments and simulation models. An overview of abiotic growth patterns is, for example, provided by Ben-Jacob (1993); interestingly a review was published by the same author (Ben-Jacob 1997) on how a similar approach could be used to model growth patterns from biology (bacteria colonies). In both modeling growth patterns from physics and bacteria colonies, coarse-grained, lattice-based, aggregation models have been applied successfully to simulate growth patterns. Moreover, DLA has been applied to 2D simulation of silicification in diatoms (Gordon and Drum 1994).

The DLA algorithm is known for producing objects with self-similarity/fractal properties. One of the experimental methods for calculation of fractal properties of materials is X-ray scattering (Schmidt et al. 1989), in which by exposing the X-ray beam to the object and measuring intensity of scattered beam as a function of scattering angle, one can understand the distribution of matter on the sample and calculate the mass and surface fractal dimensions of the sample. Based on this method, the aggregation of silica particles in solution has been studied and the result reveals fractal properties of the growth process (Knoblich and Gerber 2001). Moreover, the structure of biosilica from cell walls of diatoms shows fractal properties (Vrieling et al. 2002) (in some cases, different fractal dimensions are detected for different regions of the curve which it is believed that is an indication on hierarchical morphology of biosilica and the existence of different spatial scales).

Therefore, considering the observed fractal properties of silica particles aggregates and biosilica, we apply the DLA algorithm to describe the first steps of biosilica formation in sponges. This, of course, will be followed with post-synthesizing effects like hardening and electrostatic forces in real biosilica.

Inside the sponge's cells, one of the material transport mechanisms is diffusion. For example, inside the vesicles of sclerocytes, where the spicule starts its growth, there is silicon in the form of particles with different sizes. TEM images of primorphs from *S. domuncula* show the high electron density regions due to the existence of silica in the form of small particles inside the vesicles (Müller et al. 2005). Therefore, we use the idea of these particles in our model as the primary particles that will make the aggregate. Moreover, the organic matrix, which in this case is mainly in the form of a central protein string, will represent the initial nucleation site for aggregate formation in the model. Particles have random walks (biased by electrostatic forces due to acidic proteins) until they join the aggregate. Hence, we consider the simulation box to be the space inside the vesicle and assume silica particles to perform a random walk until they join the aggregate which is growing on a central protein string in the middle of the box. It is assumed that the box is cubic for the first approximation and the 3D mesh is based on Cartesian coordinates.

The boundary condition of the simulation is assumed to be reflective. So, the particles that reached a position outside simulation box will be moved back to a mirror reflection position inside the box and will take part in the next time step of the simulation.

There are three different cases of our particle simulation which are based on facts that have been observed in images from growth process of spicules of *S. domuncula*. First, particles (precursors nanoparticles that will join the aggregate) inside the silicalemma (vesicle membrane that surrounds first steps of spicule formation) don't show any particular distribution and they seem to be randomly located. Second, before spicules leave the cell, silicalemma grows along with the growth of spicules inside them (Müller et al. 2005). Third, based on images of the distribution of particles inside the vesicle (Müller et al. 2005), it seems that there are primary particles distributed inside the volume of vesicle during the silica deposition and not only around the boundaries. This shows that the vesicle concentrates silica itself and inside the vesicle there is the supply of nanoparticles everywhere.

To implement these three assumptions, we introduce related setups for DLA simulation. The sink for particles is always the surface of growing aggregate, and as soon as particles reach the aggregate, they stick with probability equal to one. The location of source is different in three cases; in the first one a simulation box with fixed boundaries is assumed, and the source is a number of particles with random locations on planes located close and parallel to boundaries.

In the second one the source is of the same geometrical form, but boundaries and source planes are located close to the initial template and will expand during time in such a way that they will keep a distance from aggregated particles.

In the third case, boundaries are growing like the previous case, but the source is a random distribution in the whole simulation box.

In reality it can be a combination of two of the mentioned sources, for instance, if we consider that there is some silica inside the vesicles prior to starting spicule formation and there can be more additives later on during the growth coming

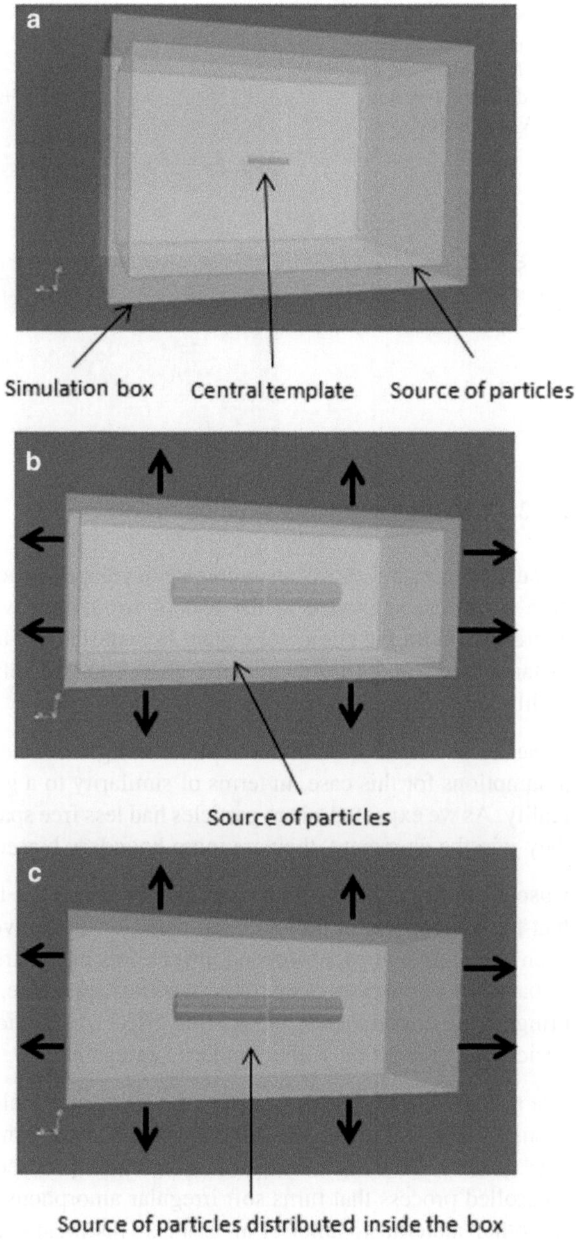

Fig. 5.2 3D geometry of the model for three cases. (**a**) Case 1: Simulation box (vesicle boundaries) has a fixed size particles are randomly distributed on planes next to boundaries. (**b**) Case 2: Simulation box grows along with aggregate growth particles are randomly distributed on planes next to boundaries. (**c**) Case 3: Simulation box grows along with aggregate growth particles are randomly distributed in the whole box

through cytoplasm and from boundaries. The geometry of the model and three setups of simulations are shown in Fig. 5.2.

For the particle simulation and fractal calculation Python and, for visualization, Mayavi2 have been used.

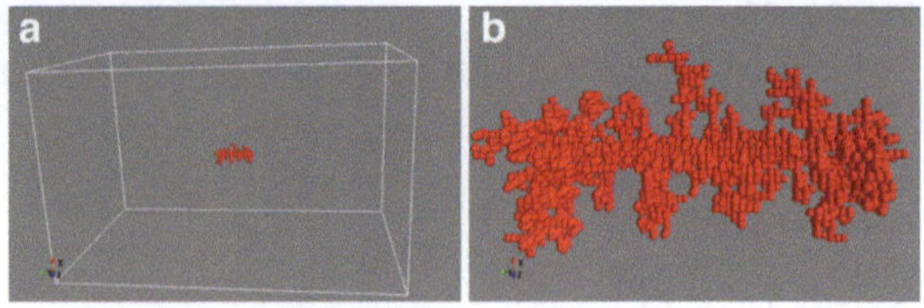

Fig. 5.3 Aggregate after joining 2000 particles in case 1. (**a**) The view from outside simulation *box*. (**b**) A closer view

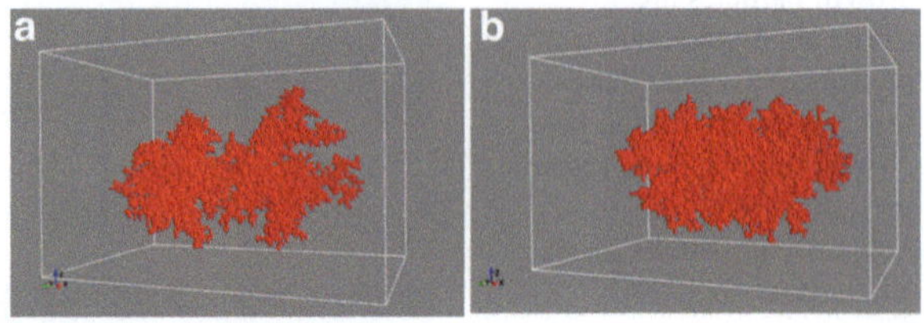

Fig. 5.4 An aggregate of 20,000 particles in setup (**a**) case 2 and (**b**) case 3

5.2.2.3 Results and Discussion

Case 1. Figure 5.3 shows an aggregate developed inside the vesicle. The snapshots here show the aggregate at the end of simulation when all particles joined the aggregate. With the choice of distant boundaries, we allow particles to have more distance for random movements; therefore the effect of diffusion is more highlighted.

Case 2. The result of simulation at the last growth step is shown in Fig. 5.4a. The assumptions for this case, in terms of similarity to a growing vesicle, are closer to reality. As we expected, since particles had less free space to travel randomly before they join the aggregate, they are more boundary biased compared to case 1.

Case 3. In Fig. 5.4b the final aggregate of case 3 has been shown. It demonstrates that the assumption of a distributed particle source over the whole volume results even in a more compact form and fewer voids in the structure. Also, since a number of particles can appear close to the growing aggregate, they can stick to it quickly, bringing the importance to the guiding effect of organic template and boundaries of vesicle in competition with a random aggregate.

The surface and structure of most of the silica in spicules, as it is shown in Fig. 5.1, is smooth (though there are some areas inside the central canal where biosilica is not smooth). However, biosilica in spicules proceeds an aging and hardening controlled process that turns soft irregular amorphous silica to a much harder and smoother material (Müller et al. 2011). This means that there is an intermediate state of aggregates where the material is quickly joined to the scaffold and is more randomly distributed. This is similar to the process of silicatein filaments formation in vitro (Murr and Morse 2005; Schloßmacher et al. 2011) where the oligomers of silicatein first form a complex random structure with fractal properties with a fractal dimension close to fractal dimension of a 2D DLA object and afterwards, they

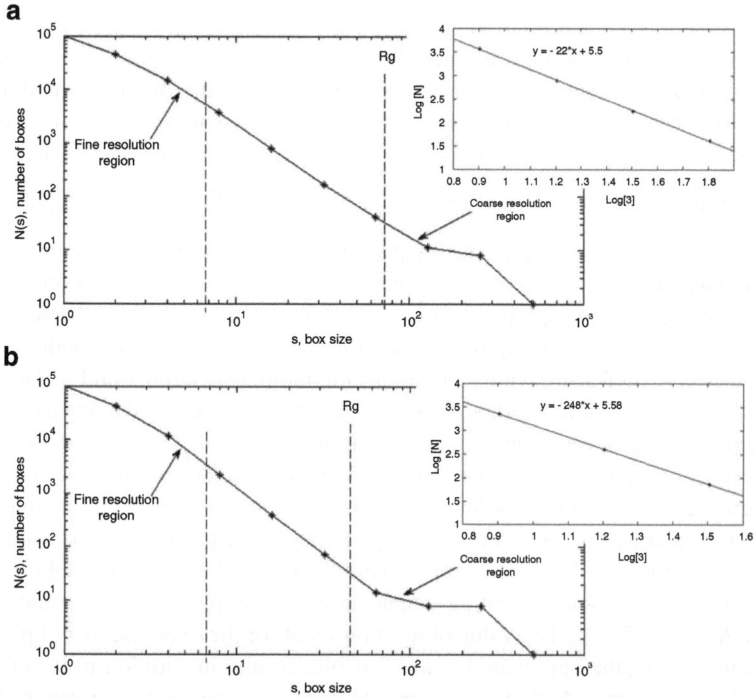

Fig. 5.5 Logarithmic representation of number of boxes versus the scale for calculation of fractal dimension of the aggregates of 100 k particles (**a**) in case 2 and (**b**) in case 3. R_g is the radius of gyration for which the bigger values of box sizes do not count. Smaller inner figures are linear fit of the *middle region* of box-counting method. The value for the slope of the *line* is shown in the equation

continue their self-assembly until reaching the filament structures. The reason for later steps from fractal to filament shape has been suggested to be the anisotropic surface of silicatein oligomers and different interactions like hydrophobic, ionic, and van der Waals interactions (Murr and Morse 2005).

A method to quantify the complex-shaped geometry of the aggregates is computation of fractal dimensions or Hausdorff dimension (Mandelbrot 1983, Bradbury and Reichelt 1983; Morse et al.1985; Kaandorp 1991; Kaandorp et al. 2011) which quantifies the self-similarity properties. In order to calculate fractal dimension, different methods, like box counting, radius of gyration, and correlation function, have been introduced. One of the most common ones is box-counting method in which dividing the space in successive decreasing scales, *s*, in each step the number of boxes containing some part of the pattern, *N*, is calculated. So if there is a self-similarity in the pattern, it will present the box-counting dimension, D_Box, defined as $N(s) = s^{-D\_Box}$. Figure 5.5 is the result of this analysis on aggregates showed in Fig. 5.4 (the same analysis can be done on the aggregate in Fig. 5.3, but since case 2 and 3 are the most realistic ones, we only show their curves). A structure is a fractal if there is a linear relation between log N and log s. The data points shown in Fig. 5.5 are not all meaningful. Since the box counting is only valid where the box size is not bigger than the entire object or it is not of the order or smaller than resolution of the object, to be more specific, for very small boxes or boxes bigger than the radius of gyration, R_g, of the aggregate box, counting is not valid. The radius of gyration around the center of mass, x_0, is defined as $R_g^2 = \frac{1}{M}\int_V d^3x\rho(x)(x - x_0)^2$. Here, the value of the density, $\rho(x)$, is equal to 1 where the particle exists and 0 where it is empty. In Fig. 5.5 the inner images show the linear regression analysis performed for the appropriate region of data in order to calculate the line slope which corresponds to D_Box, the fractal dimension by box-counting method. Since the object is a 3D fractal, we usually expect a number between 2 and 3. The other point about box-counting method, especially in 3D, is that we need many particles in the aggregate and several runs of the simulation, in order to have a good estimation of fractal dimension. For example, in Fig. 5.5 the fractal calculations have been done on aggregates of 100,000 particles. For 20,000 particles, there won't be enough valid points for calculation and hence the slope is not accurate. The D_box values for 3 simulation runs are 2.2 $\pm$ 0.05 for the case 2, and for the case 3 it is 2.48 $\pm$ 0.05. The radius of gyration for the aggregate in case 2 is $R_g = 71.5 \pm 1$ and in the case 3, $R_g = 45.5 \pm 1$. The value of R_g shows that for the same number of particles (the same mass), the aggregate in case 2 is bigger, and in addition to lower value for D_Box, we conclude that this aggregate which has the source for new particles only close to the boundaries forms a looser structure with many branchings. In another words, when source of nanoparticles is distributed in all the vesicle volume, it makes the young silica aggregate to form a more compact structure compared to the case 2 where particles start their random walks only next to the membrane, allowing a more diffusive motion.

The theoretical fractal dimension for a 3-dimensional ideal DLA object is known to be approximately 2.5. However, we should keep in mind that this value is for a very large number of particles. The other point is that box-counting method does not necessarily give us the exact Hausdorff dimension; instead it is an estimation of the mathematical definition of Hausdorff dimension (Theiler 1990). Although, it still can be used for the comparison of aggregates.

As we have mentioned earlier in this section, the X-ray scattering of materials is an experimental way to measure the self-similarity of their patterns. The fractal dimension that has been calculated for silica nanoparticles with this method is 2.58 (Vrieling et al. 2002). This value is still different from our simulation, but given the different methods for measuring fractal dimensions and their sensitivity to parameters like size of the aggregates, this difference is expected. Also the results of fractal dimension calculations show that in the third case, D_f is higher and this declares a more compact structure as we get closer to three which is the full space.

As we have mentioned before, the mature spicules of *S. domuncula* have a compact structure inside (Fig. 5.1b, d). The results presented here are the preliminary outputs of this model, but still they can lead us to the fact that the biological conditions assumed for cases 2 and 3 are closer to the real biosilica. They provide a fractal form of the aggregate which is closer to a compact structure, and also this highlights the effect of confined space (shape of membrane) as a control parameter on morphology. However, it is very important to notice there are several post-synthesizing factors in biosilica growth which make the silica harder. One of them is the interaction forces between proteins and silica particles that usually carry a negative electrical charge, which can modify the particle paths. Another factor is the surface migration which is the movement of a particle inside the aggregate. This can happen because if we consider a potential for absorption/connection to the aggregate, particles can move slowly towards a direction in which they can find a lower energy configuration [e.g., Cronemberger and Sampaio (2006), Cronemberger et al. (2010)]. Moreover, it has been shown that biosintering and hardening of silica due to additional release of water molecules from the young silica can cause the compactness of silica in later stages (Müller et al. 2011). The present model has the possibility for adding these factors to spicule growth. For example, biosintering and surface migration can be introduced to the model by assigning a sticking probability or surface potential to the particles inside the aggregate and allowing them to follow their secondary changes due to these effects.

5.3 Biosilicification in Diatoms

5.3.1 About Diatoms

Diatoms are unicellular algae which is one of the largest groups of silicifying organisms. They have walls made of amorphous silica deposited in a delicately nanopatterned form, which is why they attract increasing interest since a long time ago from a material science perspective (Schultze 1863; Gordon et al. 2009).

Figure 5.6 shows the asexual cell cycle of a typical diatom. Before a cell divides, new silica valves for the next generation will be formed inside the cell. This happens inside a specialized vesicle called silica deposition vesicle (SDV) (Drum and Pankratz 1964). After this event, daughter cells will separate and continue growing. This includes the growth in cell material and also the silica walls around the cell, called girdle bands, until the cell reaches the adult size.

The silica wall architecture is a species-specific characteristic of diatoms (Round et al. 1990), and this shows that the synthesis of silica is highly genetically controlled, in addition to chemical and physical controls. Since the entire genome of some species of diatoms has been sequenced [*T. pseudonana*: Armbrust et al. (2004) and *P. tricornutum*: Bowler et al. (2008)], there have recently been greater insights into genetic control (Sumper and Brunner 2006, 2008); it is believed that such control takes place mainly through two processes: first, via special biomolecules that regulate the uptake and transport of silicon and, second,

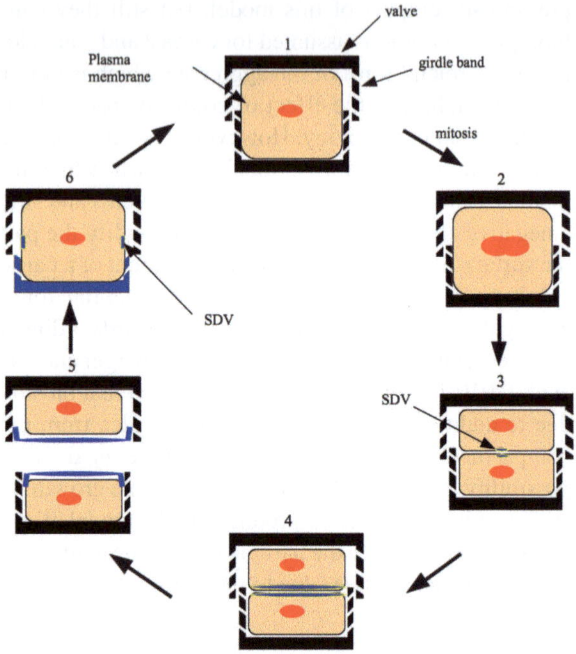

Fig. 5.6 The cell cycle of a typical diatom. *1* An adult diatom. *2* DNA synthesis has occurred but the cell does not divide immediately. *3* and *4* Deposition of new valves has started inside the SDV. *5* After the valves are complete, the cell will divide. *6* Daughter cells will continue growing along with the deposition of silica in girdle bands [picture after Kröger and Poulsen (2008)]

via an organic matrix (mainly of proteins and polyamines) which plays the role of a structure-forming scaffold (Mann 1993; Gröger et al. 2008a).

The role of active transporters has been more discussed (Thamatrakoln and Hildebrand 2006). After the cell uptakes silicon (mainly in the form of monosilicic acid), it is supposed that cell accumulates silicon in a storage pool (Hildebrand 2000) and then transports it through cytoplasm to reach the SDV, the location of new synthesizing walls. Interestingly, if we consider all of the intracellular silicon pool in the form of monosilicic acid, considering the small volume of the cell, it should have a very high concentration, much higher than the stability of monosilicic acid which is around 2 mM at pHs below 9 (Iler 1979). It has been an open question how the cell can maintain this concentration without deposition. There are different scenarios for explaining storage and transport of silicon in the pool. One of them assumes that silicic acid connects to some kind of organic molecule and thus makes a solvable silicon pool (Azam 1974), and therefore it turns the silicic acid into another chemical form. This explanation is in agreement with the uptake behavior of diatoms after different starvation conditions (Thamatrakoln and Hildebrand 2008). Another method assumes special silicon transport vesicles (STV) which transport silicic acid from the cell membrane and release their content into SDV by fusing to its membrane (Schmid and Schulz 1979). The third explanation is that oligomerization does indeed occur inside the cell as soon as there is some monosilicic acid available inside the cytoplasm, and it generates precursors for later deposition inside the SDVs. This explanation is based on NMR data from silicon pools. Recently, ^{29}Si NMR chemical shift technique has been a powerful way for understanding the forms of silicic acid based on their connections to other molecules. With this data it has been shown that the majority of silicon in the entire cell is in some polymerized form (Gröger et al. 2008b). However, this data does not necessarily exclude the possibility of the first explanation (because this method cannot distinguish between free monosilicic acid and its attachment to an organic molecule), but it shows clearly that intracellular silicon is mostly in the form of oligomers.

Finally, after the nutrient (oligomers of silicic acid and pre-synthesized silica) is provided for SDV, it will be deposited assisting by organic matrix and will result in the complex silicon structure of the daughter cells walls.

5.3.2 Synthesis of Silica in Diatoms: Reaction–Diffusion Model

Several models have been proposed to explain synthesis of silica in diatoms. In most of these models, the goal is to describe morphology of silica inside the SDV, including DLA growth models (Gordon and Drum 1994; Parkinson et al. 1999), biomolecule-mediated formation and control of aggregate formation (Vrieling et al. 2002; Spinde et al. 2011), and self-assembly of organic matrices which in turn controls silica deposition (Sumper 2002, 2004). There have been few computational models though for describing transport and pre-synthesis of intracellular silicon, mainly in cell cytoplasm. They usually calculate the temporal changes of silicon amount or the timing of silicon synthesis and its relation to other nutrients regarding to the cell cycle (Brasser et al. 2012; Flynn and Martin-Jézéquel 2000). To the best of our knowledge, there is no spatial-temporal simulation of silicon transport and polymerization in diatom cytoplasm.

In our model, we focus on the uptake and transport of silicic acid through the cell membrane and cytoplasm—the step that is vital in providing the material for the special silica depositing compartment and, consequently, vital for the cell division. The diffusion-reaction model used here consists of several compartments and includes polymerization reactions and mass transport by silicon transporters. In this way we can compare the kinetics of silica condensation in a cell with in vitro chemical experiments and obtain some insights into the biocontrol of silicification.

In order to describe temporal changes during the silica uptake and synthesis and to look at the cell compartments based on in vivo images of diatoms, we need to use a spatiotemporal model. This will lead us to a partial differential system of equations which need to be solved on different boundary conditions. Therefore, the best approach for this model is the finite element method.

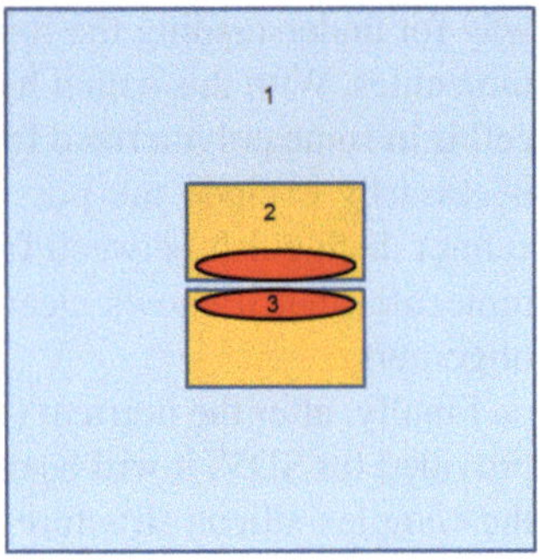

Fig. 5.7 2D geometry of the model: domain 1 represents seawater surrounding the diatom. Domain 2 and 3 are the intercellular space and SDV, respectively

5.3.2.1 Simulation Setup

The species *Thalassiosira pseudonana* is our model organism. This diatom has a cylindrical shape, very close to a cylindrical symmetry around an axis which passes through the center of its valves. Therefore, it is a good approximation to build the model in two dimensions.

The simulation box includes one diatom in the center of an area of seawater. Figure 5.7 shows the geometry of the model. For diatoms, only the most important compartments for silicification have been considered in the model, including SDVs and the two-section cell membrane.

The old silicon wall around diatom is not considered in the model as they have a rough structure compared to cell membranes. Therefore, there are not obstacles or boundaries. Although wall structure can potentially have an effect on the water flow and distribution of silicon around the membrane, but since the diffusion time scale is very short compared to other time scales of the process, diffusion is fast enough to provide enough silicic acid for uptake. Therefore, diffusion is not a limiting factor, and hence the water flow through the wall structure very likely doesn't have a big effect on uptake process.

Our model concerns the period of valve formation in diatom. It does not include all of the cell cycle. The reason for this choice is that during this time, the most important event regarding silicification occurs. Another reason is that the geometry of the cell does not change significantly during this time (see Fig. 5.6). Moreover, it has been measured that most of the silicon uptake happens during this time (Thamatrakoln and Hildebrand 2008). So, for the purpose of simplicity, we chose this period of time while keeping in mind that it still covers the most significant event in silicification. This period involves the G2 and M phases, which in the case of *T. pseudonana* is approximately 3.5 h, 35 % of cell cycle.

The boundary representing the seawater should be far enough to make sure there is no significant gradient of either species concentrations in the model in its proximity, after we run the simulation and generate results. Otherwise boundaries should be moved further until we make sure that the effect of truncation on the solution is minimal and we have smooth concentration profiles. We found that locating boundaries representing the seawater in ten times the size of diatom is a good choice and is also in agreement with the previous models regarding diatoms (Wischmeyer et al. 2003).

5.3.2.2 Diffusion–Reaction System

There are several domains in the model. The first domain represents the seawater (region 1 in Fig. 5.7) that is the environment in which chemical species are distributed by diffusion, and there is also reaction between H^+ and OH^-. Since silicic acid concentration is far below its solubility (Hildebrand 2000), we do not consider any polymerization reactions in this domain. We also do not take the charged species of monosilicic acid into account in silicon uptake in the first approximation, because based on a theoretical study, the uncharged species (H_4SiO_4) is the most uptake one by diatoms (Wischmeyer et al. 2003). In the next step, we can consider negatively charged species of silicic acid in seawater and their contribution to the whole silicon uptake by diatoms.

The second domain (region 2 in Fig. 5.7) is the intracellular space. In this domain, there are diffusion and chemical reactions (mainly oligomerization of silicic acid). This is the domain which includes the so-called silicon pool. As mentioned in the Introduction section, there have been a number of studies on the transport and storage of silicon in this domain. This model is based on the NMR data of silicon pool (Gröger et al. 2008b) and assumes that a chain of polymerization reactions occurs inside the cell, and we try to understand the kinetics of these reactions and intracellular distribution of chemical components.

Deposition of silica is a complex process and it contains different stages. At ambient temperatures the process begins with the condensation of monosilicic acid to form disilicic acid. Then this process will continue with monomer addition to form higher oligomers. The first steps of polymerization reactions have the general form as follows (Perry and Keeling-Tucker 2000):

$$\begin{array}{l}
\mathrm{Si(OH)_4 + Si(OH)_3O^- \rightarrow (OH)_3Si - O - Si(OH)_3 + OH^-}\\
\mathrm{Si(OH)_3O^- + Disilicic\,acid \rightarrow Trisilicic\,acid + OH^-}\\
\mathrm{Si(OH)_4 + (Disilicic\,acid)^- \rightarrow Trisilicic\,acid + OH^-}\\
\mathrm{Si(OH)_3O^- + Trisilicic\,acid \rightarrow Tetrasilicic\,acid + OH^-}\\
\mathrm{Disilicic\,acid + Disilicic\,acid \rightarrow Tetrasilicic\,acid + OH^-}\\
\mathrm{Si(OH)_4 + (Oligomer)^- \rightarrow Oligomer + OH^-}
\end{array}$$

In summary, in solutions with concentrations that exceed the solubility of amorphous silica, monosilicic acid continues condensation reactions leading to the formation of nanocolloidal silica, which can then (given appropriate conditions) precipitate.

Polymerization of silica has been studied in many chemical experiments [e.g., Iler (1979), Perry and Keeling-Tucker (2000)], where it takes place in a controlled environment, a solution with specific concentration, pH, temperature, salinity, etc. Different kinetics models have been published recently [e.g., Icopini et al. (2005)].

Moreover, there is evidence that additives to a solution have a significant impact on the kinetics of the process (Perry and Keeling-Tucker 2000). In fact, the intracellular space includes a variety of different molecules, compartments and chemical components, or possible protein catalyzers. Therefore, the results from this model can lead us to an interesting comparison between biokinetics of silicon synthesis in diatoms and the kinetics of the most similar chemical equivalent (similar pH, temperature, and range of concentration). This will show the effect of a biocontrol on polymerization and transport of silica.

The third domain (region 3 in Fig. 5.7) lies in SDVs where silicon deposition and new valves formation occur. For our current study, this domain is not considered to be active in numerical solutions. However, we have a boundary condition which is an influx of all the silicic acid species. Thus, the membrane of SDVs only acts as a sink of silicon and the rate is an input for the model. Inside the SDVs, silicon deposits in such a way that it makes the chemical species to be separated from the solution, causing increased absorption of those chemical species. On the other hand, the present model does not take into account the mechanism by which the absorption or the deposition event inside the SDVs occurs. Instead, it applies an absorption rate as an input, which comes from a time average of silicon input that SDVs need to make silica walls.

5.3.2.3 System of Equations

For the current model we consider that we only have five chemical components present:

$$c_1 = [H_4SiO_4],\ \ c_2 = [H_3SiO_4{}^-],\ \ c_3 = [H_6Si_2O_7],\ \ c_4 = [H^+],\ \ c_5 = [OH^-]$$

Domain 1. Equations for the first domain have the form,

$$\frac{\partial c_i}{\partial t} = D_i \nabla^2 c_i; \quad i = 1,2,3$$

$$\frac{\partial c_i}{\partial t} = D_i \nabla^2 c_i + \text{Reaction}(c_i, c_k); \quad i,k = 4,5.$$

Domain 2. Diffusion-reaction system of equation has the following form:

$$\frac{\partial c_i}{\partial t} = D_i \nabla^2 c_i + \text{Reaction}(c_i, c_k); \quad i,k = 4,5.$$

Considering chemical reactions as follows

$$H_2O \rightleftarrows H^+ + OH^-$$
$$H_4SiO_4 \rightleftarrows H_3SiO_4^- + H^+$$
$$H_4SiO_4 + H_3SiO_4^- \rightleftarrows H_6Si_2O_7 + OH^-$$

The system of equations has the form of

$$\frac{dc_1}{dt} = D_1\nabla^2 c_1 - k_1^+ c_1 + k_1^- c_2 c_4 - k_2^+ c_1 c_2 + k_2^- c_3 c_5$$
$$\frac{dc_2}{dt} = D_2\nabla^2 c_2 + k_1^+ c_1 - k_1^- c_2 c_4 - k_2^+ c_1 c_2 + k_2^- c_3 c_5$$
$$\frac{dc_3}{dt} = D_3\nabla^2 c_3 + k_2^+ c_1 c_2 - k_2^- c_3 c_5$$
$$\frac{dc_4}{dt} = D_4\nabla^2 c_4 + k_w^+[H_2O] - k_w^- c_4 c_5 + k_1^+ c_1 - k_1^- c_2 c_4$$
$$\frac{dc_5}{dt} = D_5\nabla^2 c_5 + k_w^+[H_2O] - k_w^- c_4 c_5 + k_2^+ c_1 c_2 - k_2^- c_3 c_5.$$

This is a nonlinear second-order partial differential system of equations describing chemical species rates (first-order time derivative). The diffusion term makes the equations spatial, causes reactions to act as source/sink terms, and also couples equations together. Since multipliers of concentrations appear in reactions terms, they also make the system of equations nonlinear.

The system of equations was solved with the finite elements method and using COMSOL Multiphysics application.

5.3.2.4 Parameters and Experimental Data

A typical cell size that has been used in the model is approximately 5 by 5.5 μm in a 2D cross section. We found that a square of 60 μm is a good choice for boundaries representing the seawater (see also simulation setup section). Other values used for this model are listed in Table 5.1. The parameters of the model are based on the experimental data (references in the table). The value for *Tr* is the transport rate of silicic acid through the cell membrane and has been applied in the model as Neumann boundary condition. This value was calculated from 3D (in vivo) uptake data and assuming that cell membrane uptakes silicic acid uniformly in all the surface. With this assumption and with reduction from surface boundaries integrals to linear boundaries integrals, we have calculated the uptake rate for our 2D model.

Table 5.1 Parameters of model

Name	Value	Description
C10	100e-6 M	C1 in seawater boundary[a]
C20	3.18e-6 M	C2 in seawater boundary[a]
C30	0	C3 in seawater boundary[a]
C40	10e-8 M	C4 in seawater boundary[a]
C50	2.75e-6 M	C5 in seawater boundary[a]
D	1e-9 m^2/s	Order of diffusion constant[a]
Tr	9.42e-8 mol/s	Calculated from 3D diatom uptake data[b] to the 2D model

[a]Wischmeyer et al. (2003)
[b]Thamatrakoln and Hildebrand (2008)

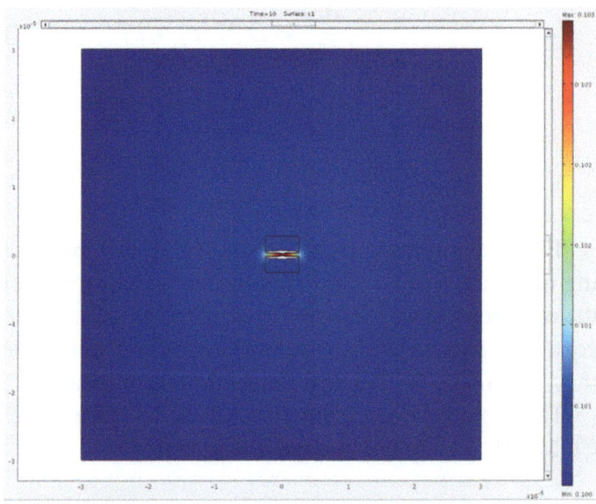

Fig. 5.8 Simulation of spatial distribution of chemical component within a diatom: distribution of [H_4SiO_4]

5.3.2.5 Results

The simulations were performed until the equilibrium was reached. The concentration profiles for five types of chemical agents are shown in the following figures. Figure 5.8 shows the distribution of concentration of monosilicic acid in the neutral form. The diatom cell acts as a sink of [H_4SiO_4], but in some areas, because of the geometry of the problem, especially centric area of diatoms, there is some local accumulation of monosilicic acid.

Figure 5.9 is a closer view to the cell boundaries together with the gradients of concentration shown as a vector field by arrows.

Concentration profiles of [$H_3SiO_4^-$] and [$H_6Si_2O_7$] are shown in Figs. 5.10 and 5.11. For [$H_3SiO_4^-$], monosilicic acid in a negatively charged form, there is some outflow, because it will be produced during polymerization reactions. [$H_6Si_2O_7$], disilicic acid, is confined to the cell membrane because the boundaries are assumed to be nontransparent for polymerized species.

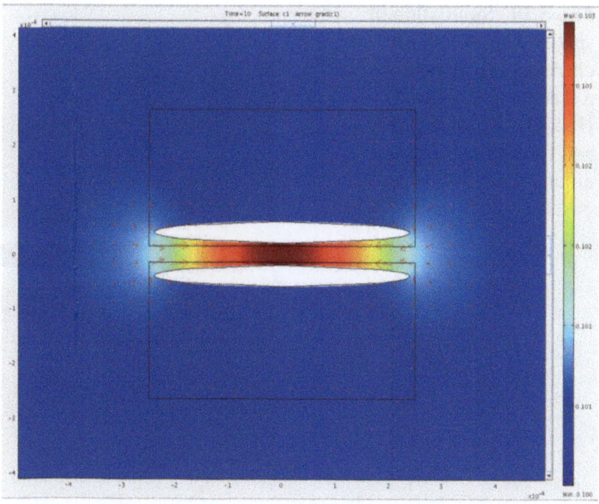

Fig. 5.9 Simulation of spatial distribution of chemical component within a diatom: distribution of [H_4SiO_4] around the cell and its gradient as *arrows*

Here, instead of showing [H^+] and [OH^-] concentrations, we look at pH profile which is a more meaningful quantity. Figure 5.12 shows that the current system of reactions causes inside of the cell to be more acidic, especially around the SDV, which is in agreement with experimental observations (Vrieling et al. 1999).

5.3.2.6 Conclusion and Discussion

One of the important and leading factors in biomineralization is the role of material transport and the rate of providing chemical agents for the deposition event. The diffusion-reaction model is one promising approach for studying transport of nutrients in a cell compartmental analysis. It results in spatial-temporal information of concentrations, and, moreover, properties of membranes can be simulated by controlling the boundary conditions in the model.

As it was shown in Figs. 5.8 and 5.9, the area between two newly forming valves where seawater can penetrate has a high concentration of some chemical species, such as monosilicic acid, due to the geometry of the problem. This can be related to the difference of the silica structure between the two sides of silica valves (proximal and distal surfaces). The interior layer of the silica valve has a smoother surface than the outer layer (Hildebrand 2008), and this can be explained by an abundance of nutrients around the outer surface.

The pH values calculated by the model (Fig. 5.12) show the acidic property of the solution inside the cell and specially around deposition area, SDV. This agrees with experimental data that silicification in diatoms happens in an acidic environment (Vrieling et al. 1999). The low pH would lead the reactions towards gel formation (Iler 1979); however, the existence of organic molecules like sillafins which are most active in acidic condition will cause the formation of silica spheres (Vrieling et al. 1999). In our model the pH value can also be used as a tuning

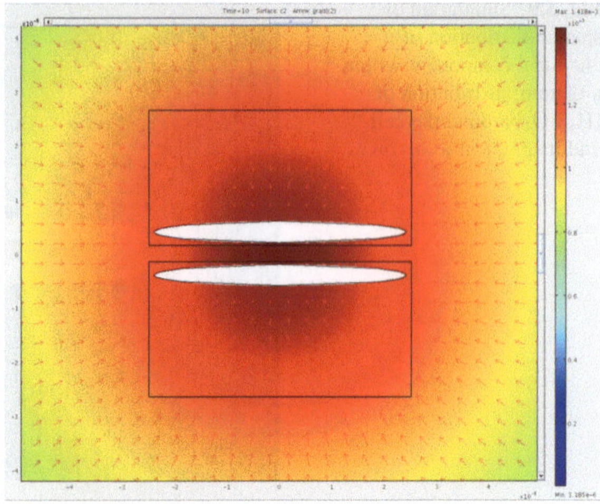

Fig. 5.10 Simulation of spatial distribution of chemical component within a diatom: distribution of $[H_3SiO_4^-]$ concentration and its gradient (*arrows*) around the cell

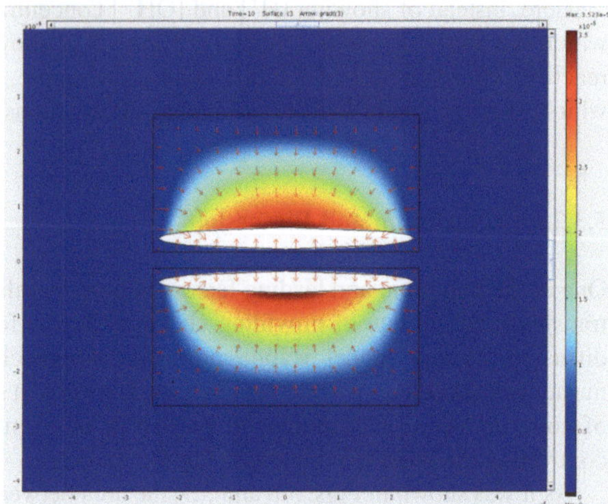

Fig. 5.11 Simulation of spatial distribution of chemical component within a diatom: distribution of $[H_6Si_2O_7]$ concentration and its gradient (*arrows*) around the cell

parameter in the model. Here, the area around SDV is only slightly acidic, and this is the effect of simplification in the reaction terms. Inside the silicon pool there is a chain of polymerization reactions. In the present model, the polymerization is considered only up to disilicic acid formation, and this already provides an indication of how the polymerization can change the equilibrium and together with diffusion change the pattern of mass distribution. Adding more reactions is part of a future work. However, it should be noticed that introducing more reactions will result in a more complex numerical system with several coupled PDEs and also reaction kinetic constants that should be estimated by experimental data or be considered as free parameters of the model, adding then more accuracy to the model.

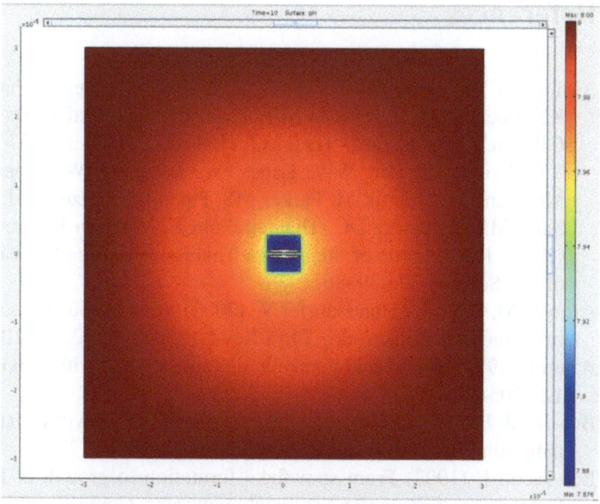

Fig. 5.12 Simulation of spatial distribution of chemical component within a diatom: pH of the environment around and inside the diatom cell

Another control factor in our model is fluxes through membranes. In the present work, the flux of monosilicic acid towards the cell has been selected based on an experimental measurement of an average of the uptake rate by diatoms. This can also be considered differently, for example, by a Michaelis-Menten saturation equation or an equation for exchange pumps in the model, in order to study its effect on all distributions and rates [e.g., Cui and Kaandorp (2006)]. Regarding this consideration, there are studies on *T. pseudonana* silicon transporters activities and their regulatory level during the cell cycle (Thamatrakoln and Hildebrand 2007). The result of this type of study in our model can be compared to the protein data, and this can be a good combination of genetic, physical, and chemical mechanisms in biosilicification.

In summary, we have presented a spatiotemporal model for simulation of silica synthesis intracellularly in diatoms. In this model the kinetics of silica polymerization reactions and diffusion as the mean of transport has been studied. Two cell compartments here are considered but the focus is on transport of silica in the cytoplasm or the so-called silicon pool. Since the synthesis of silica is a very complex process, we applied some simplifications, including considering only the most important oligomers of silica for the first approximation. The results of our model which are the concentration profiles of different chemical components show some agreements with the experimental in vivo evidences, especially about the pH of the environment around SDV. In addition to understanding the mechanisms of silica transport and synthesis, this type of model gives us the information about the nature of silica being uptake by SDV (the form of precursors) and the rate by which cell can provide them for deposition of silica in SDV. Having this information, the simulations for pattern formation (like the modified DLA model explained in the first section) can be more realistic and result in morphologies closer to the real biosilica structures.

References

Armbrust EV, Berges JA, Bowler C, Green BR, Martinez D, Putnam NH, Zhou S, Allen AE, Apt KE, Bechner M, Brzezinski MA, Chaal BK, Chiovitti A, Davis AK, Demarest MS, Detter JC, Glavina T, Goodstein D, Hadi MZ, Hellsten U, Hildebrand M, Jenkins BD, Jurka J, Kapitonov VV, Kröger N, Lau WW, Lane TW, Larimer FW, Lippmeier JC, Lucas S, Medina M, Montsant A, Obornik M, Parker MS, Palenik B, Pazour GJ, Richardson PM, Rynearson TA, Saito MA, Schwartz DC, Thamatrakoln K, Valentin K, Vardi A, Wilkerson FP, Rokhsar DS (2004) The genome of the diatom Thalassiosira pseudonana: ecology, evolution, and metabolism. Science 306:79–86

Arrayás M, Ebert U, Hundsdorfer W (2002) Spontaneous branching of anode-directed streamers between planar electrodes. Phys Rev Lett 88:174502. doi:10.1103/PhysRevLett.88.174502

Azam F (1974) Silicic acid uptake in diatoms studied with [68Ge] germanic acid as tracer. Planta 121:205–212

Ben-Jacob E (1993) From snowflake formation to growth of bacterial colonies. Part I. Diffusive patterning in azoic systems. Contemp Phys 34:247–273

Ben-Jacob E (1997) From snowflake formation to growth of bacterial colonies II: cooperative formation of complex colonial patterns. Contemp Phys 38(3):205–241. doi:10.1080/001075197182405

Bensimon D (1986) Stability of viscous fingering. Phys Rev A 33:1302–1308. doi:10.1103/PhysRevA.33.1302

Bowler C, Allen AE, Badger JH, Grimwood J, Jabbari K, Kuo A, Maheswari U, Martens C, Maumus F, Otillar RP, Rayko E, Salamov A, Vandepoele K, Beszteri B, Gruber A, Heijde M, Katinka M, Mock T, Valentin K, Verret F, Berges JA, Brownlee C, Cadoret JP, Chiovitti A, Choi CJ, Coesel S, De Martino A, Detter JC, Durkin C, Falciatore A, Fournet J, Haruta M, Huysman MJ, Jenkins BD, Jiroutova K, Jorgensen RE, Joubert Y, Kaplan A, Kröger N, Kroth PG, La Roche J, Lindquist E, Lommer M, Martin-Jézéquel V, Lopez PJ, Lucas S, Mangogna M, McGinnis K, Medlin LK, Montsant A, Oudot-Le Secq MP, Napoli C, Obornik M, Parker MS, Petit JL, Porcel BM, Poulsen N, Robison M, Rychlewski L, Rynearson TA, Schmutz J, Shapiro H, Siaut M, Stanley M, Sussman MR, Taylor AR, Vardi A, von Dassow P, Vyverman W, Willis A, Wyrwicz LS, Rokhsar DS, Weissenbach J, Armbrust EV, Green BR, Van de Peer Y, Grigoriev IV (2008) The Phaeodactylum genome reveals the evolutionary history of diatom genomes. Nature 456:239–244

Bradbury RH, Reichelt RE (1983) Fractal dimension of a coral reef at ecological scales. Mar Ecol Prog Ser 10:169–171

Brasser H, van der Strate H, Gieskes W, Krijger G, Vrieling E, Wolterbeek H (2012) Compartmental analysis suggests macropinocytosis at the onset of diatom valve formation. Silicon 4:39–49. doi:10.1007/s12633-010-9059-2

Brener E, Müller-Krumbhaar H, Temkin D (1996) Structure formation and the morphology diagram of possible structures in two-dimensional diffusional growth. Phys Rev E Stat Phys Plasmas Fluids Relat Interdiscip Topics 54:2714–2722. doi:10.1103/PhysRevE.54.2714

Cronemberger CM, Sampaio LC (2006) Growth of fractal electrodeposited aggregates under action of electric and magnetic fields using a modified diffusion-limited aggregation algorithm. Phys Rev E Stat Nonlin Soft Matter Phys 73:041403. doi:10.1103/PhysRevE.73.041403

Cronemberger C, Sampaio LC, Guimarães AP, Molho P (2010) Model for the growth of electrodeposited ferromagnetic aggregates under an in-plane magnetic field. Phys Rev E Stat Nonlin Soft Matter Phys 81:021403. doi:10.1103/PhysRevE.81.021403

Cui J, Kaandorp JA (2006) Mathematical modeling of calcium homeostasis in yeast cells. Cell Calcium 39(4):337–348. doi:10.1016/j.ceca.2005.12.001

Drum RW, Pankratz HS (1964) Post mitotic fine structure of Gomphonema parvulum. J Ultrastruct Res 10:217–223

Flynn KJ, Martin-Jézéquel V (2000) Modelling Si–N limited growth of diatoms. J Plankton Res 22:447–472

Gordon R, Drum RW (1994) The chemical basis of diatom morphogenesis. Int Rev Cytol 150:243–372

Gordon R, Losic D, Tiffany MA, Nagy SS, Sterrenburg FAS (2009) The glass menagerie: diatoms for novel applications in nanotechnology. Trends Biotechnol 27:116–127

Gröger C, Lutz K, Brunner E (2008a) Biomolecular self-assembly and its relevance in silica biomineralization. Cell Biochem Biophys 50:23–39

Gröger C, Sumper M, Brunner E (2008b) Silicon uptake and metabolism of the marine diatom *Thalassiosira pseudonana*: solid-state 29Si NMR and fluorescence microscopic studies. J Struct Biol 161(1):55–63. doi:10.1016/j.jsb.2007.09.010

Harding JH, Duffy DM, Sushko ML, Rodger PM, Quigley D, Elliott JA (2008) Computational techniques at the organic-inorganic interface in biomineralization. Chem Rev 108 (11):4823–4854. doi:10.1021/cr078278y

Harrison FW, Davis DA (1982) Morphological and cytochemical patterns during early stages of reduction body formation in *Spongilla lacustris* (Porifera: Spongillidae). Trans Am Microsc Soc 101:317–324

Hildebrand M (2000) Silicic acid transport and its control during cell wall silicification in diatoms. In: Bäuerlein E (ed) Biomineralization—from biology to biotechnology and medical applications. Wiley, Weinheim, pp 170–188

Hildebrand M (2008) Diatoms, biomineralization processes, and genomics. Chem Rev 108:4855–4874

Icopini GA, Brantley SL, Heaney PJ (2005) Kinetics of silica oligomerization and nanocolloid formation as a function of pH and ionic strength at 25°C. Geochim Cosmochim Acta 69 (2):293–303. doi:10.1016/j.gca.2004.06.038

Iler RK (1979) The chemistry of silica. Wiley, New York, NY

Kaandorp JA (1991) Modelling growth forms of the sponge Haliclona oculata (Porifera, Demospongiae) using fractal techniques. Mar Biol 110(2):203–215

Kaandorp JA, Filatov M, Chindapol N (2011) Simulating and quantifying the environmental influence on coral colony growth form, Coral Reefs: an ecosystem in transition. Springer, New York, NY, pp 177–185

Knoblich B, Gerber TH (2001) Aggregation in SiO2 sols from sodium silicate solutions. J Non Cryst Solids 283:1–3

Krasko A, Batel R, Schröder HC, Müller IM, Müller WEG (2000) Expression of silicatein and collagen genes in the marine sponge *Suberites domuncula* is controlled by silicate and myotrophin. Eur J Biochem 267:4878–4887

Kröger N, Poulsen N (2008) Diatoms-from cell wall biogenesis to nanotechnology. Annu Rev Genet 42:83–107. doi:10.1146/annurev.genet.41.110306.130109

López-Tomàs L, Claret J, Sagués F (1993) Quasi-two-dimensional electrodeposition under forced fluid flow. Phys Rev Lett 71:4373–4376. doi:10.1103/PhysRevLett.71.4373

Mandelbrot BB (1983) The fractal geometry of nature. W. H. Freeman, San Francisco, CA

Mann S (1993) Molecular tectonics in biomineralization and biomimetic materials chemistry. Nature 365:499–500

Mann S (2009) Self-assembly and transformation of hybrid nano-objects and nanostructures under equilibrium and non-equilibrium conditions. Nat Mater 8:781–792. doi:10.1038/nmat2496

Morse DR, Lawton JH, Dodson MM, Williamson MH (1985) Fractal dimension of vegetation and the distribution of arthropod body lengths. Nature 314(6013):731–733

Müller WEG, Rothenberger M, Boreiko A, Tremel W, Reiber A, Schröder HC (2005) Formation of siliceous spicules in the marine demosponge Suberites domuncula. Cell Tissue Res 321:285–297

Müller WEG, Wang X, Wiens M, Schloßmacher U, Jochum KP, Schröder HC (2011) Hardening of bio-silica in sponge spicules involves an aging process after its enzymatic polycondensation: evidence for an aquaporin-mediated water absorption. Biochim Biophys Acta 1810 (7):713–726

Murr MM, Morse DE (2005) Fractal intermediates in the self-assembly of silicatein filaments. Proc Natl Acad Sci USA 102:11657

Niemeyer L, Pietronero L, Wiesmann HJ (1984) Fractal dimension of dielectric breakdown. Phys Rev Lett 52:1033–1036. doi:10.1103/PhysRevLett.52.1033

Park SS, Durian DJ (1994) Viscous and elastic fingering instabilities in foam. Phys Rev Lett 72:3347–3350. doi:10.1103/PhysRevLett.72.3347

Parkinson J, Brechet Y, Gordon R (1999) Centric diatom morphogenesis: a model based on a DLA algorithm investigating the potential role of microtubules. Biochim Biophys Acta 1452:89–102

Perry C, Keeling-Tucker K (2000) Biosilicification: the role of the organic matrix in structure control. J Biol Inorg Chem 5:537–550. doi:10.1007/s007750000130

Round F, Crawford R, Mann D (1990) The diatoms. Cambridge University Press, Cambridge

Schloßmacher U, Wiens M, Schröder HC, Wang X, Jochum KP, Müller WEG (2011) Silintaphin-1—interaction with silicatein during structure-guiding bio-silica formation, the. FEBS J 278:1145–1155. doi:10.1111/j.1742-4658.2011.08040.x

Schmid A-MM, Schulz D (1979) Wall morphogenesis in diatoms: deposition of silica by cytoplasmic vesicles. Protoplasma 100:267–288

Schmidt PW, Höhr A, Neumann HB, Kaiser H (1989) Small-angle x-ray scattering study of the fractal morphology of porous silicas. J Chem Phys 90:5016–5023

Schröder HC, Krasko A, Batel R, Skorokhod A, Pahler S, Kruse M, Müller IM, Müller WEG (2000) Stimulation of protein (collagen) synthesis in sponge cells by a cardiac myotrophin-related molecule from *Suberites domuncula*. FASEB J 14:2022–2031

Schröder HC, Boreiko A, Korzhev M, Tahir MN, Tremel W, Eckert C, Ushijima H, Müller IM, Müller WEG (2006) Co-Expression and functional interaction of silicatein with galectin: matrix-guided formation of siliceous spicules in the marine demosponge *Suberites domuncula*. J Biol Chem 281:12001–12009

Schröder HC, Natalio F, Shukoor I, Tremel W, Schloßmacher U, Wang XH, Müller WEG (2007) Apposition of silica lamellae during growth of spicules in the demosponge *Suberites domuncula*: biological/biochemical studies and chemical/biomimetical confirmation. J Struct Biol 159:325–334

Schultze M (1863) Die Structur der Diatomeenschale verglichen mit gewissen aus Fluorkiesel künstlich darstellbaren Kieselhäuten. In: Weber CO (ed) Verhandlungen des naturhistorischen Vereines der preussischen Rheinlande und Westphalens. Max Cohen & Sohn, Bonn, pp 1–41

Shimizu K, Cha J, Stucky GD, Morse DE (1998) Silicatein alpha: cathepsin L-like protein in sponge biosilica. Proc Natl Acad Sci USA 95:6234–6238

Spinde K, Pachis K, Antonakaki I, Paasch S, Brunner E, Demadis KD (2011) Influence of polyamines and related macromolecules on silicic acid polycondensation: relevance to "soluble silicon pools"? Chem Mater 23:4676–4687. doi:10.1021/cm201988g

Sumper M (2002) A phase separation model for the nanopatterning of diatom biosilica. Science 295:2430–2433

Sumper M (2004) Biomimetic patterning of silica by long-chain polyamines. Angew Chem Int Ed Engl 43:2251–2254

Sumper M, Brunner E (2006) Learning from diatoms: nature's tool for the production of nanostructured silica. Adv Funct Mater 16:17–26

Sumper M, Brunner E (2008) Silica biomineralization in diatoms: the model organism Thalassiosira pseudonana. Chembiochem 9:1187–1194

Thamatrakoln K, Hildebrand M (2006) Analysis of *Thalassiosira pseudonana* silicon transporters indicates distinct regulatory levels and transport activity through the cell cycle. Eukaryot cell 6(2):271–279. doi:10.1128/EC.00235-06

Thamatrakoln K, Hildebrand M (2007) Analysis of *Thalassiosira pseudonana* silicon transporters indicates distinct regulatory levels and transport activity through the cell cycle. Eukaryot Cell 6(2):271–279

Thamatrakoln K, Hildebrand M (2008) Silicon uptake in diatoms revisited: a model for saturable and nonsaturable uptake kinetics and the role of silicon transporters. Plant Physiol 146 (3):1397–1407. doi:10.1104/pp. 107-107094

Theiler J (1990) Estimating fractal dimension. J Opt Soc Am A 7(6):1055–1073

Vrieling EG, Gieskes WWC, Beelen TPM (1999) Silicon deposition in diatoms: control by the pH inside the silicon deposition vesicle. J Phycol 35:548–559

Vrieling EG, Belen TPM, van Santen RA, Gieskes WWC (2002) Mesophases of (bio)polymer-silica particles inspire a model for silica biomineralization in diatoms. Angew Chem Int Ed Engl 41:1543–1546

Wiens M, Bausen M, Natalio F, Link T, Schlossmacher U, Müller WEG (2009) The role of the silicatein-alpha interactor silintaphin-1 in biomimetic biomineralization. Biomaterials 2009 (30):1648–1656

Wiens M, Schröder HC, Wang XH, Link T, Steindorf D, Müller WEG (2011) Isolation of the silicatein- interactor silintaphin-2 by a novel solid-phase pull-down assay. Biochemistry 50:1981–1990

Wischmeyer AG, Del Amo Y, Brzezinski M, Wolf-Gladrow DA (2003) Theoretical constraints on the uptake of silicic acid species by marine diatoms. Mar Chem 82(1–2):13–29. doi:10.1016/S0304-4203(03)00033-1

Witten TA Jr, Sander LM (1981) Diffusion-limited aggregation, a kinetic critical phenomenon. Phys Rev Lett 47:1400–1403. doi:10.1103/PhysRevLett.47.1400

Chapter 6
Cell Reactivity to Different Silica

Marco Giovine, Sonia Scarfì, Marina Pozzolini, Antonella Penna, and Carlo Cerrano

Contents

Abstract The interaction between mineral structures and living beings is increasingly attracting the interest of research. The formation of skeletons, geomicrobiology, the study of the origin of life, soil biology, benthos biology, human and mammalian diseases generated by the inhalation of dust and biomaterials are some examples of scientific areas where the topic has a relevance. In this chapter we focus on cell reactivity to siliceous rocks and to the various forms of silicon dioxide, in particular. The examples here reported carefully review how

---

M. Giovine (✉) • S. Scarfì • M. Pozzolini
Dipartimento di Scienze della Terra, dell'Ambiente e della Vita, University of Genova, Genoa, Italy
e-mail: mgiovine@unige.it

A. Penna
Dipartimento di Scienze Biomolecolari, Università di Urbino "Carlo Bo", Pesaro, Italy

C. Cerrano
Dipartimento di Scienze dell'Ambiente e della Vita, Università Politecnica delle Marche, Ancona, Italy

W.E.G. Müller et al. (eds.), *Biomedical Inorganic Polymers*, Progress in Molecular and Subcellular Biology 54, DOI 10.1007/978-3-642-41004-8_6,

such minerals may strongly affect different living beings, from simple ones to humans. The biomineralogy concept is explained, focusing on the effects of rocks on cell growth and development. The toxic action of silicon dioxide in mammalian lungs is the oldest evidence of crystalline silica bioactivity. More recently, we could demonstrate that crystalline silica has a deep impact on cell biology throughout the whole animal kingdom. One of the most illustrative case studies is the marine sponge *Chondrosia reniformis*, which has the amazing ability to incorporate and etch crystalline silica releasing dissolved silicates in the medium. This specific and selective action is due to the chemical reaction of ascorbic acid with quartz surfaces. One consequence of this is an increased production of collagen. The discovery of this mechanism opened the door to a new understanding of silica toxicity for animal cells and mammalian cells in particular. The presence of silica in sea water and substrates also affects processes like the settlement of larvae and the growth of diatoms. The following sections review all such aspects.

6.1 Biomineralogy: Minerals Shaping Organisms, Organism Shaping Minerals

The ways nature shapes the life on Earth are so original that man has just to observe and get inspired. Biomimetics, the science that seeks innovative technologies by emulating nature's time-tested ideas, dates back to the first man who observed in awe a natural phenomenon and tried imitation. Among the early scholars of such a science, the best known is likely Leonardo da Vinci who, in his *Codex of Flight*, describes the flight of birds and draws detailed sketches of several flying machines. Nature fascinates scientists and engineers with numerous examples of highly efficient building materials. These often show a complex hierarchical organisation from the nano- to the macroscopic scale. Biomineralogy is the science that studies biomineralisation processes (Bouligand 2004), i.e. the way living organisms can model the mineral world. Thanks to biomimetics and biomineralogy, biomaterials are increasingly considered a promising avenue towards a genuine "chimie douce" for the building of innovative tools for technology and biomedicine. Today, biomimetics and biomineralogy are gaining a strong foothold in the scientific arena, with exponential growth in the number of publications and active research groups in academia and, to a lesser extent yet, in industry. Botanists and zoologists working on the morphofunctional adaptations of living organisms have a "front row seat" in the discipline. They disseminate their observations to fertilise the studies on the mechanisms involved in the construction of these original natural models and on the way society can benefit from them.

The mineral and the living worlds are compatible—since the latter arose from the former—and still coexist, so that the study of mineral metabolism and of skeleton formation could provide ideas for experiments in pure inorganic chemistry. In this perspective, the concept of biomineralogy has been reinterpreted (Cerrano et al. 1999a), considering not only how living organisms can handle mineral elements to build amazing and low-energy frameworks (e.g. diatom shells, siliceous spicules, calcareous sclerites) but also how ambient mineral elements affect living organisms. This has been embedded in clinical studies since the eighteenth century but is still poorly considered by marine biologists and ecologists. If we look at the species that are able to incorporate foreign mineral particles, the relevance of the interactions between living and nonliving realms can be better realised. Several metazoans use foreign bodies to set up frameworks. However, these are usually external protective structures and no incorporation of foreign bodies inside tissues occurs. In demosponges and zoanthids, on the contrary, foreign bodies are actively incorporated in their tissues. Among Porifera, sediment incorporation is a widespread phenomenon yet without an evident phylogenetic pattern, suggesting a polyphyletic origin of this peculiar behaviour (Cerrano et al. 2007a). Among Cnidaria the incorporation of sediment is known only in the order Zoanthidea (Previati et al. 2010), the oldest hexacorals (Brugler and France 2007) probably originating from a mesopsammic lifestyle (Fujii and Reimer 2011).

The present review will describe the evolution of our understanding of such interactions starting from the first relevant evidence of effects on humans, moving to the development of new model organisms in the study of this field with implications for research in the clinical, biological, ecological and biomaterials fields.

6.2 Lung Diseases and Silica

6.2.1 Silica and Mammalian Pathologies

Silica, namely silicon dioxide, is the most abundant mineral on the Earth's crust and exists in two forms: amorphous and crystalline (also called free silica). This molecular species is formed from silicon and oxygen under conditions of increased pressure and heat; the amorphous form has no crystalline structure and shows relatively low toxicity to the lung (IARC Monographs 1987), while the crystalline form is based on a tetrahedral structure with silicon as a central atom and oxygen at the corners. Free silica has three principal polymorphs, all of them highly toxic to the lung (IARC Monographs 1987), tridymite, cristobalite and quartz, the latter being by far the most common. Quartz is abundant in most rock types, such as marbles, sandstones, flint and slate as well as in sands and soils, while cristobalite and tridymite are found in volcanic rocks. The drilling, grinding, cutting, breaking, crushing and abrasive blasting of all these materials may produce fine silica dust; indeed, because of the wide usage of quartz-containing materials, workers all over the world are exposed to respirable silica in a large variety of industries and occupations (Gamble 2011). Even if ambient dust control measures are applied to prevent lung diseases in most of developed countries, the incidence of lung pathologies continues to occur at an alarming rate also here. As such, exposure to silica and its detrimental effects on human health are a global occupational problem associated with numerous different diseases, not necessarily restricted to the pulmonary compartment, including various forms of silicosis (acute, accelerated, simple and silico-tuberculosis), industrial bronchitis, emphysema, rheumatoid complications, vascular diseases, glomerulonephritis, autoimmune systemic disorders (lupus, scleroderma) and last, but not least, bronchogenic carcinoma (Ding et al. 2002).

Due to the diversity of pathologies associated to silica exposure, it is quite unlikely that a single, common mechanism is responsible for all the possible diseases. It is generally accepted that alveolar macrophages (AMs) are a relevant cell type to study (Davis 1986). Since the main role of AMs in the lung is to clear the alveolar surface of inhaled debris, it is reasonable to assume that macrophages are the first cells of the body that will have crucial contact with silica particles. Indeed, once inhaled, silica particles are captured by AMs and the subsequent cellular and molecular cascades of events establish the conditions for the development of health complications, the severity of which is determined by the extension and intensity of respirable dust exposure. Although the precise sequence of events (from silica inhalation to development of disease) is not completely known, numerous investigators have been analysing these first fundamental steps and the following mechanisms/pathways have been proposed (Maeda et al. 2010):

1. Initial recognition of silica by cell membrane receptors on the macrophages
2. Engulfment of silica by macrophages and entrapment within lysosomes
3. Production of reactive oxygen species (ROS) in the macrophages
4. Induction of cellular and tissue damages due to the production of ROS
5. Production of various chemokines/cytokines such as tumour necrosis factor-α (TNF-α), monocyte chemoattractant protein-1 (MCP-1), macrophage inflammatory protein 1/2 (MIP1/2), IL-1β and IL-8 causing chronic inflammation and proliferation of fibroblasts
6. Apoptosis of the AM
7. Release of silica particles by the AM and repetition of similar cellular reactions by additional macrophages

If the AM survives the engulfment of silica, it will either migrate out of the lungs (to the proximal lymph nodes or out of the respiratory tract through the mucosal-ciliary escalator) or stay in the lung and move to the interstitial space becoming an activated interstitial macrophage (IM), able to directly contribute to lung disease (Migliaccio et al. 2005). Some evidence suggests that, indeed, IMs could play an important role in the development of silica-induced lung disease (Bowden et al. 1989; Adamson et al. 1991).

The following sections will briefly describe the salient features of the overall-induced silica diseases and the hypotheses currently most accredited to explain how crystalline silica (CS) is toxic to the AM, or in other words, how it all begins.

6.2.2 Silica-Induced Diseases

6.2.2.1 Silicosis

Silicosis is a type of pneumoconiosis caused by the inhalation of CS dust and characterised by chronic inflammation, progressive fibrosis and by the presence of peculiar scars in the form of nodular lesions in the upper lobes of the lungs (Davis 1986). This lung disorder, also known as Potter's rot, is one of the oldest known occupational diseases. The recognition of respiratory problems from respirable dust dates to ancient Greeks and Romans. Agricola, in the mid-sixteenth century, wrote about lung problems from dust inhalation in miners, while Bernardino Ramazzini, in 1713, noted asthmatic symptoms and sandlike substances in the lungs of stone cutters (Rosen 1943). The term silicosis comes from the Latin *silex*, or flint, and was originally coined by the Italian pathologist and prosecutor Achille Visconti in 1870 at the Ospedale Maggiore in Milan as a result of his observations during anatomical dissection studies on a flint worker with almost perfectly sclerotic lungs which presented many irregular, opaque corpuscles spread all over the tissue. The chemical analysis revealed a 7 % content of ashes in the lungs of the cadaver (referred to the dry weight of the organ), half of which consisting of silica (Rovida 1871). With industrialisation, as opposed to hand tools, came also an increased production of dust. The pneumatic hammer drill and sandblasting were introduced in the early 1900s, both significantly contributing to the increased prevalence of silicosis (Rosen 1943). Recent reports indicate that, nowadays, over 23 million workers are exposed to CS in China and over ten million in India alone (Chen et al. 2012). In the United States and Europe, the respective figures are 1.7 million and over three million (Chen et al. 2012).

The classification of silicosis is made according to the disease onset, severity and rapidity of progression. The various forms include:

1. *Chronic simple silicosis*. This is the most common form of silicosis usually resulting from prolonged exposure (10 years or more) to relatively low concentrations of silica dust. It may appear 10–30 years after the first contact of the lung with silica particles and is characterised by the formation of distinct, nodular, hyalinised lesions in the lung, frequently containing silica inclusions (Green and Vallyathan 1996). These nodular lesions are unique pathological entities of silicosis alone. The lesions are mainly present in the upper lobes and nodules can measure from 4 to 10 mm in diameter with a tendency to enlarge and become densely profuse with continued exposure. Cellular infiltrates and dust-engulfed macrophages are usually present at variable amounts in the peripheral zones. A completely developed silicotic nodule is composed of concentric fibrotic laminar layers with calcifications and minimal amounts of dust in the central part. With time and continued exposure to silica, the lesions of simple silicosis, measuring less than 1 cm, enlarge symmetrically and fuse to one another disrupting the surrounding pulmonary architecture and resulting in massive fibrosis, then called complicated or conglomerate silicosis. Because chronic silicosis is very slow to

develop, patients may not show symptoms until the overt of the disease, when a decreased lung function, with dyspnea, tachypnea, fatigue, cough, chest pain and fever, becomes obvious.

2. *Acute silicosis*. A form of silicosis that develops rapidly from a few weeks to 5 years after exposure to high concentrations of silica dust. Alveoli are usually filled with lipid-rich pulmonary oedema, also called silicolipoproteinosis, and interstitial inflammation is widespread (Green and Vallyathan 1996). The disease progresses rapidly with severe loss of pulmonary function which can be fatal.
3. *Accelerated silicosis*. It's a silicosis that develops between 5 and 10 years after exposure to concentrations of respirable silica dust higher than those leading to acute silicosis. It is morphologically very similar to the latter with alveoli showing silicoproteinosis and chronic inflammation and the rapid development of silicotic lesions (Green and Vallyathan 1996). Similarly to acute silicosis, accelerated silicosis is mostly common in occupations where silica is mechanically crushed or fractured and intensely inhaled.
4. *Silico-tuberculosis*. Patients with silicosis are at an increased risk of tuberculosis infection despite that the prevalence of this disease is now reduced in the overall population (Hnizdo and Murray 1988; Cowie 1994). It is well known that the inhalation of silica can lead to impaired cell-mediated immunity with alterations of the number of lymphocyte subsets (lower T cells and higher B cells) and of serum immunoglobulins (Green and Vallyathan 1996).

6.2.2.2 Lung Cancer

In 1987 the International Agency for Research on Cancer (IARC) decided that CS was a probable carcinogen to humans, belonging to Group 2A agents. This was based on the evaluation that there was sufficient evidence for carcinogenicity in experimental animals but limited evidence in humans (IARC Monographs 1987). Ten years later, when a second IARC working group met, CS was classified as a Group I carcinogen based on the reviewing of more scientific papers, as well as nine cohorts and mortality studies published since the earlier meeting (IARC Monographs 1997), stating that CS exposure gave sufficient evidence for carcinogenicity in humans. Actually, there was no unanimous agreement on this classification among the working group (anyway not an unusual situation in an IARC working group). Doubts were based on several issues including (1) the consistency of a strong association between silicosis and lung cancer and (2) the negative outcome of coal mine dust as a lung carcinogen established in the same meeting. Nevertheless, recognising a clear heterogeneity of the quartz hazard in the epidemiological and toxicological studies revised by the working group, the final evaluation included the statement "...the Working Group noted that carcinogenicity in humans was not detected in all industrial circumstances studied. Carcinogenicity may be dependent on inherent characteristics of the crystalline silica or on external factors affecting its biological activity or distribution of its polymorphs..." (IARC

Monographs 1997). Since then, and probably in response to the IARC statement, many papers have been published focusing on the differential hazards of different forms of natural or commercially available CS. These studies provide a potential explanation for the dramatic differences in the exposure response curves observed in many epidemiological studies (Donaldson and Borm 1998). Extensive research has also addressed the genotoxic mechanisms underlying the occurrence of lung cancer mainly questioning whether the tumour is a result of a direct genotoxicity of CS or of indirect genotoxicity acting through inflammation (Borm et al. 2011). Direct primary genotoxicity refers to particles that cause genetic damage in the absence of inflammation. This is by a mechanism involving a direct physical interaction or an oxidative attack by reactive oxygen species (ROS) at the particle surface on the genomic DNA or its associated components (Schins and Knaapen 2007). Indirect primary genotoxicity arises from the enhanced production of ROS and/or depletion of antioxidant defences by particle-stimulated or particle-engulfed cells. On the other hand, indirect genotoxicity is driven by inflammatory cells such as macrophages and granulocytes permanently recruited to the sites of particle deposition. This leads to chronic inflammation, persistent oxidative stress and DNA damage in a pro-survival and proliferation signal-rich environment. The genetic defects that can be accumulated in this situation facilitate the transformation of cells into malignant forms (Azad et al. 2008). An apparent agreement on the mechanism of quartz genotoxicity comes from a third IARC meeting, held in May 2009 (Straif et al. 2009). In that occasion the working group met to reassess the carcinogenicity of ten separate agents (including CS), previously classified as Group I carcinogens, and to identify additional tumour sites and mechanisms of carcinogenesis. The working group reaffirmed the carcinogenicity of "silica dust, crystalline in the form of quartz or cristobalite", indicating the lung as the sole tumour site. Moreover, in the "established mechanism events", the working group indicated an "impaired particle clearance leading to macrophage activation and persistent inflammation" without mentioning any direct genotoxicity mechanism, as such supporting the idea that lung cancer caused by quartz in vivo is a secondary, inflammation-driven event.

6.2.2.3 Dysregulation of the Immune System

There is evidence that silica exposure is also associated to the development of autoimmune diseases such as scleroderma (systemic sclerosis), rheumatoid arthritis, chronic renal disease and systemic lupus erythematosus (Cooper et al. 2002). Indeed, it is well known that silicotic patients often suffer complications from the above-mentioned autoimmune diseases (Hess 2002). More in general, silica is considered one of the most important environmental substances, like vinyl chloride and epoxy resins, that give rise to autoimmune diseases in exposed hosts (Hess 2002; Cooper et al. 2009). An increase of serum polyclonal immunoglobulins associated with high levels of circulating rheumatoid factor, anti-nuclear antibodies and immune complexes is often linked to some of these autoimmune conditions (Craighead et al. 1988). CS, as well as amorphous silica, shares such common pathogenic responses, but little insight into the underlying mechanism has been provided. It is generally believed that the effect of silica on the human immune system results, in part, from the potential adjuvant activity of silica and, additionally, from the increasing extracellular presence of various autoantigens such as RNA, DNA and organelles released by AMs undergoing apoptosis after silica engulfment. Accordingly, it would derive from the increased opportunity for reactive T-lymphocytes to meet circulating autoantigens favouring the development of a detrimental, self-reactive environment in the organism (Hamilton et al. 2008; Thakur et al. 2008).

6.2.2.4 Toxicity of Crystalline Silica in Alveolar Macrophages

Research is currently strongly focused on CS toxicity to the AM. Indeed, most researchers agree that the initial toxicity of silica upon binding and phagocytosis by resident macrophages is a crucial step towards the development of further diseases.

The literature can be divided into three main threads that support as many hypotheses developed to explain how CS particles are toxic to the macrophage cell (1) the free radical hypothesis, (2) the receptor-mediated hypothesis and (3) the lysosomal permeability hypothesis (Hamilton et al. 2008).

1. *The free radical hypothesis*. There are many possible sources of free radicals resulting from silica internalisation, i.e. particle-derived ROS, cell-derived ROS and reactive nitrogen species (RNS), as well as ROS arising from the interaction of cell-derived and particle-derived free radicals such as peroxynitrite and superoxide anion (Fubini and Hubbard 2003). Upon silica phagocytosis, AMs respond with a respiratory burst that produces superoxide anions, hydrogen peroxide and hydroxyl radicals, all potentially toxic to the AM and the surrounding lung tissue being able to damage all types of cellular macromolecules (Valko et al. 2007). Silanol groups, on silica surface, are also thoroughly studied for their ability to generate silicon-based free radicals from freshly fractured silica (Vallyathan et al. 1988) and because they can form hydrogen bonds with oxygen and nitrogen groups on biological membranes, affecting their structure and integrity (Fubini and Hubbard 2003). Furthermore, it is well documented that water reacts with freshly fractured silica producing all types of ROS also thanks to the presence of traces of iron catalysing the Fenton reaction (Fubini and Hubbard 2003). Researchers sustaining the free radical hypothesis (Fig. 6.1, panel a) maintain that ROS may be mediators of silica-induced responses from macrophages and lung epithelial cells such as acute cell injury and proliferation, inflammation and fibrosis, and also DNA modifications (Ding et al. 2002), although some admit that ROS are not completely responsible for silica toxicity (Fubini and Hubbard 2003).
2. *The receptor-mediated hypothesis*. Some researchers believe that the specific receptors involved in the first critical step of particle recognition and internalisation (upon physical contact with the particle surface) are the ultimate

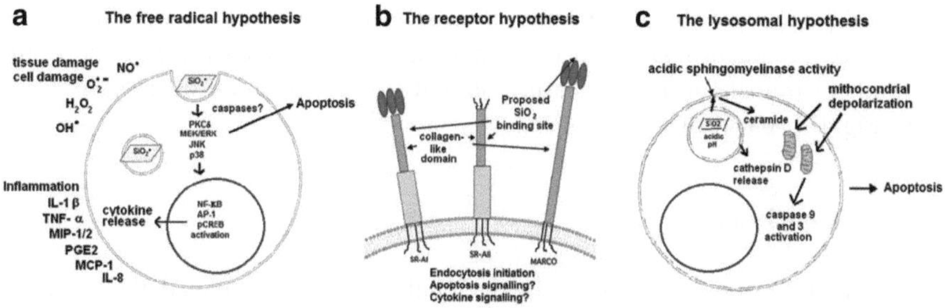

Fig. 6.1 Schematic representation of the three main hypotheses to explain the toxicity of CS to alveolar macrophages. Panel (**a**) The free radical hypothesis: overview of the reported effects of silica phagocytosis, the following respiratory burst, inflammatory pathway activation and consequential cell and tissue damage. Panel (**b**) The receptor-mediated hypothesis: schematic representation of class A scavenger receptors involved in the binding of CS. The extracellular domains proposed for silica binding are indicated for SR-AI, SR-AII and MARCO. Panel (**c**) The lysosomal permeability hypothesis: diagram of the events leading to apoptosis (following lysosomal leakage of cathepsin D in CS-engulfed alveolar macrophages), production of ceramide, mitochondrial depolarisation and cell death

mediators of silica toxicity by inducing an intracellular signalling pathway leading to apoptosis of the AM (Palecanda and Kobzik 2001). It has been described that these silica-binding receptors are also involved in normal immune function and inflammation regulation, the disruption of which can occur by excessive exposure to silica (Gough and Gordon 2000). The most accredited silica-binding receptors belong to the scavenger receptor (SR) family (Murphy et al. 2005), which are known to bind a wide variety of ligands. The SR typically associated with silica binding in the AM are the SR-AI, -AII and MARCO, where the collagen-like domain (present in all mentioned protein structures) has been implicated as the putative binding site for CS (Fig. 6.1, panel b). There are a few reports that address the potential signalling pathway elicited by SR-A and MARCO upon CS binding, some authors describing heat shock proteins and GAPDH recruitment at the level of the intracellular SR domain (Nakamura et al. 2002), while others pointing at tyrosine phosphorylation of the intracellular SR domain followed by activation of PKC (Hsu et al. 1988). Although it is not clear how these receptors can initiate different signalling pathways discriminating between various ligands, there is evidence that such phenomenon can actually occur (Hsu et al. 2001), triggering different responses that lead to the activation of macrophages or, alternatively, to apoptosis.

3. *The lysosomal permeability hypothesis*. Immediately after encountering silica, the AM initiates a yet partially undeciphered engulfment process. Silica particles are then entrapped in the lysosomal compartment where the acidic pH activates strong digestive enzymes, eventually unable to disrupt silica particles. Failure of silica destruction can lead to the loss of lysosomal membrane integrity, and some investigators believe that the failed attempt at silica digestion in the AM leads to the activation of caspases and apoptosis (Thibodeau et al. 2003, 2004). This

specific apoptotic signalling pathway caused by silica would involve lysosomal leakage of cathepsin D with activation of an acidic sphingomyelinase and would result in mitochondrial depolarisation and in the activation of caspase 3 and 9 (Fig. 6.1, panel c).

6.2.2.5 Conclusion

The wide range of diseases caused by the inhalation of CS testifies that CS toxicity is indeed a very complex process. Several biological rationales, summarised in Fig. 6.1, have been presented trying to describe silica binding and toxicity to the AM. It is important to underline that none of the hypotheses presented is mutually exclusive, as there may be more than one mechanism of silica-induced cell death, or the different toxicity mechanisms described may work in tandem to produce their detrimental effects on the exposed subjects. This heterogeneity of mechanisms might help explain why there is a diversity of pathologies linked to silica exposure and inhalation. Furthermore, not all silica-exposed individuals develop a disease. Therefore, there must also be some interaction with genetic and epigenetic factors and possibly polymorphisms that predispose groups of individuals and not others (Qu et al. 2007). Some diseases such as silicosis take years to develop, while others occur relatively rapidly after silica exposure as in the case of scleroderma and other silica-induced autoimmune disorders (Castranova et al. 1996; Cooper et al. 2002). Understanding how exactly mammalian AMs interact with inhaled silica remains a fundamental milestone to be achieved to unravel the silica-induced disease process.

6.2.3 *Mammalian Macrophages, Crystalline Silica, and Ascorbic Acid*

6.2.3.1 Cytotoxic Potential of Ascorbic Acid-Pretreated Quartz

Several physico-chemical features have been suggested as the cause of silica toxicity, among which are particles' shape, the distribution of silanols at the surface of the crystal—and consequently its hydrophilic nature—and the generation of free radicals and ROS (Ding et al. 2002; Castranova 2004). The cellular response to crystalline silica dust is a complex matter regulated by multifactorial events (Hamilton et al. 2008). Consequently, various experimental models have been proposed to explain the molecular mechanisms of quartz toxicity in detail.

Almost two decades ago, Bavestrello et al. (1995) reported that ascorbic acid (AA) is able to partially dissolve the surface of quartz, highly increasing the concentration of soluble silica in the surrounding medium. Later, Fenoglio et al. (2000) demonstrated that during this peculiar chemical reaction, while AA progressively disappears, important modifications of the quartz surface occur, leading to an increased production of free hydroxyl radicals and hydrogen peroxide.

These findings are relevant to mammalian quartz toxicity: by reacting with AA, quartz could deprive the alveolar epithelium of one of its most effective antioxidant defences, while the surface modifications induced by AA increase the concentration of particle-derived ROS in the alveolar space—one of the mechanisms proposed for quartz fibrogenicity and carcinogenicity (Ding et al. 2002). Furthermore, AA-driven quartz dissolution is specific for crystalline silica, as the amorphous silica particulate is not modified at its surface by ascorbate treatment and does not produce hydroxyl radicals, possibly explaining the dramatic difference of toxicity between amorphous and crystalline silica in the lung (Fenoglio et al. 2000).

This prompted Giovine and collaborators to further investigate the cytotoxicity of AA-treated quartz (QA) particles on the murine macrophage cell line RAW 264.7, a cell model widely used for molecular studies on cell-particle interaction. They showed that QA determines a significantly higher cytotoxicity than untreated quartz (Q) (Giovine et al. 2003), suggesting an active role of AA as a cofactor involved in the early stages of quartz-induced pathology. These results became the basis for further research on the effect of AA on the quartz-induced inflammatory response in the same cell model.

6.2.3.2 Inflammatory Potential of Ascorbic Acid-Pretreated Quartz

As cytotoxicity is one of the many aspects of macrophage reactivity to quartz, Scarfì and collaborators investigated the inflammatory response of RAW 264.7 murine macrophages in the presence of QA compared to Q. In particular they focused on the inducible enzyme cyclo-oxygenase-2 (COX-2), one of the key molecules involved in the immediate cellular inflammatory response. The protein was studied in terms of transcription, expression and enzymatic activity (PGE_2 production) on the RAW cell line and on bronchoalveolar lavage (BAL) macrophages, of the transcription factors involved and of the possible role of ROS in triggering the COX-2 synthesis (Scarfì et al. 2007).

They found that in RAW 264.7 cells and in BAL macrophages, only QA particles, and not AA-treated amorphous silica, induced a higher transcription and biosynthesis of COX-2, as well as a greater production of PGE_2 compared to Q. This overall COX-2 increase stimulated by QA seemed to be driven by an enhanced production of hydroxyl and hydrogen peroxide radicals (Fig. 6.2), which in turn stimulated the nuclear translocation and transcriptional activity of NFκB, pCREB and AP-1 transcription factors. These results could be particularly relevant because the over-expression of COX-2 seems strictly related not only to inflammation development but also to cancer progression (Hinz and Brune 2002; Brown and DuBois 2004). Interestingly, the enhancing effect of QA over Q on the synthesis of COX-2 was even more evident in the presence of IFN-γ, a cytokine produced by activated lymphocytes, which are believed to be recruited by macrophages at a later stage of the lung inflammatory process. The result of those experiments suggested that the quartz surface modifications operated by AA could be relevant not only in the first steps of the lung inflammatory response but

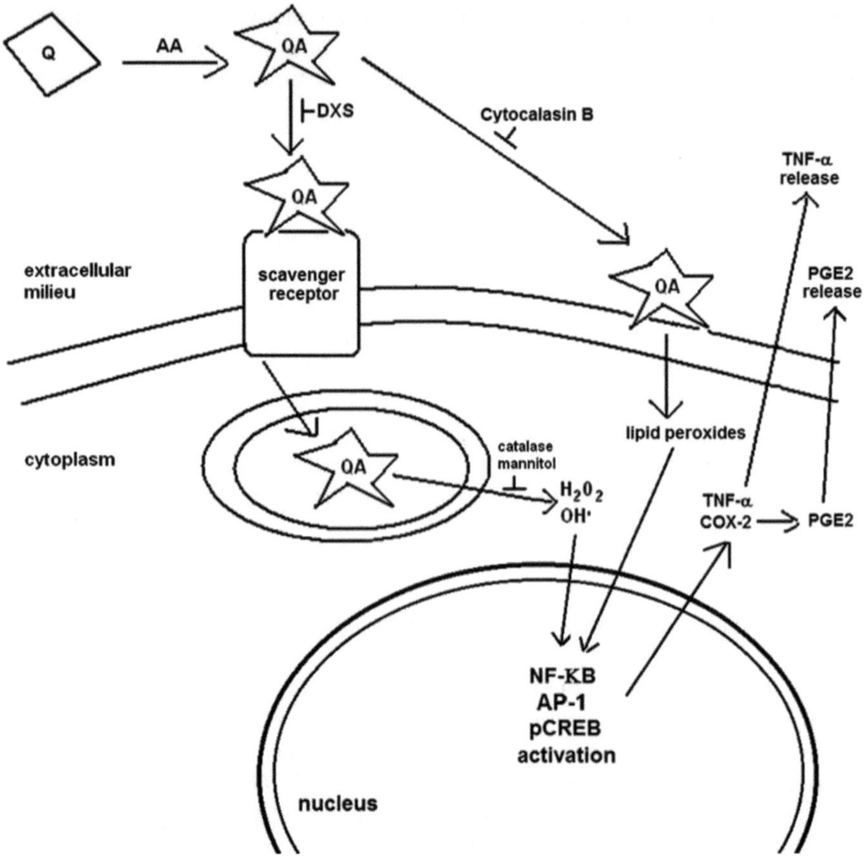

Fig. 6.2 Schematic representation of the mechanisms underlying the toxicity of ascorbic acid-modified quartz (QA) to macrophages [as proposed by Scarfì et al. (2009)], and consequent to QA endocytosis, free radicals production, TNF-α and PGE_2 release. *Q* quartz, *AA* ascorbic acid, *QA* ascorbic acid-modified quartz, *DXS* dextran sulphate, a scavenger receptor inhibitor; cytochalasin B; phagocytosis inhibitor; catalase and mannitol, inhibitors of hydrogen peroxide and hydroxyl radicals, respectively

also subsequently. The fact that macrophages are unable to dissolve the internalised quartz particles indeed prolongs the activation through multiple ingestion-reingestion cycles and exposes the same cells to cytokines produced by activated lymphocytes (Kreyling 1992). The long-lasting presence of quartz in the lung may expose the particles to AA present in the bronchoalveolar fluid, inducing the chemical modifications demonstrated by previous in vitro experiments (Fenoglio et al. 2000; Giovine et al. 2003). The resulting in vivo AA-modified quartz could eventually favour an escalation of the inflammatory response by macrophages, also stimulated by cytokines released by other cell types during the ongoing inflammation.

6.2.3.3 Fibrogenic Potential of Ascorbic Acid-Pretreated Quartz

In a further study, taking advantage of the enhanced macrophage response to AA-treated quartz (QA) compared to untreated quartz (Q), Scarfì and collaborators investigated the first steps of cell activation and the contribution of early signals directly generated by the plasma membrane to the production of TNF-α, a cytokine that is crucial to develop silicosis by activating both inflammatory and fibrogenic pathways (Scarfì et al. 2009). Results demonstrated that the secretion of TNF-α and the synthesis of mRNA were significantly increased in RAW 264.7 macrophages challenged with QA than with Q and that, as in the case of COX-2 production, the higher pro-inflammatory effect of QA over Q was even more evident in the presence of IFN-γ. To assess the direct role of the plasma membrane in the production of silica-induced TNF-α, the research group investigated the very first step of macrophage quartz phagocytosis, which is known to be mediated by scavenger receptors (Kobzik 1995). They demonstrated that membrane lipid peroxidation was significantly higher in QA- than Q-treated cells, and the use of phagocytosis inhibitors (i.e. cytochalasin D and dextran sulphate) further increased lipid peroxidation in both cell samples. These results indicated that the contact of quartz particles with the plasma membrane, which is prolonged in the absence of quartz internalisation, is responsible for lipid peroxidation (Fig. 6.2). Furthermore, the production of TNF-α increased significantly over controls after exposure to Q (and even more after exposure to QA) in cells where phagocytosis was inhibited, indicating that membrane lipid peroxides generated independently of phagocytosis could induce the production of TNF-α.

This was the first demonstration that the contact with quartz by the plasma membrane, in the absence of phagocytosis, is sufficient to trigger membrane lipid peroxidation, TNF-α release and cell death. This appeared to be more detrimental to macrophage survival than particle phagocytosis itself and to increase the fibrogenic potential of crystalline silica by means of an enhanced release of its principal mediator.

As such, an impairment of macrophage phagocytosis could exacerbate lung disease in silica-exposed individuals. This might be the case of alcoholic subjects, where pulmonary macrophage function is compromised resulting in defective phagocytosis (Brown et al. 2007). Additionally, it would imply that alcohol abuse or other pathological conditions impairing macrophage phagocytosis could be a risk factor for silica-induced lung disease and a detrimental condition facilitating complications by means of the prolonged macrophage contact with unphagocytised particles.

6.2.3.4 Conclusion

Taken altogether these studies (summarised in Fig. 6.2) demonstrate that reaction of quartz with AA in the lung is indeed a possible event. Contrary to common belief, this reaction does not neutralise the hazard of quartz as it would be expected given the well-known antioxidant properties of the vitamin. Rather, it increases the toxicity and inflammatory/fibrotic potential of crystalline silica in the lung. The subsequent events, then, lead to higher macrophage cell death and enhanced release of important mediators such as PGE_2 and TNF-α, the pro-inflammatory cytokines currently believed to be mainly responsible of lung fibrosis, and eventually cancer, after exposure to crystalline silica (Piguet et al. 1990; Brown and DuBois 2004).

6.3 The Effects of Minerals on Animal Growth and Development

Minerals in the aquatic environment may affect the growth and development of benthic animals, as suggested by the following examples (1) the development of thicker primmorphs of *Petrosia ficiformis* on quartz rock than on marble suggests that different minerals (here calcium carbonate vs. silica) elicit different responses in sponges (Pozzolini et al. 2010), (2) evidence regarding the positive effects of quartz on collagen production suggests that the thickness of the ectosome in *Chondrosia reniformis* may be enhanced by quartz dissolution (see Sect. 6.4.2.5) and (3) a long-term field experiment on the gamma (massive) stage of the boring sponge *Cliona viridis* (=*nigricans*) maintained on gravels with different mineralogy evidenced a negative effect of quartz on sponge growth compared to a calcareous substrate (Cerrano et al. 2007b).

Similarly, if the interaction with different mineral particles can cause a response in a single species discriminating calcium carbonate from silica, which are the effects of mineral substrates on benthic assemblages? Can the mineralogical features of substrates affect larval settlement and survival?

These considerations are not new to environmental scientists but, as reported in a recent review on the topic (Davis 2009), it is difficult to disentangle the properties of natural mineral substrata that may affect the structure of benthic communities, both on soft and hard bottoms. Owing to the rapidity of biofilm formation, it has always been assumed that larvae interact with surface biofilms, rather than with natural rock surfaces (Wahl 1989; Hadfield 2011). Therefore, until recently authors have not considered the crystallographic structure of sand or rock as a relevant abiotic factor in the structuring of benthic assemblages. In addition, the texture and the crystallographic structure of sand grains, and the roughness in case of rock surfaces, can have confounding effects. This is enhanced by the effects of the different biofilms that may be specific to soft (Manini and Luna 2003; Van Colen et al. 2009) and hard substrates (Doiron et al. 2012; McDougald et al. 2012), whether they are biogenic or not (e.g. Gil-Turnes et al. 1989), and may alter the biological response to minerals (Faimali et al. 2004; Totti et al. 2007). Field experiments indicate that microflora or bioorganic films can play an important role in modifying the patterns of settlement of sessile invertebrates (Todd and Keough 1994; Keough and Raimondi 1995, 1996).

While unravelling the specific role of the mineralogical composition of rocks in benthic ecology is far from easy, scientific evidence increasingly indicates that it may therefore affect the settlement of larvae (Cerrano et al. 1999a) and algal propagules, which may occur by direct as well as indirect mechanisms and be facilitating, inhibiting or neutral to the various organisms. Understanding how larvae respond to such cues across a variety of spatial and temporal scales is a real challenge for marine ecologists. Different rock types can have both physical (roughness) and chemical effects on settling organisms. Calcareous rocks can be eroded by bioborers while quartz rocks (e.g. granite) and those rich in silicates cannot. Generally, pitted substrates are easier to colonise as they offer a wide range of microhabitats, thus enhancing biodiversity. Siliceous rocks cannot be bioeroded and only their surfaces are colonised. However, the production of basal collagen or of a calcareous sheet during larval settlement provides for a physical barrier that enables benthic organisms to compete for space by avoiding several effects derived from substrate features.

Regarding the chemical effects, the ability of a particular rock type to sequester/partition solutes and the scale of its physico-chemical heterogeneity (Holmes et al. 1997) must be considered. Bavestrello and co-authors (2000) provided evidence that quartz-rich rocks and sands negatively affect species richness and epibenthic assemblages. Laboratory experiments revealed that the planulae of a hydroid tended to avoid quartz sands and preferentially settled on marble ones; however, no difference in metamorphosis was recorded (Cerrano et al. 1999a).

The formation of silicon-based and OH radicals in the surrounding water may affect community structure at the fish, algal and sessile invertebrate levels (Guidetti et al. 2004). Cattaneo-Vietti et al. (2002) also studied non-granitic shores detecting statistically significant differences between epibenthic assemblages developed on different substrates, with higher total abundance for sandstone and basalt than serpentine rocks. While leaving many questions open, these works highlight that the effects of rock type on the structure of benthic communities still deserve experimental attention.

6.4 Reactivity of Marine Sponges to Silica Particles

6.4.1 Historical Overview

The incorporation of foreign materials has been described in sponges, many of which appeared to be rich in spongin fibres or collagen. This peculiar behaviour was originally observed 140 years ago by Haeckel (1872). In the following years other authors remarked this behaviour, describing different examples of inorganic particles embedded into spongin skeletons. Specifically, the regular presence of inorganic particles inside the skeletal fibres was used as a taxonomic tool to classify

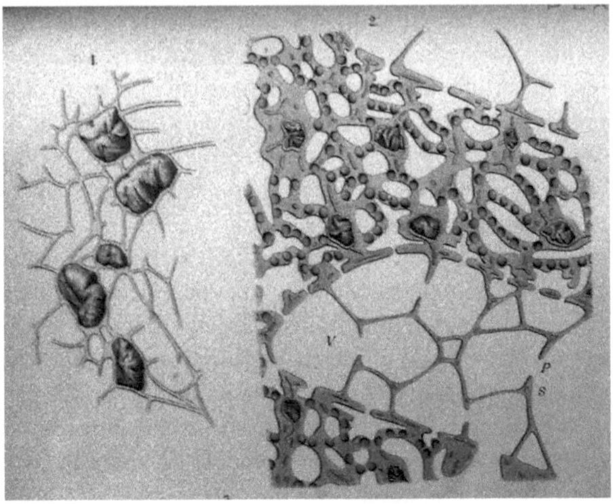

Fig. 6.3 Details of skeleton organisation of *Dysidea villosa* (Lendenfeld 1886), where sand grains are embedded within spongin fibres [partially reproduced by Von Lendenfeld (1889), Table 10, Figs. 1 and 2]

some horny sponges (Von Lendenfeld 1889; de Laubenfels 1950). However, whether the sponges actively select such material is controversial (Haeckel 1872; Schulze 1879; Von Lendenfeld 1889; Sollas 1908; Shaw 1927; Teragawa 1986a, b). Haeckel (1872) suggested a selective uptake. Schulze (1879), on the contrary, considered it passive, since the type of foreign matter incorporated depends on the features of the sponge surface, such as the ectosome texture or the presence of a thin mucous layer. The active selection hypothesis was also supported by Von Lendenfeld (1889), who observed that the incorporation is species specific and stable for the same species from different locations. In his "Monograph of Horny Sponges", a widely documented description of the selective incorporation of sand grains in some species or of foreign siliceous spicules in others is provided, and their peculiar embedment within sponging fibres is detailed. At pages 768–769 of the volume, the author underlined the structural role of the selected inorganic material incorporated, which is organised differently in the different species (see an example in Fig. 6.3). In all cases, Lendenfeld remarked that inorganic materials are firmly associated with spongin fibres and evidenced a different organisation of spongin in the same species grown in the presence of different amounts of suitable sand grains or foreign spicules (Von Lendenfeld 1889). These observations anticipated some ideas resumed and expanded more than 100 years later (Bavestrello et al. 2003).

The occurrence of sand grains or foreign spicules in horny sponges has been a matter of scientific interest also at the beginning of twentieth century. Minchin (1900) thought that sand grains adhering to the surface of sponges were surrounded by the growing points of spongin fibres and were therefore incorporated as the fibres grew upwards. Sollas (1908) studied the incorporation process in *Liosina paradoxa* (=*Migas porphyrion*), suggesting another way by which the inclusion of sand may take place: pseudopodia-like extensions of surface cells surround sand grains, and the inclusion of sand and foreign bodies is due to the action of amoebocytes, also responsible for their conveyance to and orientation in the spongin fibres. In the same report it is shown that some species include only sand in their fibres while closely related ones incorporate only foreign spicules, underlining the highly selective action of sponge cells towards foreign materials. In the technical description of the procedures used to obtain good histological preparations, Sollas advised to pretreat sponge materials with hydrofluoridric acid before cutting slices. This determined the almost complete removal of sand grains (Sollas 1908). This again remarked for the first time a selective incorporation of siliceous sands through amoeboid movements of cells: a similar behaviour is reported for mammalian lung macrophages after the inhalation of silica dust (see Sect. 6.2.1).

In 1927 Shaw continued this research line. In her "*Note on the Inclusion of Sand in Sponges*", she confirmed the prevalence of the phenomenon in species characterised by almost total absence of endogenous spicules, counterbalanced by a high production of spongin fibres. In the same report the author evidenced the absence of detectable amounts of sand granules inside sponges of the genera *Chondrosia* and *Chondrilla*, cited as a peculiarity. Now we know that also *Chondrosia* species, when living in environments rich in siliceous sand, incorporate and select large amounts of foreign material as described below (see Sect. 6.4.2.2). *Chondrilla* species produce their own spicules. In the conclusive remarks, and on the basis of histological observations of fixed materials, the author supported the idea that sponges actively select sand grains. She agreed on the hypothesis previously suggested by Sollas to explain how sand is incorporated and underlined the inverse correlation between the abundance of foreign material and the amount of endogenous spicules inside sponges (Shaw 1927). To our best knowledge, apart from some additional (taxonomical) remarks by De Laubenfels (1950), no further research has been carried out on the topic until the late 1980s (Teragawa 1986a, b; Cerrano et al. 2007a). Teragawa worked on the keratose demosponge *Dysidea etheria* de Laubenfels, a species building a spongin skeleton that can be filled with a wide variety of foreign particles. In a first work based on histological techniques, Teragawa showed a remarkable interaction of dermal cells with particles, suggesting a direct role of the same cells in moving, internalising and orienting sand grains (Teragawa 1986a). In a second contribution, she used *D. etheria* to perform relevant experiments directly demonstrating particle transport and incorporation during the formation of sponge skeleton. The loose construction and the relative transparency of the sponge species she used allowed for the direct observation of the interactions and movements of foreign particles between the outer surface and the underlying distal skeletal components in the living animals, using a time lapse cinemicrography technique. The results clearly demonstrated an active role of the sponge in the selection and incorporation process. To perform her experiments, the author used sand grains of different typology and dimension, revealing different reactions of the sponge surface towards particulates, depending on size and abundance. As a whole, the smallest particles are inhaled by the aquiferous system; the others are retained by the surface but only ca. 20 % are definitively incorporated into fibres (Teragawa 1986b). To better evaluate the contribution of sand to skeleton formation, two different types of sand grains were used: siliceous sand, as a reference, and carbonatic sand, as test grains. The results showed how the incorporation of sand grains positively contributes to build up spongin fibres (Teragawa 1986b), but the experimental design applied did not allow the unequivocal discrimination between the specific contributions of siliceous and calcareous grains. The latter consideration is pivotal in the interpretation of current research that considers sponges, especially collagen-rich demosponges, as an evolutionary cornerstone for the production of the proteins of the extracellular matrix in the animal kingdom. Recently, Cerrano and co-authors reviewed how sponges incorporate foreign matter by comparing the strategies of species living in soft and hard bottoms. They showed that in both habitats, sponges are able to incorporate sand grains or foreign spicules, often discriminating carbonate materials from silica (according to the species). However, if the animal or parts of it get detached from the substrate, they invariably lose the ability to select particles until they find a new settlement place (Cerrano et al. 2007a).

6.4.2 *Present Time* Chondrosia reniformis *Nardo 1847: A Case Study*

6.4.2.1 The Sponge

Chondrosia reniformis (Nardo 1847) is a marine sponge that lacks calcite or siliceous spicules. Its body consists of a collagenous mesohyl, which is located between the external epithelium and the internal epithelia that line the inhalant and exhalant canals (Wilkie et al. 2006). Although the mesohyl confers the whole animal a cartilaginous texture, it is also plastic in terms of shape and mechanical properties. For example, within 40–60 h after excision, fragments of *C. reniformis* (fragmorphs) undergo rounding off of cut surfaces and marked bending (Nickel and Brümmer 2003). Whole sponges or tissue explants can lose rigidity under continuous compression (Garrone et al. 1975; Garrone 1978). The animal, in its natural environment, may lose its adhesion substrate and slowly elongate under gravity (Fig. 6.4) until its eventual separation from the still-attached portion, a process that may be regarded as a form of opportunistic asexual reproduction (Sarà and Vacelet 1973; Bonasoro et al. 2001; Zanetti 2002). This plasticity is mainly due to the extraordinary attributes of *C. reniformis* collagen, which has a behaviour similar to the contractile collagens described in echinoderms (Wilkie et al. 2006). The structure and features of the collagen from *C. reniformis* (Heinemann et al. 2007), the sponge's amazing ability to incorporate and select foreign inorganic materials (Bavestrello et al. 1995) and the relationship between the incorporation of particles and the biosynthesis of collagen (Pozzolini et al. 2012) make this species a unique model to study the evolution of the extracellular matrix.

Fig. 6.4 *C. reniformis* in its natural environment, here displaying the peculiar elongation process typical of the species

6.4.2.2 Incorporation of Foreign Material

In horny sponges, foreign inorganic particles are embedded in the spongin fibres and in the ectosome, while in *C. reniformis*, they form a layer underneath the sponge surface (Bavestrello et al. 1996). The cells of the exopinacoderm are responsible for their subsequent incorporation. In *C. reniformis*, particles that settle on the mucous ectosome (Fig. 6.5) are translocated, at varying rates, to special areas where they are quickly engulfed (Bavestrello et al. 1996). In this sponge the incorporation of foreign materials is a complex phenomenon involving the passive settling of suspended sediment on the sticky sponge surface (Fig. 6.5), the active selection of particles by pinacocytes and their eventual engulfment (Bavestrello et al. 1996, 1998a). The availability of particles depends on environmental conditions. As shown comparing two locations in the Ligurian Sea, the incorporation of foreign material by *C. reniformis* depends on the rate of sedimentation (Cerrano et al. 1999b). The rate of incorporation also depends on the amount of sediment suspended in the water column, directly related to the intensity of water movement. The size of the particles incorporated depends on environmental conditions, too. When the sea is rough, only fine particles (about 10 μm in diameter) can adhere to the mucous surface of *C. reniformis*, while coarser particles (about 30 μm) may be collected only during calm sea conditions (Cerrano et al. 1999b).

6.4.2.3 Selection

Field observations and laboratory experiments indicate that *C. reniformis* selects and incorporates only siliceous materials, in particular quartz particles (crystalline silica) and sponge spicules (amorphous biogenic silica) from the sediment, even though carbonate particles are often dominant in the same sediments. Laboratory experiments demonstrated that the cells of the ectosome are the main actors in this

Fig. 6.5 *C. reniformis* in its natural environment. Note the foreign particles settled on its mucous ectosome

process. Here siliceous particles are incorporated while carbonate ones are agglutinated and drop off the sponge surface (Bavestrello et al. 1998b). Nevertheless, a striking finding was the remarkable ability of *C. reniformis* to discriminate between crystalline (quartz) and amorphous biogenic silica (foreign spicules) (Bavestrello et al. 1998a). When crystalline quartz particles settle on the sponge, ectosome cells contract uniformly, creating a ruffled surface that remains throughout the process of quartz incorporation. Indeed, when exopinacocytes interact with crystalline quartz, they contract and become globular, creating a slightly raised layer that contrasts with the flattened epithelium of undisturbed sponges. Quartz particles are gradually engulfed by pinacocytes that stretch to form a rim around them (Bavestrello et al. 2003). Despite this reaction, a gap is maintained between the exopinacocytes and the foreign grains. As the grains sink into the collagenous cortical matrix of the sponge, the cellular ring shrinks until the grain is completely incorporated. Eventually, the exopinacoderm returns to its original state (Bavestrello et al. 2003). In contrast, siliceous spicules elicit a motile response in exopinacocytes. SEM analysis showed that these cells, without losing their epithelial configuration, release an organic matrix that covers the spicules. This process creates a coat towards which other exopinacocytes protrude long and thin pseudopodia. As time goes by, other exopinacocytes line up and cover the spicules that become completely incorporated (Bavestrello et al. 2003).

The selection and incorporation of foreign inorganic matter in *C. reniformis* relies on the ability of exopinacocytes to contract and to move (Bavestrello et al. 1998a). The peculiar ability of this sponge to recognise and select mineralogical species (calcite, quartz, opal, chalcedony, etc.) also depends on the life cycle phase and the anatomical region concerned. Usually, only the upper surface recognises and eliminates calcium carbonate, while the lower surface, which settles on every kind of substratum, does not. During fragmentation, the exclusive affinity for silica is lost by the sponge propagules, allowing the incorporation of calcareous particles as well (Bavestrello et al. 1998b). This behaviour is shared by other sponge species (see Sect. 6.4.1) and is probably related to the anchoring strategy of sponge fragments (Cerrano et al. 2007a).

6.4.2.4 The Destiny of Incorporated Particles: The Dissolution of Quartz

What happens after particles have been incorporated by the sponge ectosome is remarkably different, depending on the cellular reactivity towards crystalline or amorphous structure and on the final destiny of the particles incorporated: whereas foreign spicules remain unaltered within the sponge tissue, quartz particles are quickly etched and show typical corrosion patterns at the SEM analysis (Bavestrello et al. 1995). The experimental administration of quartz sand to several specimens of *C. reniformis* showed that their size became uniform (about 30 μm) after a week, irrespective of the initial size. The etching of quartz particles by *C. reniformis* is the first report of such an activity in the animal kingdom (Bavestrello et al. 1995). It is remarkable that only crystalline quartz is dissolved. Not only opal biominerals like sponge spicules, but also inorganic amorphous minerals remain unaltered. This is particularly evident for benthic specimens in the volcanic Aeolian Archipelago (Tyrrhenian Sea), where the only available siliceous material is volcanic glass. In this area, *C. reniformis* incorporates several fragments of such amorphous material, which is not etched inside sponge tissues, so grain size is unusually large (about 60 μm) (Bavestrello 1996). The ability to etch quartz sand is ascribed to a specific chemical reaction. Previous evidence that several polyvalent organic acids are able to dissolve quartz (Bennet 1991) suggested the involvement of ascorbic acid in the quartz etching process seen in sponges (Bavestrello et al. 1995). Ascorbic acid is the reducing agent in proline hydroxylation during collagen synthesis; its concentration in *C. reniformis* ranges from 1–10 μg/g of wet weight (Cerrano et al. 1999c), the same order of magnitude observed in plants (Loewus 1980). Ascorbic acid is continuously released from the ectosome, especially when the sponge is disturbed (Bavestrello et al. 1995). Its extrusion may also be deduced by the presence of calcium oxalate crystals on the sponge surface (Cerrano et al. 1999c). In slightly alkaline environments, like sea water (pH 7.8–8.0), ascorbic acid is indeed spontaneously oxidised to oxalic acid (Cerrano et al. 1999c). Fenoglio et al. (2000) observed that ascorbic acid dissolves quartz more rapidly than it does with amorphous silica. The author suggested that ascorbic acid interacts via hydrogen bonding with the surface of silica, followed by the attack on the silicon atom and the removal of the silicate from the crystal. This peculiar reaction is favoured by the specific distance between surface silanols (Si-OH) in crystalline quartz but not in amorphous silica (Fenoglio et al. 2000). The dissolved silica seems to form a complex with an unidentified ascorbate derivative, which could be similar to the hexacoordinate organosilicon complexes that are normally generated by catechol reaction with quartz (Iler 1979). Ascorbic acid, like catechol, exhibits a planar 1.2 diols moiety. Moreover, evidence of organosilicon complexes in the presence of diols in neutral solutions has been reported (Kinrade et al. 2001).

In both fresh and sea water, the rate of silica dissolution as $Si(OH)_4$ is greatly affected by crystallinity, and it is generally accepted that the most soluble forms of silica are amorphous, while crystalline forms are characterised by a very low solubility (Kamatani 1971). Inside *C. reniformis* the dynamic of dissolution is remarkably different: amorphous hydrated silica is engulfed and "protected" against the slow dissolution operated by sea water, while the almost insoluble quartz grain is incorporated in a sort of "chamber" inside the ectosome (Bavestrello et al. 2003). Here, it is dissolved by ascorbic acid, which selectively reacts with crystalline silica rather than with the amorphous biogenic structure (Fenoglio et al. 2000).

The chemical study of this process evidenced an unexpected but extremely interesting contribution to the modification of quartz surface features generated by ascorbic acid. Indeed, once pretreated with vitamin C, quartz grains show relevant alterations in the silanol population, specifically a reduced number of free silanols (Fenoglio et al. 2000). The same scientific study also evidenced a remarkable increase of $^{\bullet}OH$ in the presence of hydrogen peroxide in an aqueous suspension of quartz pretreated with ascorbic acid, while radical production was not detected with amorphous silica (Fenoglio et al. 2000). This may explain why the sponge interacts in a different way with quartz sand or opal spicules, suggesting a new perspective in the earlier molecular events of lung injuries generated by the inhalation of quartz dust in human being, as reviewed above (see Sect. 6.2.3).

6.4.2.5 Biological Meaning of Quartz Etching

Although the process of selection, incorporation and modification of foreign materials has been clearly documented for different sponges, its adaptive function has not been fully understood yet. Sponges seem to incorporate foreign materials to strengthen their structure. Therefore, in a high-sedimentation environment, sponges select the settled material at the epithelial level, thus avoiding the incorporation of high amounts of calcite (often the most abundant fraction in the sediment). In the specific case of *C. reniformis*, while opal remains unaltered, quartz is rapidly dissolved and expelled as pellets. One consequence of quartz dissolution is an increased concentration of dissolved silica in the intercellular medium and in the nearby water environment (Bavestrello et al. 1995). Recent studies on this peculiar sponge definitively demonstrated a positive role of dissolved silica in the expression the collagen gene (Pozzolini et al. 2012). The authors identified one of the most represented collagen genes in *C. reniformis* (COLch), describing it for the first time at the molecular level. Its full length cDNA codes for a nonfibrillar collagen type, characterised by two noncollagenic and three collagenic domains (Fig. 6.6). On these grounds, COLch proves to belong to the short-chain spongin-like collagen family, with a high structural similarity to type IV collagen, even if phylogenetic analysis shows it evolved independently from a common ancestor (Pozzolini et al. 2012). This suggests a specific role by COLch in mediating the attachment

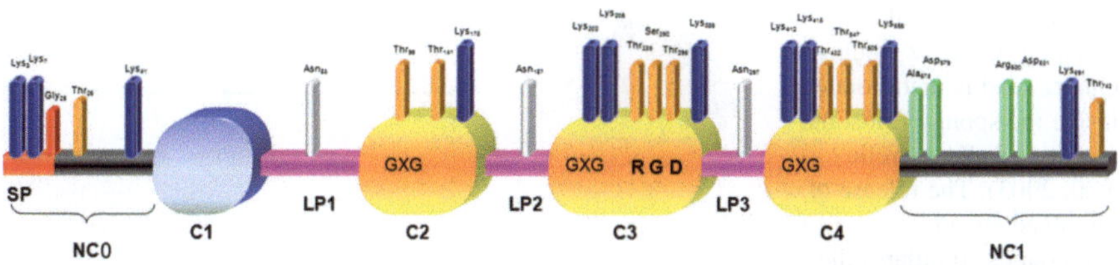

Fig. 6.6 Deduced organisation of COLch amino acid sequence domains. The non-collagenous regions at the N terminus and C terminus of the protein (NC0 and NC1, respectively) are marked in *dark grey*. In NC0, the 25-amino acid signal peptide (SP) is in *red*, and the cleavage site is between glycine$_{25}$ and threonine$_{26}$. In NC1, the four amino acid residues involved in the two putative C-proteinase cleavage sites, Ala$_{578}$–Asp$_{579}$ and in Arg$_{630}$–Asp$_{631}$, are indicated in *green*. The collagen-like domain (C1) is indicated as a *blue block* while the three conventional collagen domains (C2, C3 and C4) are indicated as *yellow blocks*. The GXG triplet imperfection is shown inside each of them; additionally, the RGD motif is shown in C3. The three linker peptides (LP1, LP2 and LP3) are indicated as *pink lines*. The potential glycosylation sites on lysine residues are marked in *blue*, the potential glycosylation sites on threonine and serine are marked in *yellow* and the potential glycosylation sites on asparagine are marked in *white*

to proper substrata by sponges (Exposito et al. 2002) as well as in reinforcing the ectosome structure. Using the in situ hybridisation technique, Pozzolini and co-authors confirmed this hypothesis. The results demonstrated that the mRNA encoding for COLch is mainly expressed in the peripheral areas of the sponge, both in the upper and in the basal region. In the same work, the authors unequivocally demonstrated the role of soluble silicon compounds in the up-regulation of the expression of this spongin-related gene. They used two different compounds: Na_2SiO_3, which generates the same anion normally found in sea water, and Na-hexafluorosilicate (Krasko et al. 2000). Both of them activated the transcription of COLch mRNA in fragmorphs model (Pozzolini et al. 2012). These findings are relevant to the molecular biology research about the proteins of the extracellular matrix. Indeed, it is widely known that silicon compounds are pivotal in the production of collagen in many living beings, from sponges (Krasko et al. 2000) to vertebrates (Carlisle 1972, 1980; Maehira et al. 2008). However, to our knowledge, there is no strong evidence yet to support a direct role of these compounds in the regulation of gene expression. The etching process has a physiological role in *C. reniformis*. The incorporation of crystalline silica causes an increase of dissolved silicon concentration and consequently of the up-regulation of collagen gene expression (Fig. 6.7).

Figure 6.8 shows how the interaction with different silica dusts in *C. reniformis* results in a clear modification of the body shape. Fragmorphs of *C. reniformis* (Pozzolini et al. 2012) were treated for 10 days with MinUSil (a well characterised quartz dust from US Silica Company), with Aerosil (amorphous silica dust from Degussa) or with *P. ficiformis* minced spicules. At the end of the treatment, the external epithelium was partially restored under each condition, but the morphology of the respective fragmorphs was significantly different. The samples treated

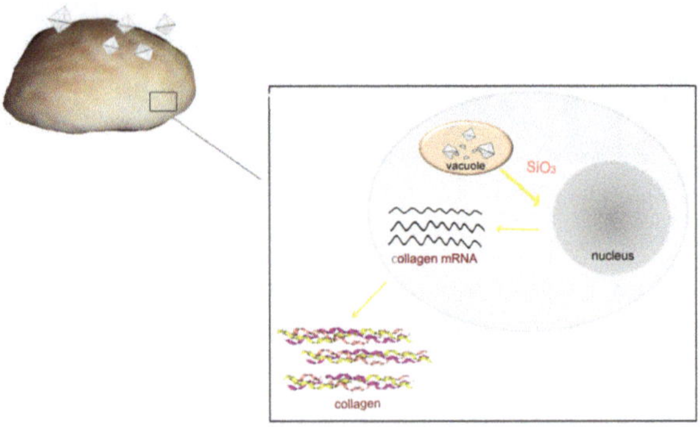

Fig. 6.7 Biological meaning of quartz etching. Quartz sand is incorporated inside the sponge ectosome and etched (Bavestrello et al. 2003). The release of dissolved silicon compounds stimulates the expression of COLch gene, with the consequent production of new fibres

with quartz dust were larger than those treated either with amorphous silica dust or minced spicules that were much more rounded. The difference may be due to the latter models trying to minimise their surface/volume ratio hence exposing the least surface to the powder. Yet, the influence of different silica dust on the organisation of collagen fibres in *C. reniformis* is still to be elucidated at the ultrastructural and molecular levels.

6.5 Different Silica Affect Diatoms Growth

Diatoms are the dominant species in many ecosystems and account for up to 35 % of total primary production in oligotrophic oceans and up to 75 % in coastal waters (Tréguer et al. 1995). Jin et al. (2006) estimated their global contribution to net primary production at 15 % and to carbon export at 40 %. Moreover, diatoms are the most important group of organisms affecting silicon balance in the marine ecosystem as they form polymerised frustules (Jézéquel et al. 2000; Ragueneau et al. 2000). Silicon constitutes the diatom cell wall and its availability is a key factor in regulating diatom growth (Shipe and Brzezinski 1999). Silicon enters the oceans by river run-off (either dissolved, in monomeric or oligomeric form, or as particulate solids of varying size and crystallinity) and by the dissolution of particulate silica present in the lithogenic and biogenic marine environment. Not much is known on Si biomineralisation processes (Finkel and Kotrc 2010), while the biochemical pathways involved in silicification and their regulation are still being investigated (Hildebrand 2008; Ramanathan et al. 2011). Moreover, little information is available about the possible influence of different silica sources on the growth of marine diatoms. This has an ecological relevance for Si-biogeochemical patterns in coastal sea water, where dissolved silica may be depleted during diatom blooms and under grazing pressure, while siliceous lithogenic material and biogenic silica from settled diatoms are abundant (Schultes et al. 2010).

Fig. 6.8 Effect of different silica dusts on the body shape of *C. reniformis*. Comparison between day 0 (a) and day 10 (b). Control fragmorph from *C. reniformis* (1), quartz dust (2), amorphous silica (3), *P. ficiformis* minced spicules (4)

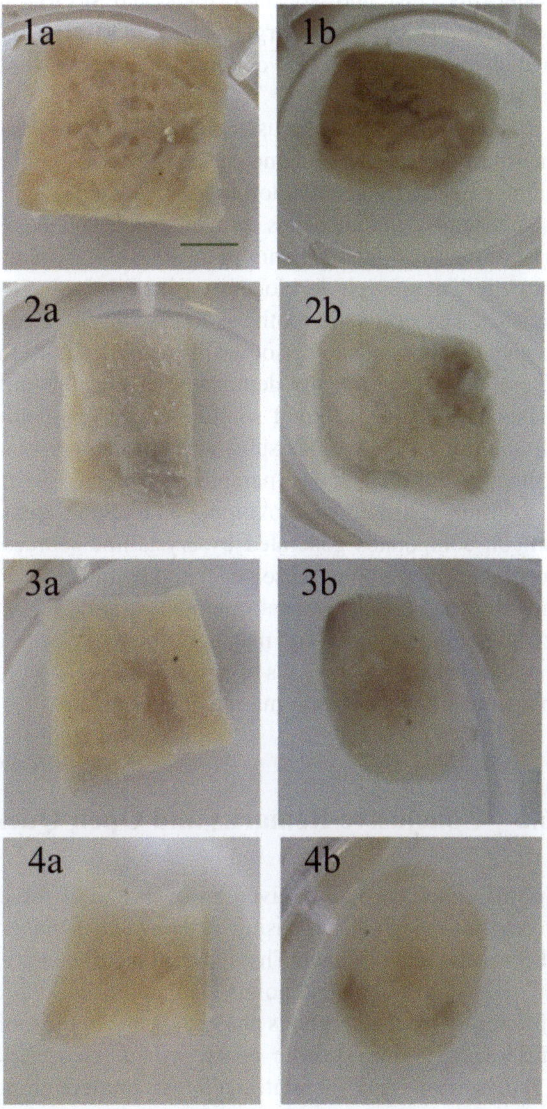

The marine silica cycle is dominated by the production and subsequent dissolution of biogenic silica in the water column. The complex chemical balance of different silica compounds in aqueous solutions with respect to the source type (i.e. biogenic silica or quartz) and the physico-chemical features of the aqueous environment are known (Dove and Rimstidt 1994; Dove et al. 2008). In sea water, dissolved silica occurs in different chemical forms $SiO_x(OH)_{x2}$ and concentrations depending on the polymorphs from which it originates. At the natural pH of sea water (pH = 8), undissociated silicic acid $Si(OH)_4$ is the dominant form (97 % of total dSi), the remainder largely being $SiO(OH)_3^-$. However, little is known on the actual bioavailability to diatoms of different silicon compounds (Demarest et al. 2009).

The group of Penna focused on the role of silica sand in regulating the growth of marine diatoms. They demonstrated that cell culture media containing different lithogenic and biogenic/amorphous silica sands (such as crystalline quartz—natural quartz sand—diatomaceous earth or sponge spicules) significantly affect the growth rate of marine diatoms (Penna et al. 2003). In particular, quartz sand induced higher growth than amorphous biogenic silica, likely due to surface-driven uptake mechanisms. The original hypothesis to explain these results was that diatoms excrete organic compounds to dissolve silica and make it bioavailable (Penna et al. 2003). Present-day experiments demonstrated that the differences are due to different bioavailability of different soluble silicon compounds generated by spontaneous dissolution of the various substrates (Penna, personal communication). More specifically, the analysis of parameters typical for diatoms (such as cell growth, maximum cell density, biovolume and silicon uptake) suggested that the soluble silicon compounds generated by crystalline sources in sea water are highly bioavailable compared to those generated by biogenic and amorphous materials (Data not shown, Penna, personal communication). These findings are of considerable ecological importance and may contribute to clarify anomalous spatial and temporal distributions of siliceous organisms with respect to the presence of lithogenic or biogenic silica sources in marine environments.

6.6 Conclusive Remarks and Open Questions

In this chapter we have given an overview of the complex interactions between rocks and living cells, focusing on case studies that range from human pathology to the aquatic environment. The apparent heterogeneity of these examples is actually solved by several shared points (1) in aqueous solutions, crystalline silica behaves differently from amorphous phases; (2) one of the molecular keys at the basis of this difference is the specific reactivity of ascorbic acid with crystalline silica; (3) dissolved silica stimulates collagen production in marine sponges; and (4) different silica substrates affect the growth of diatoms in different ways. These points remark the deep interaction between living beings and silica. However, little is still known about the molecular mechanism at the basis of such processes. The main open question is related to the chemistry of silica in aqueous solution. The generally accepted idea is that in aqueous solution, silica generates orthosilicic acid although with different kinetics and yields depending on several factors. Yet, the specific reactivity of ascorbic acid with crystalline silica and the findings related to diatoms are not consistent with this assumption. "Silicon bioavailability" is the correct perspective hereby introduced and "bioavailable silicon" is likely the result of peculiar chemical species not yet identified.

Acknowledgements The research leading to these results has received funding from the European Union Seventh Framework Programme (FP7/2007-2013) under grant agreement no. KBBE-2010-266033 to MG and from University of Genova Funding (PRA 2012) to SS.

References

Adamson IY, Letourneau HL, Bowden DH (1991) Comparison of alveolar and interstitial macrophages in fibroblast stimulation after silica and long or short asbestos. Lab Invest 64:339–344

Azad N, Rojanasakul Y, Vallyathan V (2008) Inflammation and lung cancer: roles of reactive oxygen/nitrogen species. J Toxicol Environ Health B Crit Rev 11:1–15

Bavestrello G (1996) Particolarità nella formazione dello scheletro osservato in alcune specie di poriferi delle isole Eolie. In: Faranda FM, Povero P (eds) Caratterizzazione ambientale marina del sistema Eolie e dei bacini limitrofi di Cefalù e Gioia (EOCUMM95). G. Lang Artigrafiche, Genova, pp 317–322

Bavestrello G, Benatti U, Cattaneo-Vietti R, Cerrano C, Giovine M (1995) Quartz dissolution by the sponge *Chondrosia reniformis* (Porifera, Demospongiae). Nature 378:374–376

Bavestrello G, Cerrano C, Cattaneo-Vietti R, Sarà M, Calabria F, Cortesogno L (1996) Selective incorporation of foreign material in *Chondrosia reniformis* (Porifera, Demospongiae). Ital J Zool 63:215–220

Bavestrello G, Arillo A, Calcinai B, Cerrano C, Lanza S, Sarà M, Cattaneo-Vietti R, Gaino E (1998a) Siliceous particles incorporation in *Chondrosia reniformis* (Porifera, Demospongiae). Ital J Zool 65:343–348

Bavestrello G, Benatti U, Calcinai B, Cattaneo-Vietti R, Cerrano C, Favre A, Giovine M, Lanza S, Pronzato R, Sarà M (1998b) Body polarity and mineral selectivity in the demosponge *Chondrosia reniformis*. Biol Bull 195:120–125

Bavestrello G, Bianchi CN, Calcinai B, Cattaneo-Vietti R, Cerrano C, Morri C, Puce S, Sarà M (2000) Bio-mineralogy as a structuring factor for rocky bottom communities. Mar Ecol Prog Ser 193:241–249

Bavestrello G, Benatti U, Cattaneo-Vietti R, Cerrano C, Giovine M (2003) Sponge cell reactivity to various forms of silica. Microsc Res Tech 62:327–335

Bennet PC (1991) Quartz dissolution in organic-rich aqueous systems. Geochim Cosmochim Acta 55:1781–1797

Bonasoro F, Wilkie IC, Bavestrello G, Cerrano C, Candia Carnevali MD (2001) Dynamic structure of the mesohyl in the sponge *Chondrosia reniformis* (Porifera, Demospongiae). Zoomorphology 121:109–121

Borm PJ, Tran L, Donaldson K (2011) The carcinogenic action of crystalline silica: a review of the evidence supporting secondary inflammation-driven genotoxicity as a principal mechanism. Crit Rev Toxicol 41(9):756–770

Bouligand Y (2004) The renewal of ideas about biomineralisations. CR Palevol 3:617–628

Bowden DH, Hedgecock C, Adamson IY (1989) Silica-induced pulmonary fibrosis involves the reaction of particles with interstitial rather than alveolar macrophages. J Pathol 158:73–80

Brown JR, DuBois RN (2004) Cyclooxygenase as a target in lung cancer. Clin Cancer Res 10:4266s–4269s

Brown LA, Ping XD, Harris FL, Gauthier TW (2007) Glutathione availability modulates alveolar macrophage function in the chronic ethanol-fed rat. Am J Physiol Lung Cell Mol Physiol 292: L824–L832

Brugler MR, France SC (2007) The complete mitochondrial genome of the black coral *Chrysopathes formosa* (Cnidaria: Anthozoa: Antipatharia) supports classification of antipatharians within the subclass Hexacorallia. Mol Phylogenet Evol 42:776–788

Carlisle EM (1972) Silicon: an essential element for the chick. Science 178:619–621

Carlisle EM (1980) Biochemical and morphological changes associated with long bone abnormalities in silicon deficiency. J Nutr 110:1046–1056
Castranova V (2004) Signaling pathways controlling the production of inflammatory mediators in response to crystalline silica exposure: role of reactive oxygen/nitrogen species. Free Radic Biol Med 37(7):916–925
Castranova V, Vallyathan V, Wallace WE (1996) Silica and silica-induced lung diseases. CRC, Boca Raton, FL
Cattaneo-Vietti R, Albertelli G, Bavestrello G, Bianchi CN, Cerrano C, Chiantore M, Gaggero L, Morri C, Schiaparelli S (2002) Can rock composition affect sublittoral epibenthic communities? Mar Ecol 23:65–77
Cerrano C, Arillo A, Bavestrello G, Benatti U, Calcinai B, Cattaneo-Vietti R, Cortesogno L, Gaggero L, Giovine M, Puce S, Sarà M (1999a) Organism-quartz interactions in structuring benthic communities: towards a marine bio-mineralogy? Ecol Lett 2:1–3
Cerrano C, Bavestrello G, Cattaneo-Vietti R, Giovine M, Benatti U, Sarà M (1999b) Incorporation of inorganic matter in *Chondrosia reniformis* (Porifera, Demospongiae): the role of water turbulence. Mem Queensl Mus 44:85–90
Cerrano C, Arillo A, Bavestrello G, Benatti U, Bonpadre S, Cattaneo-Vietti R, Gaggero L, Giovine M, Leone L, Lucchetti G, Sarà M (1999c) Calcium oxalate production in the marine sponge *Chondrosia reniformis*. Mar Ecol Prog Ser 179:297–300
Cerrano C, Calcinai B, Di Camillo CG, Valisano L, Bavestrello G (2007a) How and why do sponges incorporate foreign material? Strategies in Porifera. In: Custódio MR, Lôbo-Hajdu G, Hajdu E, Muricy G (eds) Porifera research: biodiversity, innovation and sustainability. Museu Nacional, Série Livros 28, Rio de Janeiro
Cerrano C, Sambolino P, Calcinai B, Azzini F, Bavestrello G (2007b) Growth of the massive morph of *Cliona nigricans* (Schmidt, 1862) (Porifera, Clionaidae). Ital J Zool 74:13–19
Chen W, Liu Y, Wang H (2012) Long-term exposure to silica dust and risk of total and cause-specific mortality in Chinese workers: a cohort study. PLoS Med 9(4):e1001206
Cooper GS, Miller FW, Germolec DR (2002) Occupational exposures and autoimmune diseases. Int Immunopharmacol 2:303–313
Cooper GS, Makris SL, Nietert PJ, Jinot J (2009) Evidence of autoimmune-related effects of trichloroethylene exposure from studies in mice and humans. Environ Health Perspect 117:696–702
Cowie RL (1994) The epidemiology of tuberculosis in gold miners with silicosis. Am J Respir Crit Care Med 150:1460–1462
Craighead JE, Kleinerman J, Abraham JL, Gibbs AR, Green FHY, Harley RA, Ruettner JR, Vallyathan V, Juliano EB (1988) Diseases associated with exposure to silica and non-fibrous silicate minerals. Arch Pathol Lab Med 112:673–720
Davis GS (1986) The pathogenesis of silicosis: state of the art. Chest 89:166S–169S
Davis AR (2009) The role of mineral, living and artificial substrata in the development of subtidal assemblages. In: Wahl M (ed) Marine hard bottom communities. Springer, New York, NY
De Laubenfels MW (1950) The Porifera of Bermuda Archipelago. Trans Zool Soc Lond 27:1–154
Demarest MS, Brzezinski MA, Beucher C (2009) Fractionation of silicon isotopes during biogenic silica dissolution. Geochim Cosmochim Acta 73:5572–5583
Ding M, Chen F, Shi X, Yucesoy B, Mossman B, Vallyathan V (2002) Diseases caused by silica: mechanisms of injury and disease development. Int Immunopharmacol 2:173–182
Doiron K, Linossier I, Fay F, Yong J, Wahid EA, Hadjiev D, Bourgougnon N (2012) Dynamic approaches of mixed species biofilm formation using modern technologies. Mar Environ Res 78:40–47, http://dx.doi.org/10.1016/j.marenvres.2012.04.00
Donaldson K, Borm PJ (1998) The quartz hazard: a variable entity. Ann Occup Hyg 42:287–294
Dove PM, Rimstidt JD (1994) Silica-water interactions. In: Heaney P, Prewitt C, Gibbs G (eds) The silica polymorphs, vol 29, Reviews in mineralogy. Mineralogical Society of America, Washington, DC, pp 259–308

Dove PM, Han N, Wallace AF, De Yoreo JJ (2008) Kinetics of amorphous silica dissolution and the paradox of the silica polymorphs. Proc Natl Acad Sci USA 105:9903–9908

Exposito J, Cluzel C, Garrone R, Lethias C (2002) Evolution of collagen. Anat Rec 268:302–316

Faimali M, Garaventa F, Terlizzi A, Chiantore M, Cattaneo-Vietti R (2004) The interplay of substrate nature and biofilm formation in regulating *Balanus amphitrite* Darwin, 1854 larval settlement. J Exp Mar Biol Ecol 306:37–50

Fenoglio I, Martra G, Coluccia S, Fubini B (2000) Possible role of ascorbic acid in the oxidative damage induced by inhaled crystalline silica particles. Chem Res Toxicol 13:971–975

Finkel ZV, Kotrc B (2010) Silica use through time: macroevolutionary change in the morphology of the diatom frustule. Geomicrobiol J 27:596–608

Fubini B, Hubbard A (2003) Reactive oxygen species (ROS) and reactive nitrogen species (RNS) generation by silica in inflammation and fibrosis. Free Radic Biol Med 34:1507–1516

Fujii T, Reimer JD (2011) Phylogeny of the highly divergent zoanthid family Microzoanthidae (Anthozoa, Hexacorallia) from the Pacific. Zool Scr 40:418–431

Gamble JF (2011) Crystalline silica and lung cancer: a critical review of the occupational epidemiology literature of exposure response studies testing this hypothesis. Crit Rev Toxicol 41(5):404–465

Garrone R (1978) Phylogenesis of connective tissue: morphological aspects and biosynthesis of sponge intercellular matrix, vol 5, Frontiers of matrix biology. Karger, Basel

Garrone R, Huc A, Junqua S (1975) Fine structure and physicochemical studies on the collagen of the marine sponge *Chondrosia reniformis* Nardo. J Ultrastruct Res 52:261–275

Gil-Turnes MS, Hay ME, Fenical W (1989) Symbiotic marine bacteria chemically defend crustacean embryos from a pathogenic fungus. Science 246:116–118

Giovine M, Pozzolini M, Fenoglio I, Scarfi S, Ghiazza M, Benatti U, Fubini B (2003) Crystalline silica incubated in ascorbic acid acquires a higher cytotoxic potential. Toxicol Ind Health 18:249–255

Gough PJ, Gordon S (2000) The role of scavenger receptors in the innate immune system. Microbes Infect 2:305–311

Green FHY, Vallyathan V (1996) Pathologic responses to inhaled silica. In: Castranova V, Vallyathan V, Wallace WE (eds) Silica and silica-induced lung diseases. CRC, Boca Raton, FL, pp 39–59

Guidetti P, Bianchi CN, Chiantore MC, Schiaparelli S, Morri C, Cattaneo-Vietti R (2004) Living on the rocks: substrate mineralogy and the structure of subtidal rocky substrate communities in the Mediterranean Sea. Mar Ecol Prog Ser 274:57–68

Hadfield MG (2011) Biofilms and marine invertebrate larvae: what bacteria produce that larvae use to choose settlement sites. Ann Rev Mar Sci 3:453–470

Haeckel E (1872) Die Kalkschwämme. G. Reiser, Berlin

Hamilton RE Jr, Thakur SA, Holian A (2008) Silica binding and toxicity in alveolar macrophages. Free Radic Biol Med 44:1246–1258

Heinemann S, Ehrlich H, Douglas T, Heinemann C, Worch H, Schatton W, Hanke T (2007) Ultrastructural studies on the collagen of the marine sponge *Chondrosia reniformis* Nardo. Biomacromolecules 8:3452–3457

Hess EV (2002) Environmental chemicals and autoimmune disease: cause and effect. Toxicology 181–182:65–70

Hildebrand M (2008) Diatoms, biomineralization processes, and genomics. Chem Rev 108:4855–4874

Hinz B, Brune K (2002) Cyclooxygenase-2 10 years later. J Pharmacol Exp Ther 300:367–375

Hnizdo E, Murray J (1988) Risk of pulmonary tuberculosis relative to silicosis and exposure to silica dust in South Africa gold miners. Occup Environ Med 55:496–502

Holmes SP, Sturgess CJ, Davies MS (1997) The effect of rock-type on the settlement of *Balanus balanoides* (L) cyprids. Biofouling 11:137–147

Hsu H, Hijjar DP, Khan KM, Falcone DJ (1988) Ligand binding to scavenger receptor-A induces urokinase-type plasminogen activator expression by a protein kinase C delta-dependent. J Biol Chem 273:1240–1246

Hsu H, Chiu SL, Wen MH, Chen KY, Hua KF (2001) Ligands of macrophage scavenger receptor induce cytokine expression via differential modulation of protein kinase signalling pathways. J Biol Chem 276:28719–28730

IARC (1987) Silica, some silicates, vol 42, IARC monographs on the evaluation of carcinogenic risk of chemicals to humans. IARC, Lyon

IARC (1997) Silica, some silicates, coal dust and para-aramid fibrils, vol 68, IARC monographs on the evaluation of carcinogenic risk of chemicals to humans. IARC, Lyon

Iler RK (1979) The chemistry of silica. Plenum, New York, NY

Jézéquel MV, Hildebrand M, Brzezinski MA (2000) Silicification in diatoms. Implication for growth. J Phycol 36:821–840

Jin X, Gruber N, Dunne JP, Sarmiento JL, Armstrong RA (2006) Diagnosing the contribution of phytoplankton functional groups to the production and export of particulate organic carbon, $CaCO_3$, and opal from global nutrient and alkalinity distributions. Global Biogeochem Cy. doi:10.1029/2005GB002532

Kamatani A (1971) Physical and chemical characteristics of biogenous silica. Mar Biol 68:91–96

Keough MJ, Raimondi PT (1995) Responses of settling invertebrate larvae to bioorganic films: effects of different types of films. J Exp Mar Biol Ecol 185:235–253

Keough MJ, Raimondi PT (1996) Responses of settling invertebrate larvae to bioorganic films: effects of large-scale variation in films. J Exp Mar Biol Ecol 207:59–78

Kinrade SD, Schach AS, Hamilton RJ, Knight CT (2001) NMR evidence of pentaoxo organosilicon complexes in dilute neutral aqueous silicate solutions. Chem Commun (Cambr) 7:1564–1565

Kobzik L (1995) Lung macrophage uptake of unopsonized environmental particulates. Role for scavenger-type receptors. J Immunol 155:367–376

Krasko A, Lorenz B, Batel R, Schröder HC, Müller IM, Müller WEG (2000) Expression of silicatein and collagen genes in the marine sponge *Suberites domuncula* is controlled by silicate and myotrophin. Eur J Biochem 267:4878–4887

Kreyling WG (1992) Intracellular particle dissolution in alveolar macrophages. Environ Health Perspect 97:121–126

Loewus FA (1980) L-ascorbic acid: metabolism, biosynthesis, function. In: Press J (ed) The biochemistry of plants, vol 3. Academic, New York, NY

Maeda M, Nishimura Y, Kumagai N, Hayashi H, Hatayama T, Katoh M, Miyahara N, Yamamoto S, Hirastuka J, Otsuki T (2010) Dysregulation of the immune system caused by silica and asbestos. J Immunotoxicol 7(4):268–278

Maehira F, Iinuma Y, Eguchi Y, Miyagi I, Teruya S (2008) Effect of soluble silicon compound and deep sea water on biochemical and mechanical properties of bone and the related gene expression in mice. J Bone Miner Metab 26:446–455

Manini E, Luna GM (2003) Influence of the mineralogical composition on microbial activities in marine sediments: an experimental approach. Chem Ecol 19:399–410

McDougald D, Rice SA, Barraud N, Steinberg PD, Kjelleberg S (2012) Should we stay or should we go: mechanisms and ecological consequences for biofilm dispersal. Nat Rev Microbiol 10 (1):39–50

Migliaccio CT, Hamilton RF Jr, Holian A (2005) Increase in a distinct pulmonary macrophage subset possessing an antigen-presenting cell phenotype and in vitro APC activity following silica exposure. Toxicol Appl Pharmacol 205:168–176

Minchin EA (1900) In: Ray Lancaster E (ed) A treatise on zoology, p 42

Murphy JE, Tedbury PR, Homer-Vanniasinkam S, Walker JH, Ponnambalam S (2005) Biochemistry and cell biology of mammalian scavenger receptors. Atherosclerosis 182:1–15

Nakamura T, Hinagata J, Tanaka T, Imanishi T, Wada J, Kodama T, Doi T (2002) HSP90, HSP70 and GAPDH directly interact with the cytoplasmic domain of macrophage scavenger receptors. Biochem Biophys Res Commun 290:858–864

Nickel M, Brümmer F (2003) In vitro sponge fragment culture of *Chondrosia reniformis* (Nardo, 1847). J Biotechnol 100:147–159

Palecanda A, Kobzik L (2001) Receptors for unopsonized particles: the role of alveolar macrophage scavenger receptors. Curr Mol Med 1:589–595

Penna A, Magnani M, Fenoglio I, Fubini B, Cerrano C, Giovine M, Bavestrello G (2003) Marine diatom growth on the different forms of particulate silica: evidence of cell/particle interaction. Aquat Microb Ecol 32:299–306

Piguet PF, Collart MA, Grau GE, Sappino AP, Vassalli P (1990) Requirement of tumor necrosis factor for development of silica-induced pulmonary fibrosis. Nature 344:245–247

Pozzolini M, Valisano L, Cerrano C, Menta M, Schiaparelli S, Bavestrello G, Benatti U, Giovine M (2010) Influence of rocky substrata on three-dimensional sponge cells model development. In Vitro Cell Dev Biol Anim 46(2):140–147

Pozzolini M, Buzzone F, Berilli V, Mussino F, Cerrano C, Benatti U, Giovine M (2012) Molecular characterization of a nonfibrillar collagen from the marine sponge *Chondrosia reniformis* Nardo 1847 and positive effects of soluble silicates on its expression. Mar Biotechnol 14 (3):281–293

Previati M, Palma M, Bavestrello G, Falugi C, Cerrano C (2010) Reproductive biology of *Parazoanthus axinellae* (Schmidt 1862) and *Savalia savaglia* (Bertoloni 1819) (Cnidaria, Zoanthidea) from the NW Mediterranean coast. Mar Ecol Evol Perspect 31:555–565

Qu Y, Tang Y, Cao D, Wu F, Liu J, Lu G, Zhang Z, Xia Z (2007) Genetic polymorphisms in alveolar macrophage response-related genes, and risk of silicosis and pulmonary tuberculosis in chinese iron miners. Int J Hyg Environ Health 210:679–689

Ragueneau O, Tréguer P, Leynaert A, Anderson RF, Brzezinski MA, DeMaster DJ, Dugdale RC, Dymond J, Fischer G, François R, Heinze C, Maier-Reimer E, Jézéquel MV, Nelson DM, Quéguiner B (2000) A review of the Si cycle in the modern ocean: recent progress and missing gaps in the application of biogenic opal as paleoproductivity proxy. Global Planet Change 26:317–365

Ramanathan R, Campbell JL, Soni SK, Bhargava SK, Bansal V (2011) Cationic amino acids specific biomimetic silicification in ionic liquid: a quest to understand the formation of 3-D structures in diatoms. PLoS One 6:e17707

Rosen G (1943) The history of miners' diseases: a medical and social interpretation. Schuman's, New York, NY, pp 459–476

Rovida CL (1871) Un caso di silicosi del polmone con analisi chimica. Ann di Chim Appl alla Med 13:102

Sarà M, Vacelet J (1973) Écologie des démosponges. In: Grassé PP (ed) Traité de Zoologie, Tome III: spongiaires. Masson et Cie, Paris

Scarfì S, Benatti U, Pozzolini M, Clavarino E, Ferraris C, Magnone M, Valisano L, Giovine M (2007) Ascorbic acid pre-treated quartz enhances cyclooxygenase-2 expression in raw 264.7 murine macrophages. FEBS J 274:60–73

Scarfì S, Magnone M, Ferraris C, Pozzolini M, Benvenuto F, Benatti U, Giovine M (2009) Ascorbic acid pre-treated quartz stimulates TNF-α release in RAW 264.7 murine macrophages through ROS production and membrane lipid peroxidation. Respir Res 10:25. doi:10.1186/1465-9921-10-25

Schins RP, Knaapen AM (2007) Genotoxicity of poorly soluble particles. Inhal Toxicol 19 (S1):189–198

Schultes S, Lambert C, Pondaven P, Corvaisier R, Jansen S, Ragueneau O (2010) Recycling and uptake of $Si(OH)_4$ when protozoan grazers feed on diatoms. Protist 161:288–303

Schulze FE (1879) Untersuchungenüber den Bau und die Entwicklung der Spongien. VI. Die Gattung Spongelia. Z Wiss Zool 32:117–157

Shaw ME (1927) Note on the inclusion of sand in sponges. Ann Mag Nat Hist 19:601–609

Shipe RF, Brzezinski MA (1999) A study of Si deposition study in *Rhizosolenia* (Bacillariophyceae) mats using novel ^{32}Si autoradiographic method. J Phycol 35:995–1004
Sollas IBJ (1908) The inclusion of foreign bodies by sponges, with a description of a new genus and species of Monaxonida. Ann Mag Nat His 1:395–401
Straif K, Benbrahim-Tallaa L, Baan R, Grosse Y, Secretan B, El Ghissassi F, Bouvard V, Guha N, Freeman C, Galichet L, Cogliano V (2009) A review of human carcinogens—part C: metals, arsenic, dusts and fibers. Lancet Oncol 10:453–454
Teragawa CK (1986a) Sponge dermal membrane morphology: histology of cell mediated particle transport during skeletal growth. J Morphol 190:335–348
Teragawa CK (1986b) Particle transport and incorporation during skeleton formation in a keratose sponge *Dysidea etheria*. Biol Bull 170:321–334
Thakur SA, Hamilton RE Jr, Holian A (2008) Role of scavenger receptor: a family in lung inflammation from exposure to environmental particles. J Immunotoxicol 5:151–157
Thibodeau M, Giardina C, Hubbard AK (2003) Silica-induced caspase activation in mouse alveolar macrophages is dependent upon mitochondrial integrity and aspartic proteolisis. Toxicol Sci 76:91–101
Thibodeau M, Giardina C, Knecht DA, Helble J, Hubbard AK (2004) Silica-induced apoptosis in mouse alveolar macrophages is initiated by lysosomal enzyme activity. Toxicol Sci 80:34–48
Todd CD, Keough MJ (1994) Larval settlement in hard substratum epifaunal assemblages: a manipulative field study of the effects of substratum filming and the presence of incumbents. J Exp Mar Biol Ecol 181:159–187
Totti C, Cucchiari E, De Stefano M, Pennesi C, Romagnoli T, Bavestrello G (2007) Seasonal variations of epilithic diatoms on different hard substrates, in the northern Adriatic Sea. J Mar Biol Assoc UK 87:649–658
Tréguer P, Nelson DM, Van Bennekom AJ, DeMaster DJ, Leynaert A (1995) The silica balance in the world ocean: a re-estimate. Science 268:375–379
Valko M, Leibfritz D, Moncol J, Cronin MT, Mazur M, Telser J (2007) Free radicals and antioxidants in normal physiological functions and human disease. Int J Biochem Cell Biol 39:44–84
Vallyathan V, Shi XL, Dalal NS, Irr W, Castranova V (1988) Generation of free radicals from freshly fractured silica dust: potential role in acute silica-induced lung injury. Am Rev Respir Dis 138:1213–1219
Van Colen C, Lenoir J, De Backer A, Vanelslander B, Vincx M, Degraer S, Ysebaert T (2009) Settlement of *Macoma balthica* larvae in response to benthic diatom films. Mar Biol 156:2161–2171
Von Lendenfeld R (1889) A monograph on the horny sponges. Trübner, London
Wahl M (1989) Marine epibiosis. I. Fouling and antifouling: some basic aspects. Mar Ecol Prog Ser 58:75–89
Wilkie IC, Parma L, Bonasoro F, Bavestrello G, Cerrano C, Carnevali MD (2006) Mechanical adaptability of a sponge extracellular matrix: evidence for cellular control of mesohyl stiffness in *Chondrosia reniformis* Nardo. J Exp Biol 209(22):4436–4443
Zanetti G (2002) Dinamismo strutturale nella spugna Chondrosia reniformis (Porifera, Demospongiae): osservazioni in habitat. Dissertation, University of Milan

Chapter 7
Bioactive Poly(Arsenic) Compounds

Ines Mancini and Andrea Defant

To my father Giacomo (1926–2001), with love and gratefully.

Contents

Abstract An overview of the biological activities of arsenic compounds containing more than one arsenic atom in their molecular structure is presented. This contribution covers the literature of the last 10–12 years concerning the in vitro and in vivo studies on arsenic species. They include inorganic oxides and sulfides, already employed for a long time in traditional Chinese medicine and currently investigated against hematological or solid malignancies, with arsenic trioxide clinically used in the treatment of acute promyelocytic leukemia. Chemical and biological aspects on the marine product arsenicin A, representing the first and only organic polyarsenical isolated from Nature, have also been reviewed, pointing out

I. Mancini (✉) • A. Defant
Laboratorio di Chimica Bioorganica, Dipartimento di Fisica, Università di Trento,
Via Sommarive 14, 38123 Povo-Trento, Italy
e-mail: ines.mancini@unitn.it

W.E.G. Müller et al. (eds.), *Biomedical Inorganic Polymers*, Progress in Molecular and Subcellular Biology 54, DOI 10.1007/978-3-642-41004-8_7,

the characterization of its $C_3H_6As_4O_3$ molecular structure by experimental and theoretical vibrational spectroscopies, the potent antimicrobial activities, and the promising perspectives as an antitumor agent.

Abbreviations

AML	Acute myeloid leukemia
APCI	Atmospheric pressure chemical ionization
APL	Acute promyelocytic leukemia
ATL	Adult T-cell leukemia
Bcl-xl	B-cell lymphoma-extra large
CML	Chronic myelogenous leukemia
DFT	Density functional theory
EI	Electron impact
FDA	Food and drug administration
Hh	Hedgehog
HTLV-I	Human T-cell leukemia virus type I
HPV	Human papillomavirus
IR	Infrared
JNK	Jun N-terminal kinase
MS	Mass spectrometry
NMR	Nuclear magnetic resonance
PML	Promyelocytic leukemia
RAR	Retinoic acid receptor
ROS	Reactive oxygen species
SAR	Structure-activity relationship
SDS	Sodium dodecyl sulfate
SUMO	Small ubiquitin-like modifier
SUV	Small unilamellar vesicles
TLC	Thin layer chromatography
UV	Ultraviolet

7.1 Arsenic and Its Compounds: An Introduction

Arsenic is an element belonging to group V in the periodical table, with [Ar] $3d^{10}4s^24p^3$ electronic configuration and oxidation states from −3 to +5. In particular, it is present with oxidation state −3 in arsenides and +3 in most organoarsenic compounds including arsenites and arsenates(III), whereas the pentavalent state is less common. Arsenic forms covalent compounds with several elements, mainly with oxygen and sulfur in natural species (Cotton and Wilkinson 1988). It is a

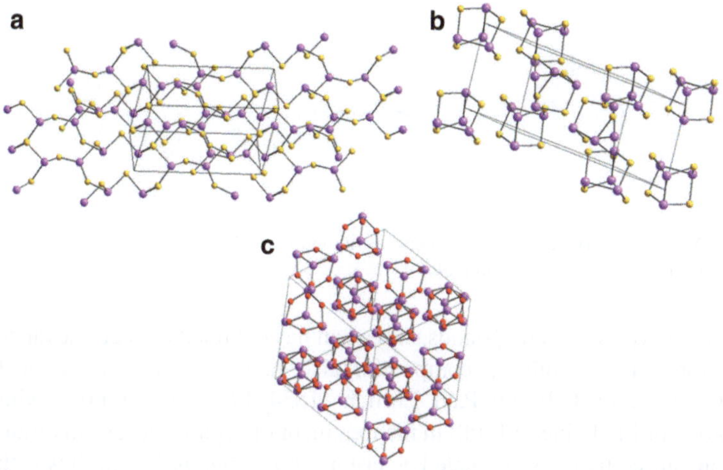

Fig. 7.1 Molecular structures of arsenic minerals (**a**) orpiment (As_2S_3), (**b**) realgar (As_4S_4), and (**c**) arsenolite (As_4O_6); arsenic (*purple*), oxygen (*red*), and sulfur (*yellow*). The corresponding crystal lattices are also depicted (data from American Mineralogist Structure Database)

widely diffused chemical element in Nature, present in rocks, soils, oceans, and the atmosphere. With an average concentration of $2 \div 3$ mg for each kilogram of the earth's crust, arsenic represents the 20th most abundant element, mainly contained in sedimentary rocks and minerals, with arsenopyrite (FeAsS) as the most abundant form (Francesconi and Kuehnelt 2002).

Regarding polyarsenic compounds (taking into account an equal or higher number than two arsenic atoms), minerals include sulfur and oxygen compounds. In the first class, there are bright yellow orpiment (As_2S_3), dimorphite (As_4S_3), alacranite (As_8S_9) whose structure consists of an ordered packing of As_4S_4 and As_4S_5 cage-like molecules, as very recently investigated by micro-Raman analysis and ab initio calculations (Pagliai et al. 2011), and red-colored realgar (As_4S_4). The second class of minerals includes arsenolite and claudetite, which are polymorphic forms of As_2O_3 (Fig. 7.1).

Both arsenic oxides and sulfides have found applications in traditional Chinese medicine and more recently in the treatment of leukemia.

Inorganic forms of arsenic are alkylated by marine organisms to give monoarsenic compounds, present in brown algae, mollusk, arthropods, and vertebrates. Arsenicals isolated from marine organisms include volatile compounds like alkyl arsines and more commonly nonvolatile methyl arsine oxides, methylarsonic acid, and dimethylarsinic acid, alongside water-soluble polar betaines, cholines, as well as carbohydrate and lipid derivatives (Francesconi and Kuehnelt). If it is true that marine sponges seldom appeared in these reports, from the poecilosclerid sponge *Echinochalina bargibanti*, a novel peculiar arsenic compound has been isolated, providing the first organic polyarsenical from Nature (Mancini et al. 2006).

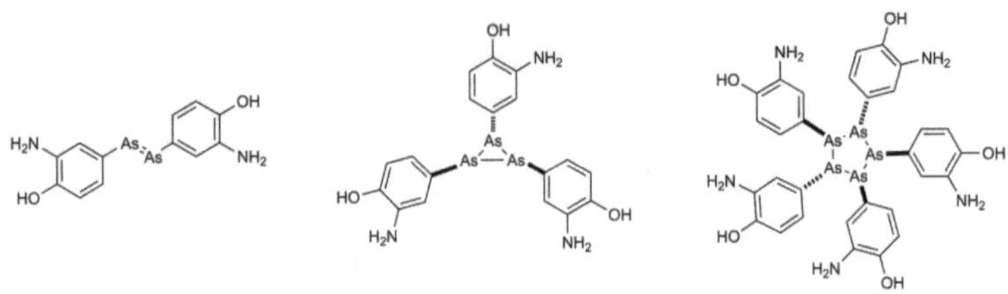

Fig. 7.2 Molecular structures for salvarsan: attributed by P. Ehrlich (*left*) and the ones recently established in solutions (*center* and *right*)

Among polyarsenic compounds, salvarsan has obtained a great relevance for its applications. This synthetic compound, also named arsphenamine or Ehrlich 606, was used in 1910 by Paul Ehrlich (1854–1915, Nobel prize winner for Physiology or Medicine in 1908) in the treatment of syphilis, a sexually transmitted infection caused by the spirochete bacterium *Treponema pallidum.* It was the most effective drug against this disease, until penicillin became available about 30 years later. For his pioneering methodology in research on clinical use of salvarsan, based on searching for magic bullets able to strike parasites and parasites only, Paul Ehrlich is regarded as chemotherapy's father (Gensini et al. 2007). Ehrlich attributed to salvarsan the structure with a central As=As bond, but the first definitive assignment is given in 2005 based on electrospray ionization mass spectrometric measurements. It was established that salvarsan in solution consists of cyclic species $(RAs)_n$, with $n = 3$ and $n = 5$, as reported in Fig. 7.2 (Lloyd et al. 2005).

7.2 The Double Face of Arsenic Compounds: Toxicity and Antitumor Activities

The effects of arsenic are paradoxical because it can act as a potent poison and environmental carcinogen and a chemotherapeutic agent. In fact it is known that chronic arsenic exposure to industrial or natural sources (e.g., food and drinking water) can cause serious poisoning (arsenicosis) and toxicity to more than 200 million people all over the world. Arsenic trioxide is readily absorbed by the digestive system, and after a long period of exposure, it can be incorporated into tissues rich in keratin (bones, muscles, skin, hair, and nails). In a recent study on a rat model, urinary level of 8-hydroxy-2′-deoxyguanosine has been reported as a valid biomarker for detecting the value of arsenic exposure (Wu and Ho 2009). In humans, prolonged and high-dose exposure to arsenic can cause the development of malignancies, severe gastrointestinal toxicities, diabetes, cardiac arrhythmias, and death. Epidemiological studies have shown a significant dose-response relationship between inorganic arsenic ingestion and cancer. On the other hand, mineral arsenicals such as orpiment, realgar, and arsenolite have long been used in traditional Chinese medicine (Liu et al. 2008); organoarsenicals were the first antimicrobial agents synthesized for the treatment of infectious diseases such as syphilis and sleeping sickness, and arsenic trioxide has a high efficiency in the therapy of acute promyelocytic leukemia (APL), a rare (about 15 % of acute myeloid leukemia) and fatal disease striking mainly young people (Au and Kwong 2008; Platanias and Robert 2009).

7.3 Arsenic Sulfides

7.3.1 Orpiment

Orpiment (As_2S_3) is a mineral present in volcanic environments. Both the crystalline and amorphous forms feature polymeric structures consisting of trigonal pyramidal As(III) centers linked by sulfide centers. Upon in vacuo heating, polymeric As_2S_3 gives a mixture of molecular species, including As_4S_6, which adopts the adamantane geometry. Orpiment has been studied for the fabrication of pho tonic crystals and used as a tanning agent. It is so insoluble that its toxicity is low. It has shown hepatotoxicity by oral administration in rats (50 mg/kg). In particular, it was responsible for decreasing cytochrome P450 and the concentration of glutathione in liver microsomes, associated with a reduction of antioxidant enzyme activities with a concurrent increase in lipid peroxidation (Singh and Sharma 1994). Later the same authors have obtained similar effects by an investigation of lipid profile in rat lungs (Singh and Sharma 1998). After oral administration in rats of orpiment at the same dosage, a kinetic and thermodynamic study was performed on enzymes associated with intestinal brush border membranes. It resulted deleterious by decreasing significantly the activities of intestinal enzymes, producing a perturbation in membrane fluidity (Singh et al. 1999). Very recently, a study on the toxicity of orpiment, realgar, and arsenopyrite has been carried out for the amphipod *Corophium volutator*, by exposure of the animals to these minerals. An arsenic accumulation deriving from minerals was observed, in line with a comparable toxicity for orpiment and realgar, higher than for arsenopyrite. For its specific sensitivity, *Corophium volutator* represents a potential bioindicator in the evaluation of toxicity for sediments containing arsenic minerals (Cui et al. 2011).

Orpiment nanoparticles, with 60 $\div$ 140 nm average diameter, have been prepared and their cytotoxic effect investigated on leukemic K562 cells, comparing antitumor effects with natural orpiment at different concentration values. The results indicated that nanoparticles gave a higher inhibition of cell proliferation, time, and were concentration dependent (Lin et al. 2007).

7.3.2 Realgar

Another arsenic sulfide is realgar (As_4S_4), a red–orange mineral used as a pigment in ancient paintings. Investigation of this compound both of natural and synthetic origin by mass spectrometry and vapor pressure technique showed that congruent sublimation gave mainly gaseous As_4S_4 together with minor degradative sulfides (Munir et al. 1971). Similarly to orpiment, realgar has a low solubility, and therefore, it is poorly absorbed from the gastrointestinal tract (Liu et al. 2008).

Traditional Chinese medicine makes use of several types of herbs and minerals, and according to the combinational principle, some of these components work as adjuvants in effects or facilitate the delivery of the principal component. Orpiment and As_2O_3 are also used, and in 2005 realgar alone was contained in 22 oral remedies approved by the Chinese Pharmacopeia Committee (Balaz and Sedlak 2010). Both traditional Chinese and Indian Ayurvedic medicines often include the combined use of realgar and cinnabar (HgS), whose toxic effects have been recently evaluated in a comparative toxicological study in mice on An-Gong-Niu-Huang Wan herbal preparation alone or combined with realgar and cinnabar. The results have indicated that taken together, realgar, cinnabar and An-Gong-Niu-Huang Wan are much less chronically nephrotoxic than common arsenic ($NaAsO_2$, Na_2HAsO_4) and mercurial compounds (CH_3Hg, $HgCl_2$) (Lu et al. 2011). A combination of plant species/minerals based on traditional Chinese medicine has been evaluated in clinical studies in order to enhance therapeutic efficiency and reduce adverse effects. As_4S_4 combined with the natural products from Chinese herbs tanshinone IIA and indirubin yielded synergy in the treatment of a murine APL model and in the induction of APL cell differentiation in vitro, with an easier transportation of As_4S_4 into APL cells (Wang et al. 2008a, b). In the brain tissue and serum of rats, realgar has also shown detoxification effects by affecting oxidative stress state with an increase of stress levels and activating some endogenous protective factors (Tang et al. 2008). ^{1}H nuclear magnetic resonance (NMR) technique was applied to the analysis of metabolic profile of urine, serum, and liver tissue of rats after administration of realgar, resulting in a reliable methodology in the study of the toxicological effects of this sample and of traditional Chinese medicine in general (Wei et al. 2009).

Bioleaching is an extraction technique using living organisms to recover specific metals, including arsenic, from their core bodies. Recently, the extraction of realgar has been successfully carried out by the Gram-negative bacterium *Acidithiobacillus ferrooxidans*, and this form of extracted realgar has resulted in a promising anti-cancer drug candidate due to its high selective affinity to tumor tissues (Zhang et al. 2010a).

It has been reported that NB4 human APL cell line and a mice model were more sensitive to As_4S_4 than As_2S_3 and As_2O_3. In addition, a clinical study on APL patients indicated that both As_4S_4 and As_2S_3 alone were very effective in complete remission induction (Lu and Wang 2002). Realgar gave a lower toxicity than As_2O_3 in the treatment of APL and human myelogenous leukemia (K562) cells became sensitive to realgar at clinically acceptable concentrations when the expression level of anti-apoptotic mitochondrial transmembrane B-cell lymphoma-extra large (Bcl-xl) protein is downregulated by transfecting bcl-xL antisense RNA vector into the cells. The results suggested an insensitivity to realgar-induced apoptosis in the case under investigation (Zhang et al. 2005).

Recent results report on the combined use of realgar and imatinib, the latter one developed by a rational drug design and approved by Food and Drug Administration (FDA) to treat ten different cancers, including chronic myelogenous leukemia (CML) and whose mesylate salt is currently marketed as Gleevec or Glivec by Novartis. The study carried out on a mouse model of CML has shown that the combined administration produces more therapeutic effects than each drug as a single agent (Zhang et al. 2009).

Although realgar has long been used in Asian traditional medical practice, its clinical use is currently not properly defined, and some mechanisms are unclear to date. In detail, its therapeutic effect has been associated with its induced cell apoptosis, but the molecular mechanism of action in APL therapy is not yet completely understood. Moreover, it is as effective as the most studied arsenic trioxide and has relatively good safety for oral administration which is very effective in the treatment not only of newly diagnosed APL but especially of refractory and relapsed patients (Qi et al. 2010). The reduced attention focused on realgar which, if compared to arsenic trioxide, is due to its poor solubility in water responsible for its poor bioavailability, which has reduced its clinical applications. In order to overcome these shortcomings, studies on dissolution methods have been recently carried out. Attention has been focused on the development and the bioactivity evaluation of some types of realgar nanoparticles, where the results were very affected by sizes. Through a comparative study of inhibitory effects carried out for realgar both in nanoparticles and raw forms on the growth of tumor-transplanted mice and human promyelocytic leukemia (HL-60) cells cultured in vitro, realgar nanoparticles showed higher antitumor activity (Xu et al. 2006). There is also evidence that size reduction of realgar particles to nano levels could effectively give an enhanced bioavailability. They have been reported in a study on nanoparticles (176 ÷ 287 nm) obtained by co-grinding realgar with polyvinylpyrrolidone or the biocompatible surfactant sodium dodecyl sulfate (SDS), giving a monodisperse suspension which exhibited significant in vitro cytotoxicity towards human ovarian and cervical cancer cell lines (Wu and Ho 2006). Another study on realgar nanoparticles prepared with the assistance of SDS has proved a stronger antiproliferation effect on human osteosarcoma (MG-63) human hepatoma carcinoma (HepG-2) cell lines than coarse realgar, showing that cytotoxicity induced by nanoparticles was typically size, concentration, and time dependent (Zhao et al. 2011). Mechanochemical solid-state properties have also been investigated for nanosized As_4S_4 particles ($d = 144$ nm) obtained by milling in the presence of SDS, which have been tested for their cytotoxic effects on some leukemia and myeloma cancer cell lines (Balaz et al. 2009). In addition, realgar nanoparticles have been studied by spectroscopic techniques in their interactions with liposomes of phosphatidylcholine-cholesterol small unilamellar vesicles (SUV) used as a model for biomimetic membranes. When nanoparticles bound to SUV, the viscosity of the membrane has been seen to increase and its fluidity decrease, followed by pore formation (Shen et al. 2009).

Some realgar nanoparticles have shown optical properties of quantum dots, devices with spatial confinement containing a tiny droplet of free electrons with properties intermediate between bulk semiconductors and discrete molecules (Wu et al. 2011). It was the case of As(II) sulfide nanocrystals prepared by a wet process which involves a dissolution of As_4S_4 in ethylenediamine to give nanoparticles with size-dependent emission from UV to the blue region. They have been used in cytotoxic tests on in vitro human ovarian, cervical, and lung fibroblast cells, and the results indicated that these nanoparticles exhibit good therapeutic efficiency (Wang et al. 2008a, b).

7.4 Arsenic(III) Oxides

7.4.1 Claudetite

Claudetite (arsenic trioxides, As_2O_3) is an amphoteric compound, weakly acidic in aqueous media and easily dissolved in basic solutions, which are responsible for its conversion into arsenites. As_2O_3 is relatively stable to oxidation, and only ozone, hydrogen peroxide, and nitric acid are able to give As(V) oxide. Two polymeric forms (claudetite I and II) are known, which both crystallized as monoclinic crystals.

Arsenic trioxides have been used in therapeutic applications for a long time. In a brief chronological sequence, it must be cited that it was used with other arsenicals in ancient Greece and Rome; As_2O_3 added to aqueous potassium bicarbonate (Fowler's solution, 1786) was empirically employed in the treatment of a series of diseases, and in 1878 it was observed reducing white blood cells in a patient affected by leucocythemia. About 50 years later, its efficiency against chronic myelogenous leukemia was detected, until the remerging interest in 1970s when Chinese physicians applied arsenic trioxide to the treatment of acute promyelocytic leukemia (APL) and its approval by FDA in 2000 for refractory APL form (Antman 2001; Miller et al. 2002). Nowadays, we can say that after more than 17 years of clinical trials conducted worldwide, the efficiency of arsenic trioxide has been demonstrated (Emadi and Gore 2010). Trisenox is the injectable form of arsenic trioxide currently used in the treatment of APL, which is a unique subtype of acute myeloid leukemia typically carrying a specific reciprocal chromosome translocation, *t*(15;17), leading to the expression of a leukemia-generating fusion protein, promyelocytic leukemia PML-RARα (RAR is for retinoic acid receptor). The aroused interest of this drug is confirmed by the high number of scientific publications reported in PUBMED: 596 references for As_2O_3 (including 28 review articles and more than 470 from 2000) and 16 references for As_4O_6.

In vitro studies showed that As_2O_3 caused apoptosis in the concentration range 0.5–2.0 μmol/L, whereas it induced a partial cell differentiation in the lower range 0.1–0.5 μmol/L (Wang 2001). In clinical use, it has shown substantial effectiveness in APL treatment as a single agent, inducing a complete remission in about 90 % of patients with both primary and relapsed APL. If severe hepatic lesions have been found in some primary cases, a few adverse effects (as gastrointestinal symptoms, liver dysfunction, skin reaction, electrocardiographic changes) resulted moderate in relapsed patients (Chen et al. 2001; Miller et al. 2002).

More recent therapeutic results have shown that orally medication by a preparation combining As_2O_3 and As_4S_4 is as effective as parenteral arsenic trioxide and has a similar toxicity. In addition, the administration of As_2O_3 together with all-*trans*-retinoic acid in patients with newly diagnosed APL has yielded more durable remission than monotherapy (Chou and Dang 2005; Jing and Waxman 2007). In the last few years, US trials have demonstrated that As_2O_3 added to standard treatment improves survival in patients with APL and that its therapeutic doses are well tolerated, with no evidence of long-term toxicity (Emadi and Gore 2010).

The potent antitumor effect has prompted investigations to elucidate the mechanisms of action. There are many pathways according to which arsenic compounds are able to cause cell alterations (Fig. 1.3), already exhaustively reviewed (Waxman and Anderson 2001; Miller et al. 2002). One effect is related to reactions of arsenic trioxide with thiols (SH) in cysteine groups close to the space inside cell proteins (Ramadan et al. 2009). Examples are given by the inhibition of tyrosine phosphatases, which play a pivotal role in modulating cell metabolism. Experimental data have shown that an alteration in the folding of PML-RARα and PML occurred, favoring an oligomerization of these proteins and promoting SUMOylation, the chemical modification of a protein after its translation by binding a small ubiquitin-like modifier (SUMO) protein (Mann and Miller 2004). Recently, the identification of PML as a direct target of arsenic trioxide has been reported, giving new insights into the specificity for APL and its mechanism of action (Zhang et al. 2010b).

A synergistic effect between As_2O_3 and all-*trans*-retinoic acid (ATRA) has been shown, inducing complete remission in patients both sensitive and resistant to retinoic acid (Shen et al. 2004), and the high rates of complete clinical remission have found supports from results reported by Nasr and coauthors (Nasr et al. 2008).

Oxidative stress is involved in another pathway, where arsenic interferes in redox equilibria, causing accumulation of reactive oxygen species (ROS) like superoxide ion generating hydrogen peroxide. These species increase mitochondrial membrane permeability, due to the alteration of membrane potential ($\Delta\Psi$) with release of cytochrome C, which activates caspases (cysteine-aspartic proteases involved in apoptosis, necrosis, and inflammation) with the final effect of inducing programmed cell death (apoptosis). Induction of apoptosis and accumulation of oxidized proteins have been observed by the treatment of cells with As_2O_3 in the presence of inhibitors of caspase (Khan et al. 2004). In addition the c-Jun N-terminal kinase (JNK) has been reported to be involved in apoptosis dependent from As_2O_3; the latter one also activates the pro-apoptotic Bcl-2–associated X protein

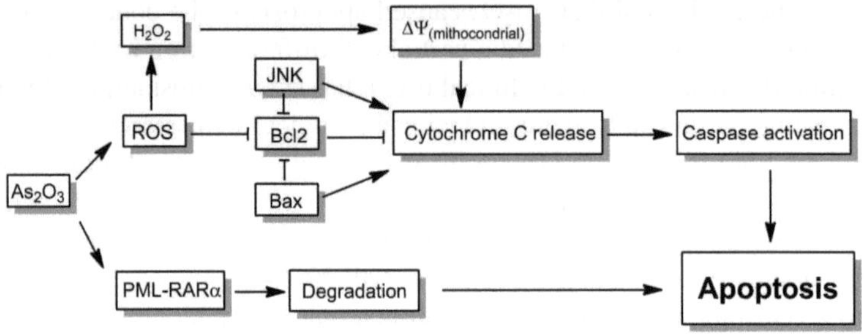

Fig. 7.3 Simplified scheme of mechanism for As_2O_3-induced apoptosis

(Bax), a member of B-cell lymphoma 2 (Bcl-2) family (Fig. 7.3) (Platanias and Robert 2009).

It is noteworthy that arsenic trioxide was dissolved in concentrated aqueous solution of sodium hydroxide, followed by adjustment to biological pH value, in most of the reported experimental procedures (Park et al. 2003a, b; Khan et al. 2004; Shen et al. 2004). However, it is known that arsenites are produced by treatment of As_2O_3 with alkaline hydroxides, so that very probably the active arsenic species could not be the starting arsenic trioxide. Aberrant hedgehog (Hh) pathway activation is implicated in promoting the growth of different cancers. Arsenic trioxide has resulted active in Hh pathway antagonism, resulting in a potential therapeutic agent in the treatment of correlated malignancies (Kim et al. 2007).

It demonstrated the effectiveness of a co-treatment with dietary known antioxidants as flavonoids, including genistein, which are able to increase apoptosis through the production of moderate oxidative stress and activation of ROS-inducible protein kinases (Sanchez et al. 2010). In addition, similarly to arsenic trioxide, realgar (As_4S_4) has shown clinical effectiveness in APL patients, and oxidative stress was associated with realgar-induced differentiation in human leukemia HL-60 cells (Wang et al. 2009). Besides inhibition of growth, promotion of differentiation, and induction of apoptosis, arsenic trioxide is also involved in blocking angiogenesis, by the inhibition of vascular endothelial growth. These effects have been observed in cultured cell lines, in animal models, and in clinical studies (Miller et al. 2002).

The study was soon extended to the potential use of As_2O_3 against other hematological and non-hematological malignancies. Arsenic trioxide has also shown to act in non-APL cells as chronic myeloid leukemia (Ito et al. 2008), myeloma cells, cells of immune origin, and human primary cells (Binet et al. 2009). If arsenic trioxide has given good results in APL inhibition where it can induce remission in more than 90 % of patients, it has instead no effectiveness in patients with non-APL acute myeloid leukemia (AML). The relationship between catalase activity and As_2O_3 sensitivity in AML has been investigated, observing that the inhibition of catalase (an alternative mechanism to convert hydrogen peroxide to water) did not adequately increase levels of reactive oxygen species to sensitize the cells to As_2O_3, so that other strategies must be taken into account for increasing As_2O_3 cytotoxicity in AML (Coe and Schimmer 2008).

A probable role of As_2O_3 has been also investigated in adult T-cell leukemia (ATL), an aggressive neoplasm of mature T-cell origin caused by human T-cell leukemia virus type I (HTLV-I). The treatment of ATL with As_2O_3 significantly inhibits the growth of HTLV-I-infected T-cell lines in clinically nontoxic concentrations (Ishitsuka et al. 2000). Arsenic trioxide induced apoptosis also in other various cancer cell lines including lymphoid malignancies, multiple myeloma, neuroblastoma, esophageal carcinoma, and gastric cancer (Wang 2001).

Multiple myeloma has remained an incurable disease, characterized by the clonal proliferation of malignant plasma cells in the bone marrow associated with bone loss, renal disease, and immunodeficiency. Therapies include thalidomide and immunomodulatory derivatives, as the first-in-class proteasome inhibitor bortezomib (known as PS-341) and As_2O_3 (Richardson et al. 2005). The effectiveness of this latter agent was confirmed in the clinical treatment of multiple myeloma; in particular, there is evidence of an immunological mechanism behind the therapeutic effects of As_2O_3, whose role in managing this disease could be enhanced if given in combination with ascorbic acid or other active chemotherapeutic agents (Hussein 2001). Besides, also in the treatment of multiple myeloma, as for APL, a connection with the sensitivity to oxidative damage has been recognized, which has allowed the investigation of pharmacological modulation of the cellular redox state for potentiation of As_2O_3 (Evens et al. 2004). Studies have been carried out to evaluate the role of As_2O_3 in the treatment of solid cancer where induction of apoptosis, inhibition of growth, and promotion of differentiation of solid tumors could be the mechanism of its actions. In particular As_2O_3 has resulted effective against cancers of digestive system (hepatoma, esophageal carcinoma, gastric cancer) and urinary and reproductive system (cervical cancer, ovarian cancer, bladder transitional carcinoma), lung cancer, breast cancer, and prostate cancer (Zhang and Ling 2003). Recent investigations have shown that it may be applied in the treatment for adenoid cystic carcinoma, an uncommon tumor of the head and neck that may occur in salivary glands which is not sensitive to conventional chemotherapeutics (Fu et al. 2010).

7.4.2 Arsenolite

Three polymorphic forms are known in the solid state: cubic As_4O_6, and two related polymeric forms, the latter ones both crystallized as monoclinic system. Arsenolite presents an unusual rigid structure with an adamantane cage, which consists of three 6-membered rings arranged in the chair configuration (Fig. 7.4) (O'Day 2006).

In vitro and in vivo tests have been carried out with the aim of comparing the activity of As_4O_6 with the more fully studied As_2O_3. Similarly to As_2O_3, tetraarsenic oxide has shown antiproliferative and apoptosis-inducing effects

Fig. 7.4 Molecular structure of tetraarsenic oxide As_4O_6 (=2,4,6,8,9,10-hexaoxa-1,3,5,7-tetraarsatricyclo [3.3.1.13,7]decane)

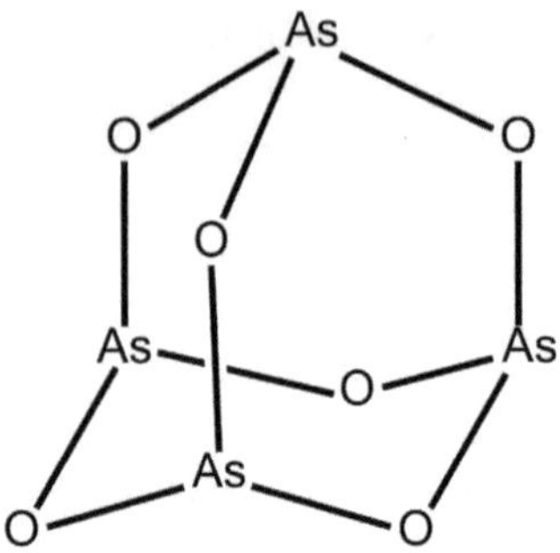

against human leukemic and solid tumor cells. In more detail, studies are carried out on the induced apoptosis in leukemic (U937) cells, provided that As_4O_6 generated ROS and induced caspase-3 activation more potently than As_2O_3 (Park et al. 2003b). Besides, another comparative evaluation of these arsenic oxides indicated that As_4O_6 could be more effective than As_2O_3 in inhibiting the growth of human papillomavirus (HPV16)-infected cervical cancer cells (Ahn et al. 2004). Tetraarsenic oxide resulted also more effective than arsenic trioxide in suppressing uterus carcinoma cells (SiHa) both in vitro and in vivo tests, also inducing apoptosis more significantly. A relevant difference between cell death pathways induced by the two arsenicals has been observed and provided important new information regarding the role of molecular network in apoptosis. The conclusion of this comparative study was that As_4O_6 is a more potent antitumor agent on human cervical cancer (Chang et al. 2007).

It is known that progressive tumor growth is dependent on angiogenesis, which corresponds to the growth of new blood vessels from preexisting vessels. Therefore, inhibiting angiogenesis represents a promising approach in antitumor therapy, and even arsenic compounds have been investigated with this purpose. The evidence that tetraarsenic oxide has effective antiangiogenic activity has been proved by in vitro studies, whereas its oral administration in rats inhibited induced new vessel formation in corneal micropocket assay and reduced substantially pulmonary metastatic nodules in mice implanted with melanoma cells (Park et al. 2003a). A comparative study on the antiangiogenic effects of tetraarsenic oxide and diarsenic oxide in the rat cornea has been conducted, and the results suggest that different mechanisms are implied when the arsenicals were used (Yoo et al. 2004). In addition, significant dose-dependent inhibition of cell proliferation was obtained in vitro when human umbilical vein endothelial cells were treated with either As_2O_3 or tetraarsenic oxide, causing a reduction of the cyclin protein levels. The inhibitory effects observed on basic fibroblast growth factor or vascular endothelial growth factor stimulating cell proliferation suggested antiangiogenic potential of these arsenic compounds. According to their results the authors propose these arsenicals as good therapeutic agents also for the inhibition of angiogenesis of endothelial cells (Woo et al. 2005).

More recently, studies on increased efficiency have been carried out combining As_4O_6 with the known antitumor drug paclitaxel, 5-fluorouracil, or cisplatin and studying its mechanism of action in the cell lines of human gastric, cervix, and head and neck tumors. In particular, tetraarsenic oxide suppressed the tubulin polymerization in the presence and absence of paclitaxel in a concentration-dependent manner, indicating a synergic interaction and good potential for the following in vivo evaluation (Chung et al. 2009). It was also reported during the first study of the enhanced activity shown by tetraarsenic oxide when combined to radiation therapy in the treatment of solid tumors, like fibrosarcoma and human squamous tumor (Park et al. 2009).

7.5 Arsenicin A: The First and Only Natural Organic Polyarsenical

7.5.1 Isolation and Structural Elucidation

Natural products have always played a relevant role in medicine, including marine metabolites which have focused attention as interesting leads in recent drug discovery, because many of them are highly active and selective. For their natural role to protect a particular organism, marine secondary metabolites have been subjected to evolutionary pressure and selected to reach optimal activity and to perform specialized functions, like defense from predators or in sexual mechanisms. Thousands of molecules have been isolated from marine organisms, many of which show a broad spectrum of biological activities (antineoplastic, cytotoxic, neurotoxic, antibiotic, antiviral, antifungal, antimitotic, and antiprotozoal). Currently, some of these compounds are involved in advanced preclinical and clinical trials, although only a few drugs from marine sources have been hitherto introduced into clinics. Regarding their molecular structure, as a rule they present a carbon skeleton rich of stereogenic centers and functional groups, including halogen atoms introduced by metabolic processes of the inorganic species present in seawaters, as a feature of marine products if compared to metabolites of terrestrial origin. However, the extremely scarce availability of these biologically active substances precludes their use in extended bioassays and therapy. Synthetic organic chemistry still represents one of the routes of choice to overcome these difficulties. It is able to produce sufficient amounts for a broad biological screening and to provide access to synthetic analogs for structure-activity relationship (SAR) studies, as well as to confirm the structures and establish the absolute configuration of natural compounds (Mancini et al. 2007).

In this framework, the new marine product arsenicin A stands as a representative example, peculiar due to its unique molecular structure.

The poecilosclerid sponge *Echinochalina bargibanti* was collected along the northeastern coast of New Caledonia at 18–25 m depth during the program "SMIB" (Substances Marines d'Interest Biologique). Raw organic extracts from dichloromethane/water partition of the residue from ethanol extraction of the sponge showed the most potent antibacterial and antifungal activities ever observed for marine organisms from New Caledonia in the ORSTOM/IRD laboratories at Noumea. In the course of a long-standing joint project among the French marine biologists in this center and the bioorganic chemists at the University of Trento, the crude extract arrived in Italy for the purification and structural elucidation of the molecule(s) responsible for the observed bioactivities. Starting with purification, the workup proved to be out of routine procedures. It was only by a bioassay-guided fractionation on silica gel of the dichloromethane extract that the fractions containing the compounds of interest could be detected. Later they were subjected to reversed-phase flash chromatography by elution with acetonitrile/water, setting apart sterols and esterified carotenoids from the compounds which showed up as bioactive and ultraviolet (UV)-active spots on thin layer chromatography (TLC). In order to avoid their oxidative degradation, acetonitrile was evaporated at room temperature, and the remaining aqueous residue extracted with ethyl acetate and evaporated also under a nitrogen atmosphere, giving a residue which was subjected to preparative HPLC purification on cyano (CN) stationary phase and UV detection, by elution with hexane/ethyl acetate to give pure arsenicin A (5 mg, 0.001 % of the sponge dry weight). The following structural assignment was not easier, and it is noteworthy that the structure simplicity of arsenicin A does not account for the efforts necessary to establish it, due to the peculiar approaches useful to arrive at the complete definition of a so outstanding molecule in the field of marine natural products.

Extensive application of mass spectrometric technique was essential for the study. Due to the low polarity of the molecule, electron impact (EI) ionization could be applied, and a typical "mass defective" signal for arsenicin A pointed to the presence of heteroatoms. Arsenic was detected by a strong signal at *m/z* 75 in the inductively coupled plasma mass spectrum. High-resolution EI-MS experiments revealed the composition $C_3H_6As_4O_3$ of the molecular ion at *m/z* 389.7177 and supported the characteristic loss of one or two formaldehyde molecules, confirmed in tandem fragmentation experiments by atmospheric pressure chemical ionization mass spectrometry (APCI(+)-MS^n) in positive ion mode.

Standard NMR technique, usually decisive in the structural assignment of organic molecules, resulted instead poorly diagnostic in this case. It was due both to the lack of reference data on similar polyarsenicals and to the few signals present in ^1H and ^{13}C NMR spectra, for the reduced presence of carbon and hydrogen atoms and mainly for the molecular symmetry of arsenicin A, which showed three CH_2 groups, two of which were magnetically equivalent. However, more than one spectrally compatible structure fitted these data and the aim of a diffraction analysis was defeated by the difficulty of obtaining suitable crystals of the available small amount of arsenicin A. It was only the synthesis of a model compound that provided the basis to discriminate the correct structure of the natural compound. Therefore, the arsenical of molecular composition $C_4H_8As_4O_4$, whose adamantane-like structure had been established by X-ray diffractometric analysis, was prepared. Its EI-MS spectrum showed the same unusual fragmentation already observed in the natural molecule, in this case by the loss of one and two acetaldehyde molecules,

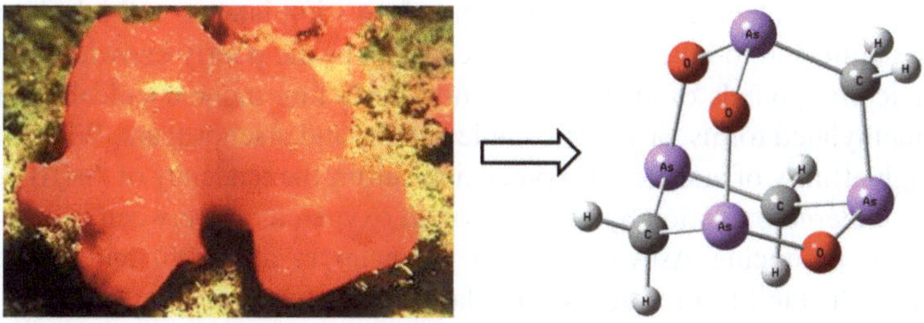

Fig. 7.5 Sponge *Echinochalina bargibanti* collected along the northeastern coast of New Caledonia (photo by ORSTOM/IRD center at Noumea) and molecular structures (energy-minimized form by DFT calculations) of the isolated metabolite arsenicin A ($C_3H_6As_4O_3$)

supporting similar molecular rearrangements based on structural similarity. Then a comparative density functional theory (DFT) simulation of infrared (IR) spectra was decisive. The nice agreement observed between the experimental and quantum chemical calculated IR spectra of the natural and synthetic compounds definitely supported the structure reported in Fig. 7.5 of the natural bioactive compound. No absolute configuration meaning could be attributed to the structure, due to the practically optical inactivity obtained in the natural compound. Arsenicin A (=2,4,6-trioxa-1,3,5,7-tetrarsa-tricyclo [3.3.1.13,7]decane) is the first organic polyarsenic compound ever found in nature (Mancini et al. 2006).

An extensive experimental and DFT-calculated vibrational analysis of arsenicin A was also carried out in our research group by exploiting the dissimilar and complementary selected rules of infrared absorption and Raman scattering transitions. In particular, Raman scattering proved to be more useful in the assignment of the adamantane structure of arsenicin A, whereas the IR spectroscopy resulted in more powerful distinguishing different atom connectivities inside the skeleton itself. It has been demonstrated that vibrational spectroscopy associated with easily accessible accurate calculations can be a very useful tool in structure elucidation in cases where MS and NMR are unable to give diagnostic indications and that this approach allows to establish a reliable methodology for the structural characterization of other polyarsenicals and in general of other molecules showing similar problems in their structural analysis (Guella et al. 2009). DFT methods were also successfully applied to the simulation of NMR spectra of polyarsenic structures previously taken into account, due to the fact that such systems are ideally suited to the application of these calculations, since they are nonpolar and rigid. In particular, this theoretical approach was able to support the assignment for arsenicin A and offers a valid tool for the investigation of other naturally occurring molecules with similar structures (Tähtineen et al. 2008).

The presence of arsenicin A was verified also in other extracts of the sponge *Echinochalina bargibanti* collected in the following years, in combination with minor related polyarsenicals also containing sulfur atoms, but it is still unknown if synthesized by sponge cells, microbial symbionts, or both in cooperation. Arsenicin A represents the first example of a marine metabolite belonging to the new biogenetic class of polyarsenicals and therefore stimulates to rethink the arsenic cycle in nature and to investigate the mechanism of the poorly known biosynthesis of organic compounds of arsenic, until now including monoarsenicals, mostly as apolar methylated forms, or water-soluble betaine and arsenosugars. What is known is the high affinity of arsenic for oxygen and sulfur, the tendency of As(III)–As(III) bonds to undergo oxidation by insertion of an oxygen atom between two As atoms rather than producing As(V) = O unit (Mancini et al. 2006) and the recently reported high stability of tetraarsenic adamantanes by electrostatic effects (Wang et al. 2010).

Recently, racemic arsenicin A has been synthesized in four steps starting from methylenebis (phenylarsinic acid) and its crystal structure described (Lu et al. 2010). Although lacking an obvious chromophore, UV spectrum showed natural metabolite, and later it was observed that the synthetic compound shows strong absorbances in the region 230–314 nm, attracting interest for a theoretical investigation. Time-dependent DFT calculations have pointed to an absorption due to the interactions between the lone pairs on the arsenic and oxygen atoms and the σ-bonding framework of the molecule (Arulmozhiraja et al. 2011).

7.5.2 Biological Activities

The bioactivities observed in the raw extracts from sponges have been found in the isolated arsenicin A, which is endowed with potent antimicrobial effects on human pathogenic strains. Antibacterial activity was evaluated on both *Staphylococcus aureus* (one of the most frequent pathogens, still associated with a high mortality) and *Escherichia coli* and antifungal activity on *Candida albicans* by using the standard microdilution plate test. Comparable or higher activities have been observed for arsenicin A when compared with the antibiotic gentamicin chosen as a control test. There is a revived interest in new active molecules due to an increasing bacterial resistance towards the currently used antibiotics and these preliminary results are of potential pharmacological interest. In fact, they recalled the known activities of inorganic forms of arsenic, which however show far weaker antibiotic activity against clinical strains of *Staphylococcus aureus* and are usually more toxic than the organic forms (Mancini et al. 2006). Antifungal activities had also been reported for synthetic adamantane cage compounds $CH_3C(CH_2AsO)_3$ and $CH_3C(CH_2AsS)_3$ towards a series of pathogenic strains, showing that these compounds resulted too toxic in mice at the concentration used in the tests (Ellermann et al. 1979).

Remembering the use of arsenic trioxide in APL treatment and the recent interest in the inhibition of solid tumors, arsenic A results also promising to be evaluated as an antileukemic agent, or in general for its cytotoxic activities, in comparison with the inorganic arsenolite (As_4O_6) showing a similar adamantane structure. These studies are now accessible through organic synthesis, which has allowed to give good amounts of racemic arsenicin A useful for extended in vitro and in vivo tests.

Profiling the cytotoxicity of arsenicin A in human carcinoma cell lines has been carried out by solubilizing the compound in water and comparing its bioactivities with the ones for arsenite as water-soluble species of As_2O_3. Arsenicin A resulted more cytotoxic than arsenite species in lung (A549) and liver (HepG2) cell lines (Moe et al. 2010).

The knowledge acquired so far and the biological potential of arsenicin A warrant deeper future investigations to verify its application in drug development. SAR studies on synthetic analogs, as well as of pure enantiomeric forms of arsenicin A should be auspicable.

7.6 Concluding Remarks

Toxicity and therapeutic properties are jointly present in a drug whose dosage is a discriminating factor in favor of positive effects. This is particularly true for paradoxical arsenic and its compounds, widely known as poisons. The aim of this review was to present and discuss the current state of the art on biological activities of polyarsenicals, both as effective and potential drugs. The emblematic example is given by arsenic trioxide, widely used in traditional Chinese medicine, approved as a therapeutic agent by FDA in 2000 and at present considered to be one of the most potent drugs in chemotherapy against APL. Similar activity has been detected for realgar (As_4S_4), but its poor water solubility responsible for its poor bioavailability has reduced its clinical applications. Studies of the development of realgar nanoparticles have been recently carried out, showing promise as a water dissolution method. Other investigations are on the activity of arsenolite (As_4O_6), especially in comparison with arsenic trioxide and on their effectiveness against other types of hematological and solid tumors. Natural products have provided new lead compounds for drug design, and among them, arsenicin A, a new tetraarsenic bioactive metabolite isolated from the marine sponge *Echinochalina bargibanti*, has been included in this overview as a unique case.

References

Ahn WS, Bae SM, Lee KH, Kim YW, Lee JM, Namkoong SE, Lee IP, Kim CK, Seo JS, Sin JI, Kim YW (2004) Comparison of effects of As2O3 and As4O6 on cell growth inhibition and gene expression profiles by cDNA microarray analysis in SiHa cells. Oncol Rep 12:573–580

Antman KH (2001) Introduction: the history of arsenic trioxide in cancer therapy. Oncologist 6:1–2. doi:10.1634/theoncologist.6-suppl_2-1

Arulmozhiraja S, Coote ML, Lu D, Salem G, Wild SB (2011) Origin of the unusual ultraviolet absorption of arsenicin A. J Phys Chem A 115:4530–4534. doi:10.1021/jp200956b

Au W-Y, Kwong Y-L (2008) Arsenic trioxide: safety issues and their management. Acta Pharmacol Sin 29:296–304. doi:10.1111/j.1745-7254.2008.00771.x

Balaz P, Sedlak J (2010) Arsenic in cancer treatment: challenges for application of realgar nanoparticles (a mini review). Toxins 2:1568–1581. doi:10.3390/toxins2061568

Balaz P, Fabian M, Pastorek M, Cholujova D, Sedlak J (2009) Mechanochemical preparation and anticancer effect of realgar As_4S_4 nanoparticles. Mater Lett 63:1542–1544. doi:10.1016/j.matlet.2009.04.008 DOI:10.1016/j.matlet.2009.04.008#doilink

Binet F, Antoine F, Girard D (2009) Interaction between arsenic trioxide and human primary cells: emphasis on human cells of myeloid origin. Inflamm Allergy Drug Targets 8:21–27, http://dx.doi.org/10.2174/187152809787582516

Chang HS, Bae SM, Kim YW, Kwak SY, Min HJ, Bae IJ, Lee YJ, Shin JC, Kim CK, Ahn WS (2007) Comparison of diarsenic oxide and tetraarsenic oxide on anticancer effects: relation to the apoptosis molecular pathway. Int J Oncol 30:1129–1135

Chen Z, Chen GQ, Shen ZX, Chen SJ, Wang ZY (2001) Treatment of acute promyelocytic leukemia with arsenic compounds: in vitro and in vivo studies. Semin Hematol 38:26–36. doi:10.1016/S0037-1963(01)90003-4 DOI:10.1016/S0037-1963%2801%2990003-4#doilink

Chou W-C, Dang CV (2005) Acute promyelocytic leukemia: recent advances in therapy and molecular basis of response to arsenic therapies. Curr Opin Hematol 12:1–6. doi:10.1097/01.moh.0000148552.93303.45

Chung WH, Sung BH, Kim SS, Rhim H, Kuh HJ (2009) Synergistic interaction between tetraarsenic oxide and paclitaxel in human cancer cells in vitro. Int J Oncol 34:1669–1679. doi:10.3892/ijo_00000298

Coe E, Schimmer AD (2008) Catalase activity and arsenic sensitivity in acute leukemia. Leuk Lymphoma 49:1976–1981. doi:10.1080/10428190802353617

Cotton FA, Wilkinson G (1988) Advanced inorganic chemistry, vol 5. Wiley, New York, NY

Cui L, Newcombe C, Urgast DS, Raab A, Krupp EM, Feldmann J (2011) Assessing the toxicity of arsenic-bearing sulfide minerals with the bio-indicator *Corophium volutator*. Environ Chem 8:52–61. doi:10.1071/EN10044

Ellermann J, Lietz M, Merbach P, Thiele G, Zoubek G (1979) Chemie polyfunktioneller liganden,53Uber methyl-substituierte triarsa-trioxa-, triarsa-trithia-und triarsa-triaza-adamantane and űber den stark aufgeweiteten adamatan-käfig im schweel-derivat. Z Naturforsch 84b:975–982

Emadi A, Gore SD (2010) Arsenic trioxide—an old drug rediscovered. Blood Rev 24:191–199. doi:10.1016/j.blre.2010.04.001 DOI:10.1016/j.blre.2010.04.001#doilink

Evens AM, Tallman MS, Gartenhaus RB (2004) The potential of arsenic trioxide in the treatment of malignant disease: past, present, and future. Leuk Res 28:891–900

Francesconi KA, Kuehnelt D (2002) Arsenic compounds in the environment. In: Frankenberger WT Jr (ed) Environmental chemistry of arsenic. Dekker, New York, NY

Fu Z-C, Zhang B, Fu Z-J (2010) As_2O_3 may be a treatment option for adenoid cystic carcinoma of salivary gland. Med Hypotheses 75:490–491. doi:10.1016/j.mehy.2010.07.001 DOI:10.1016/j.mehy.2010.07.001#doilink

Gensini GF, Conti AA, Lippi D (2007) The contributions of Paul Ehrlich to infectious disease. J Infect 54:221–224. doi:10.1016/j.jinf.2004.05.022

Guella G, Mancini I, Mariotto G, Rossi B, Viliani G (2009) Vibrational analysis as a powerful tool in structure elucidation of polyarsenicals: a DFT-based investigation of arsenicin A. Phys Chem Chem Phys 11:2420–2427. doi:10.1039/b816729j

Hussein MA (2001) Arsenic trioxide: a new immunomodulatory agent in the management of multiple myeloma. Med Oncol 18:239–242, http://dx.doi.org/10.1385/MO:18:4:239

Ishitsuka K, Hanada S, Uozumi K, Utsunomiya A, Arima T (2000) Arsenic trioxide and the growth of human T- cell leukemia virus type I infected T- cell lines. Leuk Lymphoma 37:649–655. doi:10.3109/10428190009058521

Ito K, Bernardi R, Morotti A, Matsuoka S, Saglio G, Ikeda Y, Rosenblatt J, Avigna DE, Teruya-Felstein J, Pandolfi PP (2008) PML targeting eradicated quiescent leukaemia-initiating cells. Nature 453:1072–1078. doi:10.1038/nature07016

Jing Y, Waxman S (2007) The design of selective and non-selective combination therapy for acute promyelocytic leukemia. Curr Top Microbiol Immunol 313:245–269. doi:10.1007/978-3-540-34594-7_13

Khan MAS, Oubrahim H, Stadtman ER (2004) Inhibition of apoptosis in acute promyelocytic leukemia cells leads to increases in levels of oxidized protein and LMP2 immunoproteasome. Proc Natl Acad Sci USA 101:11560–11565. doi:10.1073/pnas.0404101101

Kim J, Lee JJ, Kim J, Grdner D, Beachy PA (2007) Arsenic antagonizes the hedgehog pathway by preventing ciliary accumulation and reducing stability of the Gli2 transcriptional effector. Proc Natl Acad Sci USA 107:13432–13437. doi:10.1073/pnas.1006822107

Lin M, Wang Z, Zhang D (2007) Preparation of orpiment nanoparticles and their cytotoxic effect on cultured leukemia K562 cells. J Nanosci Nanotechnol 7:490–496. doi:10.1166/jnn.2007.145

Liu J, Lu Y, Wu Q, Goyer RA, Waalkes MP (2008) Mineral arsenicals in traditional medicines: orpiment, realgar, and arsenolite. J Pharmacol Exp Ther 326:363–368. doi:10.1124/jpet.108.139543

Lloyd NC, Morgan HW, Nicholson BK, Ronimus RS (2005) The composition of Ehrlich's Salvarsan: resolution of a century-old debate. Angew Chem Int Ed 44:941–944. doi:10.1002/anie.200461471

Lu D-P, Wang Q (2002) Current study of APL treatment in China. Int J Hematol 76:316–318

Lu D, Rae AD, Salem G, Weir ML, Willis AC, Wild SB (2010) Arsenicin A, a natural polyarsenical: synthesis and crystal structure. Organometallics 29:32–33. doi:10.1021/om90098q

Lu Y-F, Wu Q, Yan J-W, Shi J-Z, Liu J, Shi J-S (2011) Realgar, cinnabar and An-Gong-Niu-Huang Wan are much less chronically nephrotoxic than common arsenicals and mercurial. Exp Biol Med 236:233–239. doi:10.1016/j.cbi.2010.11.006

Mancini I, Guella G, Frostin M, Hnawia E, Laurent D, Debitus C, Pietra F (2006) On the first polyarsenic organic compound from nature: arsenicin A from the New Caledonian marine sponge Echinochalina bargibanti. Chem Eur J 12:8989–8994. doi:10.1002/chem.200600783

Mancini I, Defant A, Guella G (2007) Recent synthesis of marine natural products with antibacterial activities. AntiInfect Agents Med Chem 6:17–48. doi:10.1080/00397910802044249

Mann KK, Miller WH Jr (2004) Death by arsenic: implications of PML sumoylation. Cancer Cell 5:307–309, http://dx.doi.org/10.1016/S1535-6108(04)00089-3

Miller WH Jr, Schipper HM, Lee JS, Singer J, Waxman S (2002) Mechanisms of action of arsenic trioxide. Cancer Res 62:3893–3903, http://cancerres.aacrjournals.org/content/62/14/3893.full

Moe B, Chen B, Li XF, Le C (2010) Profiling the cytotoxicity of a novel polyarsenical in human carcinoma cell lines. Presented at the Faculty Research and Development Day Program, University of Alberta Canada, November 19th, 2010 Abstracts page17, http://www.pharm.ualberta.ca/~/media/University%20of%20Alberta/Faculties/Phacy%20and%20Pharmaceutical%20Sciences/Faculty%20Site/Research/Research%20Day/2010%20Research%20Day/Brochure Research_Day_2010_3.pdf

Munir ZA, Street GB, Winters HF (1971) Mass-spectrometric and vapor pressure studies on the sublimation of realgar (As_4S_4). J Chem Phys 55:4520–4527

Nasr R, Guillemin MC, Ferhi O, Soilihi H, Peres L, Berthier C, Rousselot P, Robledo-Sarmiento-M, Lallemand-Breitenbach V, Gourmel B, Vitoux D, Pandolfi PP, Rochette-Egly C, Zhu J, de Thé H (2008) Eradication of acute promyelocytic leukemia-initiating cells through PML-RARA degradation. Nat Med 14:1333–1342. doi:10.1038/nm.1891

O'Day PA (2006) Chemistry and mineralogy of arsenic. Elements 2:77–83. doi:10.2113/gselements.2.2.77

Pagliai M, Bonazzi P, Bindi L, Muniz-Miranda M, Cardini G (2011) Structural and vibrational properties of arsenic sulfides: alacranite (As_8S_9). J Phys Chem A 115:4558–4562. doi:10.1021/jp201097k

Park MJ, Park IC, Bae IJ, Seo KM, Lee SH, Hong SI, Eun CK, Zhang W, Rhee CH (2003a) Tetraarsenic oxide, a novel orally administrable angiogenesis inhibitor. Int J Oncol 22:1271–1276

Park IC, Park MJ, Woo SH, Lee HC, An S, Gwak HS, Lee SH, Hong SI, Bae IJ, Seo KM, Rhee CH (2003b) Tetraarsenic oxide induces apoptosis in U937 leukemic cells through a reactive oxygen species-dependent pathway. Int J Oncol 23:943–948

Park SG, Jung JJ, Won HJ, Kang MS, Seo SK, Choi IW, Eun CK, Ahn KJ, Park CW, Lee SW, Lew YS, Bae IJ, Choi IH (2009) Tetra-arsenic oxide (tetras) enhances radiation sensitivity of solid tumors by anti-vascular effect. Cancer Lett 277:212–217. doi:10.1016/j.canlet.2008.12.012

Platanias LC, Robert H (2009) Biological responses to arsenic compounds. J Biol Chem 284:18583–18587. doi:10.1074/jbc.R900003200

Qi J, He P, Chen W, Wang H, Wang X, Zhang M (2010) Comparative proteome study of apoptosis induced by As4S4 in retinoid acid resistant human acute promyelocytic leukemia NB4-R1 cells. Leuk Res 34:1506–1516. doi:10.1016/j.leukres.2010.03.038

Ramadan D, Rancy PC, Nagarkar RP, Schneider JP, Thorpe C (2009) Arsenic (III) species inhibit oxidative protein folding in vitro. Biochemistry 48:424–432. doi:10.1021/bi801988x

Richardson PG, Mitsiades CS, Hideshima T, Anderson KC (2005) Novel biological therapies for the treatment of multiple myeloma. Best Pract Res Clin Hematol 18:619–634, http://dx.doi.org/10.1016/j.beha.2005.01.01

Sanchez Y, Amran D, de Blas E, Aller P (2010) Arsenic trioxide as an anti-tumour agent: mechanisms of action and strategies of sensitization. J Appl Biomed 8:199–208. doi:10.2478/v10136-009-0027-3

Shen Z-X, Shi Z-Z, Fang J, Gu B-W, Li J-M, Zhu Y-M, Shi J-Y, Zheng P-Z, Yan H, Liu Y-F, Chen Y, Shen Y, Wu W, Tang W, Waxman S, de Thé H, Wang Z-Y, Chen S-J, Chen Z (2004) All-trans retinoic acid/As_2O_3 combination yields a high quality remission and survival in newly diagnosed acute promyelocytic leukemia. Proc Natl Acad Sci USA 101:5328–5335. doi:10.1073/pnas.0400053101

Shen XC, Jin T, Xie J, Liang H, Yan Y (2009) Studies on the biomimetic membrane interaction between liposome and realgar nanoparticles. Sci China B 52:1512–1518. doi:10.1007/s11426-009-0223-8

Singh P, Sharma R (1994) Effect of orpiment (As_2S_3) on cytochrome P 450, glutathione and lipid peroxide levels of rat liver. J Environ Pathol Toxicol Oncol 13:199–203

Singh P, Sharma R (1998) In vivo effects of orpiment administration on the lipid profile of rat lungs. Pollut Res 17:307–310

Singh P, Gregory MA, Sharma R, Kiran R (1999) Effects of orpiment (As_2S_3) toxicity on the kinetic and thermodynamic parameters of some brush-border membrane enzymes of rat intestine. Pollut Res 18:75–78

Tähtineen P, Saielli G, Mancini I, Bagno A (2008) Computational NMR spectroscopy of organoarsenicals and the natural polyarsenic compound arsenicin A. Chem Eur J 14:10445–10452. doi:10.1002/chem.200801272

Tang Y, Wang N, Zhang Y, Ye S, Ou W (2008) Effects of realgar on stress proteins, inflammatory mediators, and complement in brain tissue and serum of rats with inflammatory brain injury. Neural Regen Res 3:885–889

Wang ZY (2001) Arsenic compounds as anticancer agents. Cancer Chemother Pharmacol 48: S72–S76. doi:10.1007/s002800100309

Wang J, Lin M, Zhang T, Yan Y, Ho PC, Xu QH, Loh KP (2008a) Arsenic(II) sulfide quantum dots prepared by a wet process from its bulk. J Am Chem Soc 130:11596–11597. doi:10.1021/ja804436w

Wang L, Zhou GB, Liu P, Song JH, Liang Y, Yan XJ, Xu F, Wang BS, Mao JH, Shen ZX, Chen SJ, Chen Z (2008b) Dissection of mechanisms of Chinese medicinal formula realgar-indigo naturalis as an effective treatment for promyelocytic leukemia. Proc Natl Acad Sci USA 105:4826–4831. doi:10.1073/pnas.0712365105

Wang LW, Shi YL, Wang N, Gou BD, Zhang TL, Wang K (2009) Association of oxidative stress with realgar-induced differentiation in human leukemia HL-60 cells. Chemotherapy 55:460–467. doi:10.1159/000265528

Wang Y, Wu JIC, Li Q, von Ragué Schleyer P (2010) Why are some $(CH)_4X_6$ and $(CH_2)_6X_4$ polyheteroadamantanes so stable? Org Lett 12:1320–1323. doi:10.1021/ol1002187

Waxman S, Anderson KC (2001) History of the development of arsenic derivatives in cancer therapy. Oncologist 6:3–10. doi:10.1634/theoncologist.6-suppl_2-3

Wei L, Liao P, Wu H, Li X, Pei F, Li W, Wu Y (2009) Metabolic profiling studies on the toxicological effects of realgar in rats by ^{1}H NMR spectroscopy. Toxicol Appl Pharmacol 234:314–325. doi:10.1016/j.taap.2008.11.010

Woo SH, Park MJ, An S, Lee HC, Jin HO, Lee SJ, Gwak HS, Park IC, Hong SI, Rhee CH (2005) Diarsenic and tetraarsenic oxide inhibit cell cycle progression and bFGF- and VEGF-induced proliferation of human endothelial cells. J Cell Biochem 95:120–130. doi:10.1002/jcb.20329

Wu JZ, Ho PC (2006) Evaluation of the in vitro activity and in vivo bioavailability of realgar nanoparticles prepared by cryo-grinding. Eur J Pharm Sci 29:35–44. doi:10.1016/j.ejps.2006.05.002

Wu J-Z, Ho PC (2009) Comparing the relative oxidative DNA damage caused by various arsenic species by quantifying urinary levels of 8-hydroxy-2′-deoxyguanosine with isotope-dilution liquid chromatography/mass spectrometry. Pharm Res 26:1525–1533. doi:10.1007/s11095-009-9865-7

Wu JZ, Shao YB, Liu JL, Chen G, Ho PC (2011) The medicinal use of realgar (As_4S_4) and its recent development as an anticancer agent. J Ethnopharmacol 135:595–602. doi:10.1016/j.jep.2011.03.071

Xu L, Zeng F, Ye H, Yang X, Xu H (2006) Anti-tumor activity of realgar nanoparticles and the distribution of arsenic (As) in tissues of tumor-transplanted mice. Zhongguo Xinyao Zazhi 15:1845–1848, CNKI:SUN:ZXYZ.0.2006-21-013

Yoo MH, Kim JT, Rhee CH, Park MJ, Bae IJ, Yi NY, Jeong MB, Jeong SM, Nam TC, Seo KM (2004) Reverse effects of tetraarsenic oxide on the angiogenesis induced by nerve growth factor in the rat cornea. J Vet Med Sci 66:1091–1095. doi:10.1292/jvms.66.1091

Zhang C, Ling C (2003) Review of arsenic trioxide effect on solid cancer. Acad J Second Mil Med Univ 24:440–443, cnki:ISSN:0258-879X.0.2003-04-032

Zhang J, Wang JC, Han YH, Wang LF, Ji SP, Liu SX, Liu XP, Yao LB (2005) High expression of bcl-xL in K562 cells and its role in the low sensitivity of K562 to realgar-induced apoptosis. Acta Haematol 113:247–254. doi:10.1159/000084678

Zhang QY, Mao JH, Liu P, Huang QH, Lu J, Xie YY, Weng L, Zhang Y, Chen Q, Chen SJ, Chen Z (2009) A systems biology understanding of the synergistic effects of arsenic sulfide and Imatinib in BCR/ABL-associated leukemia. Proc Natl Acad Sci USA 106:3378–3383. doi:10.1073/pnas.0813142106

Zhang X, Xie Q, Wang X, Wang B, Li H (2010a) Biological extraction of realgar by and its in vitro and in vivo antitumor activities. Pharm Biol 48:40–47. doi:10.3109/13880200903029381

Zhang XW, Yan XJ, Zhou ZR, Yang FF, Wu ZY, Sun HB, Liang WX, Song AX, Lallemand-Breitenbach V, Jeanne M, Zhang QY, Yang HY, Huang QH, Zhou GB, Tong JH, Zhang Y, Wu JH, Hu HY, de Thé H, Chen SJ, Chen Z (2010b) Arsenic trioxide controls the fate of the PML-RARα oncoprotein by directly binding PML. Science 328:240–243. doi:10.1126/science.1183424

Zhao W, Lu X, Yuan Y, Liu C, Yang B, Hong H, Wang G, Zeng F (2011) Effect of size and processing method on the cytotoxicity of realgar nanoparticles in cancer cell lines. Int J Nanomedicine 6:1569–1577. doi:10.2147/IJN.S21373

Chapter 8
Biogenic Inorganic Polysilicates (Biosilica): Formation and Biomedical Applications

Heinz C. Schröder, Xiaohong Wang, Ute Schloßmacher, Matthias Wiens, and Werner E.G. Müller

Contents

H.C. Schröder (✉) • W.E.G. Müller (✉)
ERC Advanced Investigator Group, Institute for Physiological Chemistry, University Medical Center of the Johannes Gutenberg University, Duesbergweg 6, 55128 Mainz, Germany

NanotecMARIN GmbH, Duesbergweg 6, 55128 Mainz, Germany
e-mail: hschroed@uni-mainz.de; wmueller@uni-mainz.de

X.H. Wang
ERC Advanced Investigator Group, Institute for Physiological Chemistry, University Medical Center of the Johannes Gutenberg University, Duesbergweg 6, 55128 Mainz, Germany

U. Schloßmacher • M. Wiens
ERC Advanced Investigator Group, Institute for Physiological Chemistry, University Medical Center of the Johannes Gutenberg University, Duesbergweg 6, 55128 Mainz, Germany

W.E.G. Müller et al. (eds.), *Biomedical Inorganic Polymers*, Progress in Molecular and Subcellular Biology 54, DOI 10.1007/978-3-642-41004-8_8,

Abstract The siliceous sponges, the demosponges and hexactinellid glass sponges, are unique in their ability to form biosilica structures with complex architectures through an enzyme-catalyzed mechanism. The biosilica skeleton of these sponges with its hierarchically structure and exceptional opto-mechanical properties has turned out to be an excellent model for the design of biomimetic nanomaterials with novel property combinations. In addition, biosilica shows morphogenetic activity that offers novel applications in the field of bone tissue engineering and repair. In recent years, much progress has been achieved towards the understanding of the principal enzymes, the silicateins that form the sponge skeletal elements, the spicules, and their self-assembling and structure-guiding properties. The discovery of the silicatein-interacting, scaffolding proteins provided new insights in the mechanism of spiculogenesis. The now available toolbox of enzymes and proteins that are involved in biosilica formation and the biosilica material synthesized by them are of great interest for a variety of applications from nanobiotechnology to nanomedicine.

8.1 Introduction

The organization of the inorganic skeleton of the siliceous sponges (Demospongiae and Hexactinellida) is genetically fixed and architecturally complex (Schröder et al. 2008; Müller et al. 2009c); as an example, the hierarchical structure of a glass sponge, the hexactinellid *Euplectella aspergillum*, is shown in Fig. 8.1. The inorganic matrix of the spicules which build up the siliceous sponge skeleton consists of amorphous biosilica. For a long time, it remained unclear how the hard biosilica of the sponge spicules is formed. Two discoveries brought the breakthrough in the elucidation of the mechanism of spicule formation (1) the discovery that the axial filaments of the spicules mainly consist of silicatein (Shimizu et al. 1998; Cha et al. 1999), a group of proteins related to the protease cathepsin that we had identified in sponges before (Krasko et al. 1997), and (2) that this protein has biocatalytic/enzymatic activity [Krasko et al. (2000), Müller et al. (2008d); reviewed in Schröder et al. (2007a, 2008), Müller et al. (2009c, 2013b), Wang et al. (2012a)]. The silicatein-mediated biosilica formation even proceeds at silica precursor concentrations far below those at which polycondensation in silicatein-free solutions occurs.

The silicateins catalyze the synthesis of nanoscale silica particles from soluble precursors at near-neutral pH and ambient temperature, which subsequently assemble and fuse to microscopic and macroscopic spicules/skeletal structures showing a variety of morphologies. The spicules often have a lamellar organization (Müller et al. 2005, 2007e, 2010b; Woesz et al. 2006; Schröder et al. 2007b) and are provided with exceptional opto-mechanical properties, most likely caused by the hierarchical structuring of their inorganic (silica) and organic (protein) constituent materials (Mayer 2005). The assembly of the monomeric silicatein to polymeric filaments,

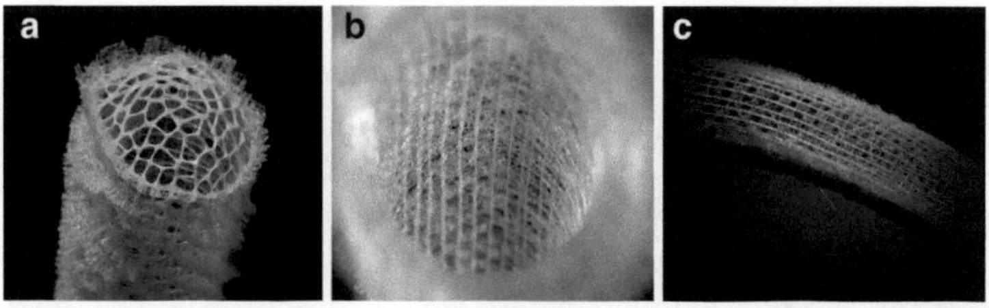

Fig. 8.1 Hierarchical structure of the biosilica skeleton of the hexactinellid glass sponge *Euplectella aspergillum*. (**a**) *Upper portion* of the skeleton with the terminal sieve plate, (**b**) interior view, (**c**) side view

acting as a template for biosilica deposition, is controlled by specific silicatein-interacting proteins (Müller et al. 2009c).

The biosilica matrix of the spicules is a hybrid material that is composed of both an inorganic (silica) component and an organic (protein) component, allowing for unusual property combinations (Müller et al. 2010b, 2013a; Wang et al. 2012d). Scanning electron microscopy (SEM) and nano-secondary ion mass spectrometry (NanoSIMS) analyses have been applied to find out if the proteinaceous material (mainly silicatein) is present within the lamellae (Müller et al. 2008f, 2010b). The studies were performed using the large (basal) spicules from the hexactinellids *Hyalonema sieboldi* (length, 30 cm) and *Monorhaphis chuni* (length of up to 3 m). The giant basal spicules of these sponges are composed of multiple (in *M. chuni* more than 500) concentrically arranged lamellae (thickness, 3–7 μm) around an axial cylinder (Fig. 8.2a; see also Fig. 8.11b, c). Spicules from the demosponge *Suberites domuncula* are much smaller (lengths <300 μm) and only consist of a solid axial-cylinder-like rod, most likely formed by fusion of the lamellae. The organic scaffold becomes visible after limited exposure of the basal spicule of *M. chuni* to hydrofluoric acid (HF) vapor (Fig. 8.2b) (Müller et al. 2008a). A high-resolution magnification (HR-SEM) of the surface of *M. chuni* lamella shows that it is composed of 5–7-nm large nanoparticles (Fig. 8.2c). Longer exposure to HF results in a progressive dissolution of the silica matrix, and an organized proteinaceous scaffold becomes uncovered (Fig. 8.2d). The element distribution of the concentrically arranged lamellae around the central axial cylinder of the giant spicule of the hexactinellid *M. chuni* has been investigated by NanoSIMS in detail (Fig. 8.2e–g) (Müller et al. 2010b). The thicker (10–30-μm) lamellae adjacent to the axial cylinder are separated from each other by 50- and 100-nm-wide gaps. These lamellae are composed of three to six sublamellae, each measuring about 5 μm (Fig. 8.2e; HR-SEM), which are not delimited by gaps and are further organized into three cylindrical slats with thickness of 1.6–1.8 μm. NanoSIMS element distribution analysis shows that the borders of the sublamellae are especially highlighted in the scan obtained from the $^{16}O/^{28}Si$ mapping (Fig. 8.2f); here the relative oxygen level is highest. The distribution patterns of C and S (Fig. 8.2g; $^{32}S/^{28}Si$ ratio) reflect a distinct zonation of the organic material, most likely silicatein (Müller et al. 2010b), within the silica matrix. These results indicate that the protein within

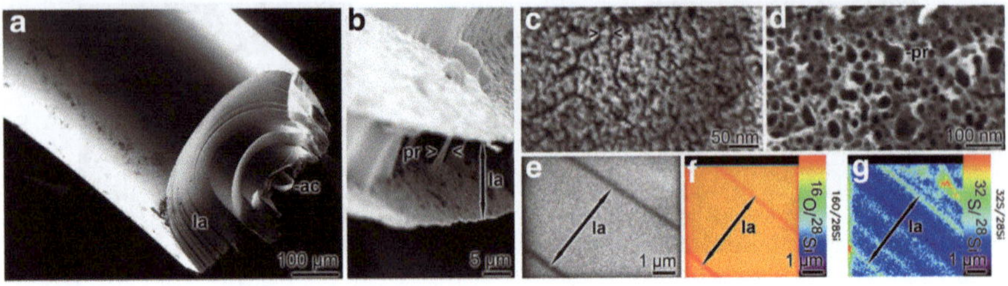

Fig. 8.2 Micro- and nanostructure of the biosilica matrix of *M. chuni*. (**a**) Giant basal spicule, after treatment at 600 °C for 20 min, resulting in a separation of the individual lamellae (la) around the axial canal (ac) (SEM). (**b**) Dissolution of lamellae after limited HF exposure. > < pr, proteinaceous bundles; la, thickness of one lamella. (**c**) Silica nanoparticles (> <) visible on the surface of a silica lamella (HR-SEM). (**d**) Proteinaceous scaffold (-pr) around the nanoparticles (SEM). (**e–g**). NanoSIMS analysis of one sublamella in a cross section through a spicule. (**e**) Thickness of the sublamella (la), 5 μm. (**f**) The $^{16}O/^{28}Si$ ratio (mapping). (**g**) $^{32}S/^{28}Si$ ratio. Pseudocolor images [Modified after Müller et al. (2008a, 2010b)]

the spicules (mainly silicatein) is not only located in the gaps between the siliceous layers (Woesz et al. 2006) but is also entrapped in the biosilica matrix (Müller et al. 2008a, f). Hence, silicatein remains inside the biosilica product after synthesis and may become active again after dissolution of the inorganic phase during maturation, growth, or self-repair of spicules (Müller et al. 2008a, 2013f).

8.2 Silicatein

Silicatein is the first enzyme that has been described to catalyze the formation of an inorganic polymer from an inorganic monomeric substrate (Müller et al. 2008d). It is related to the cathepsins but distinguished from this group of proteases by the exchange of the first amino acid in the catalytic triad, Cys, by a Ser residue which is essential for the catalytic mechanism (Shimizu et al. 1998; Krasko et al. 2000). In addition, silicateins are characterized by the presence of a Ser cluster not found in cathepsins. Silicatein cDNAs have been isolated and characterized from over 30 siliceous sponge species, not only from demosponges but also from the deep-sea hexactinellids, *Crateromorpha meyeri* (Müller et al. 2008e) and *M. chuni* (Müller et al. 2009b), the phylogenetically oldest class of the Porifera (Kruse et al. 1997, 1998). Marine demosponges (e.g., *S. domuncula*) are mostly provided with two silicatein isoforms, silicatein-α and silicatein-β (Shimizu et al. 1998; Cha et al. 1999; Krasko et al. 2000; Schröder et al. 2005b; Müller et al. 2006a, 2007b), while the freshwater demosponge *Lubomirskia baicalensis* lacks silicatein-β but has more than six silicatein-α isoenzymes (Belikov et al. 2005; Kaluzhnaya et al. 2005; Wiens et al. 2006, 2008). It should be noted that the pattern of silicatein isoforms can differ in different types of spicules; e.g., only one silicatein isoform

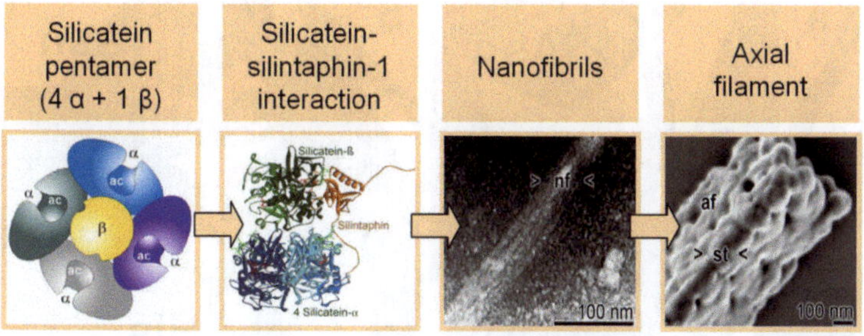

Fig. 8.3 Formation of axial filaments. From *left* to *right*: Silicatein pentamer formed by four silicatein-α and one silicatein-β molecule (ac, active center; *left*); interaction of silintaphin-1 with the silicatein pentamers during assembly of the silicatein filaments (*middle left*); alignment of the pentamers to 10-nm nanofibrils that laterally align to larger bundles (SEM; *middle right*); and axial filament (af), formed by orderly arranged strands (st) (SEM; *right*) [Modified after Müller et al. (2013a, b)]

has been found in the microscleres of the demosponge *Geodia cydonium*, in contrast to the megascleres from the same species (Müller et al. 2007f). The hexactinellid silicateins comprise a second Ser cluster, in addition to the demosponge Ser cluster, which might additionally stabilize the interaction with the surface of the formed silica product (Müller et al. 2008a). In demosponges (*S. domuncula*), the silicatein-α and silicatein-β genes comprise six exons (Schröder et al. 2005b; Müller et al. 2003). Applying homology modeling, the presumptive 3D structure of silicatein has been obtained (Schröder et al. 2008; Müller et al. 2007g).

Silicatein is the major protein component of the axial filament within the axial canal of the spicules. The two isoforms, silicatein-α and silicatein-β, which are expressed in *S. domuncula*, are able to self-assemble via fractal intermediates to long filaments (Schloßmacher et al. 2011). Thereby four silicatein-α molecules associate to tetrads, comprising a central opening into which one silicatein-β molecule is inserted (Fig. 8.3, left panel). The pentamers formed are then aligned to filaments (Müller et al. 2007a) that are stabilized by the silicatein interactor, silintaphin-1 (Fig. 8.3, middle left panel). Recently, 10-nm filaments/nanofibrils formed of only one stack of silicatein pentamers could be visualized and found having the ability to align laterally to larger bundles (Fig. 8.3, middle right panel; Müller et al. 2013a) and finally the axial filaments (Fig. 8.3, right panel). The 10-nm filaments have already the property to form a biosilica coat around them.

The silica-forming activity of silicatein could be clearly demonstrated by us; we could show for the first time that silicatein is indeed an enzyme, comprising all enzyme-characteristic properties and functions (Müller et al. 2008d; Schloßmacher et al. 2011). In addition, we could demonstrate that silicatein comprises a dual enzymatic function: it acts as both silica polymerase and silica esterase (Müller et al. 2008d). Orthosilicate is the natural substrate for the enzyme. Silicatein is also

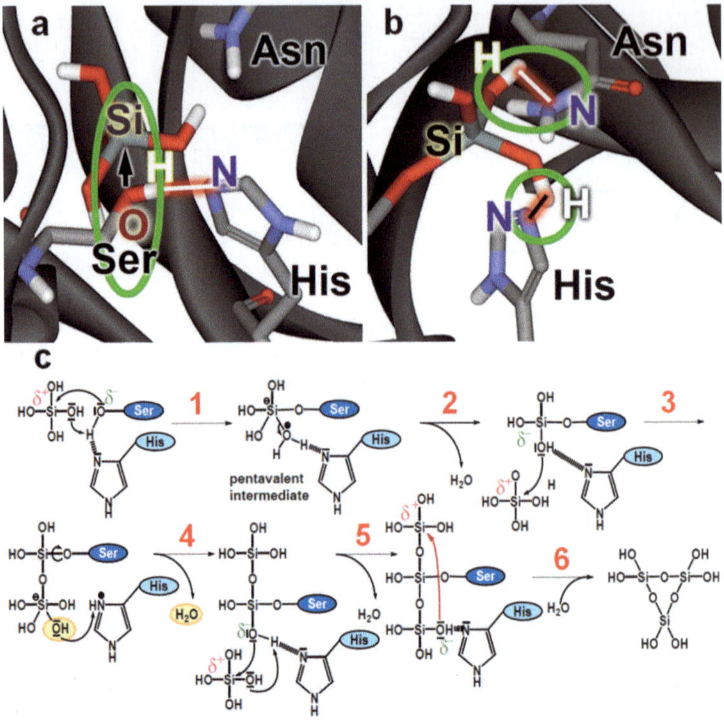

Fig. 8.4 Proposed mechanism of silicatein reaction. (**a**) and (**b**) Detail of computer model of the silicatein-α structure with the three amino acids Ser, His, and Asn of the catalytic center during the biocatalytic reaction. (**a**) Nucleophilic attack (*arrow*) of the oxygen atom of the Ser OH group at the silicon atom of the silicic acid substrate modeled in the catalytic pocket of the enzyme. The nucleophilicity of the Ser oxygen is increased by hydrogen bridge formation to the nitrogen of the His imidazole (*red-framed white bar*). (**b**) Hydrogen bridges (*red-framed white* and *black bars*) formed between the Ser-bound silicic acid and the catalytic center amino acids Asn and His. (**c**) Scheme of the catalytic reaction. Step 1: Nucleophilic attack of the negatively charged Ser oxygen at the positively charged silicon atom of the orthosilicic acid substrate molecule under formation of a pentavalent intermediate, facilitated by hydrogen bridge formation between the Ser OH and the imidazole nitrogen, and proton transfer. Step 2: Release of a water molecule from the pentavalent intermediate. Step 3: Nucleophilic attack of the oxygen of one of the OH ligands of the bound silicic acid at the silicon of a second orthosilicic acid molecule, facilitated by hydrogen bridge formation of the bound silicic acid to the imidazole nitrogen. Step 4: Release of water after proton transfer from the imidazole group. Step 5: Nucleophilic attack of the oxygen of a further OH ligand of the first silicic acid molecule, after rotation of the Si–O–C bond between this molecule and the Ser residue, at the silicon of a third orthosilicic acid. Step 6: Formation of reactive trisiloxane ring by cyclization of the enzyme-bound trisilicic acid after hydrolysis of the Si–O–C bond [Modified after Schröder et al. (2012c)]

capable of cleaving the ester-/ether-like bonds, e.g., in bis(*p*-aminophenoxy)dimethylsilane (Müller et al. 2008d), or in TEOS (Schröder et al. 2008) and in dimethoxy(dimethyl)silane (Wolf et al. 2010), allowing to determine the processivity of silicatein reaction applying EI-MS/^{29}Si DEPT NMR techniques (Müller et al. 2008d; Wolf et al. 2010). The kinetic constants of the recombinant enzyme (*S. domuncula*) have been determined (Müller et al. 2008d).

The silicateins, with the amino acids His, Asn, and Ser (instead of Cys in the cathepsins) in their catalytic center, catalyze the formation of siloxane (Si–O–Si) bonds via a nucleophilic attack (S_N2 type) of the serine hydroxy group in the catalytic site of the enzyme at the silicon atom of the orthosilicate substrate (Müller et al. 2013b). As a result, the polycondensation reaction even proceeds at low silicic acid concentrations, below 1 mM (Müller et al. 2008d; Schloßmacher et al. 2011), while this process, in the absence of silicatein, only occurs at higher concentrations (>1 mM) at neutral pH. Based on the results of modeling studies, a novel mechanism for the silicatein reaction has been proposed that includes a final cyclization step (Schröder et al. 2012c; Fig. 8.4). Other models of silicatein reaction were only able to explain the hydrolytic cleavage of nonphysiological silicatein substrates (TEOS) to silanol compounds (Cha et al. 1999) or the formation of less reactive silicic acid dimers (Fairhead et al. 2008). The initial step (nucleophilic attack of the serine oxygen at the silicon atom of the silicic acid substrate; step 1) and the covalently (serine) bound silicic acid species in the catalytic center of the enzyme are shown in Fig. 8.4a, b. The cyclic silicic acid species (trisiloxane rings) formed according to this model (Fig. 8.4c) are much more reactive than the silicic acid monomers and dimers and able to promote the further polycondensation reaction (Schröder et al. 2012c).

The silicatein molecules undergo posttranslational modifications. Analysis by 2D electrophoresis of the silicatein protein, isolated from axial filaments of demosponges and hexactinellids, revealed five phospho-isoforms (Müller et al. 2005, 2007a, g, 2008a; Wang et al. 2009a). Further posttranslational modifications are oxidations of a histidine and of a cysteine to cysteic acid, as well as methylations at the N-terminal region (Armirotti et al. 2009).

8.2.1 Autocatalytic Cleavage of Pro-silicatein

The silicatein(s) turned out to have an autocatalytic function which might act as a molecular switch for biomineral formation both in vitro and in vivo (Schröder et al. 2012a). Self-cleavage of the silicatein precursor molecule (pro-silicatein; Fig. 8.5a) into the N-terminal propeptide and the mature silicatein triggers the mature molecule not only to become enzymatically active but also to acquire structure-guiding and structure-forming properties. This autocatalytic activity can be demonstrated by expression of the gene encoding pro-silicatein as fusion protein in *Escherichia coli* together with the bacterial trigger factor, a chaperone encoded

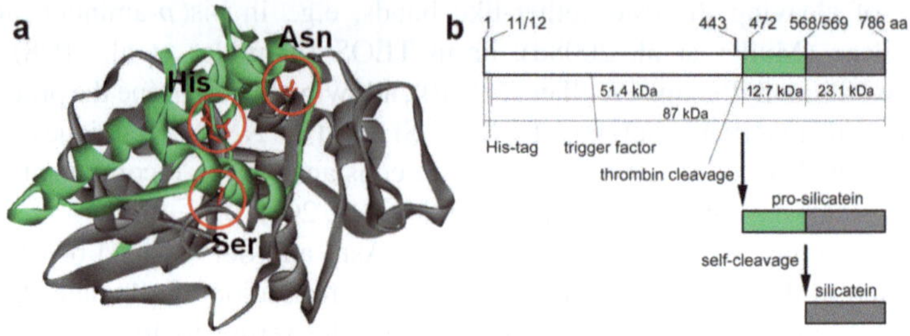

Fig. 8.5 Self-cleavage of pro-silicatein into the silicatein propeptide and the mature silicatein. (**a**) Model of pro-silicatein structure showing the propeptide (*green*) and the mature silicatein (*gray*) with the catalytic triad amino acids Ser, His, and Asn (*red*). (**b**) Construct used for the expression of recombinant silicatein-trigger factor fusion protein and cleavage products. The length of the construct (in amino acids, aa) comprising the gene for the trigger factor chaperone (*white box*) and the pro-silicatein that consist of the propeptide (*green*) and the mature silicatein (*gray*) is given. The cleavage site of the protease thrombin and the predicted autocatalytic cleavage site of silicatein are marked. The thrombin-mediated cleavage of the silicatein-trigger factor fusion protein results in the liberation of the pro-silicatein molecule which is immediately split by autocatalytic cleavage into the propeptide and the mature silicatein [Modified after Schröder et al. (2012a)]

by the pCold TF vector (Schröder et al. 2012a; Fig. 8.5b). The 87-kDa fusion protein is split with thrombin at the thrombin cleavage site within the spacer region into the trigger factor and pro-silicatein (Fig. 8.5b). This proteolytic cleavage reaction is immediately followed by an autocatalytic cleavage of the 35-kDa pro-silicatein into the propeptide and the 23-kDa mature silicatein (Fig. 8.5b), resulting (1) in the uncovering of the catalytic pocket of the silicatein molecule, enabling the enzymatic biosilica polycondensation reaction to occur, and (2), via exposure of hydrophobic patches on the surface of the molecule, in the self-assembly of the mature silicatein (via dimers, tetramers, and pentamers, consisting of four silicatein-α molecules and one silicatein-β molecule) to long insoluble filaments (axial filaments; a schematic presentation is given in Fig. 8.6). The biocatalytic activity of the mature silicatein (formation of biosilica) could be demonstrated after dissolution in urea-containing buffers (Schröder et al. 2012a).

It is assumed that the silicatein propeptide, either directly by association or indirectly by separation from the mature silicatein, controls the folding and function (structure-forming and enzymatic activity) of the mature protein (Müller et al. 2013d). The autocatalytic cleavage of pro-silicatein to assembly-competent and enzymatically active silicatein might allow the development of a bio-inspired molecular switch of the biomineralization process. Thereby the released, biocatalytically active 23-kDa mature enzyme is expected to act as a template for an organized, structure-controlled biosilica deposition, directed by the silicatein self-assembly process.

Analysis of the propeptide sequences of silicateins α and β from *S. domuncula* and *Tethya aurantium* and of cathepsin L from *G. cydonium* revealed that the positions of the basic amino acids Lys, Arg, and His are well conserved; in particular, the His sites are found to be restricted to the N-terminal α-helix of the

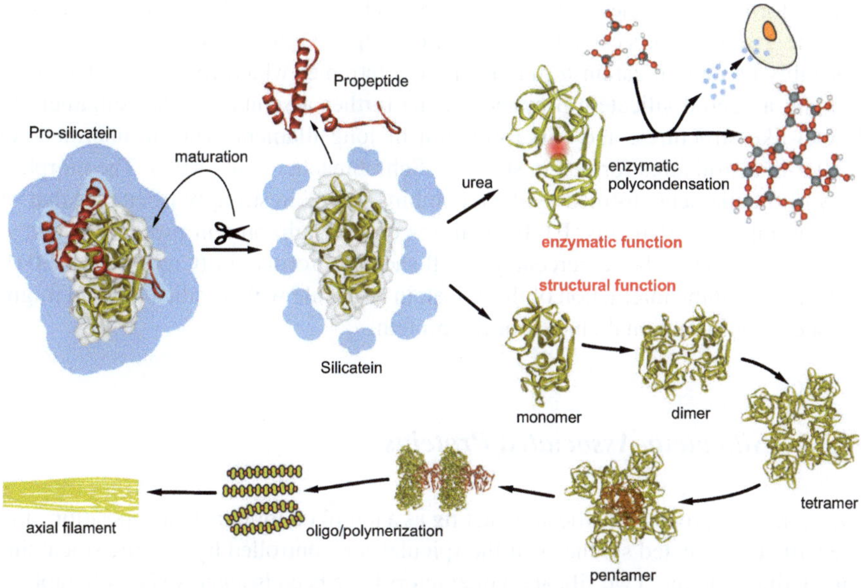

Fig. 8.6 Schematic representation of the self-assembly and mineralization process after autocatalytic cleavage of pro-silicatein to assembly-competent and enzymatically active silicatein. The self-cleavage of pro-silicatein results in the release of the N-terminal propeptide and the exposure of hydrophobic patches on the surface of the mature silicatein monomer that starts to self-assemble via dimers and tetramers and 4:1 silicatein-α/silicatein-β pentamers to long filaments. The uncovering of the catalytic pocket of the silicatein molecule enables the enzymatic reaction to occur. The water formed during the polycondensation reaction is removed by cellular uptake via aquaporin channels [Modified after Schröder et al. (2012a)]

silicatein propeptide, while in the cathepsin L propeptide, this amino acid is only present in the C-terminal β-sheet region (Müller et al. 2013d). It is assumed that the His residues form the biosilica-binding domain of the propeptide (Müller et al. 2013d). On the other hand, the silicatein interactor silintaphin-2 comprises clusters of negatively charged amino acids in the N-terminal part of the protein containing 6 Glu residues and 8 Asp residues (Wang et al. 2013c). A synthetic peptide spanning the Glu-/Asp-rich region of silintaphin-2 significantly increases the silica polycondensation reaction, suggesting that the acidic side chains favor the polycondensation reaction (see Sect. 8.2.3.2).

8.2.2 Silicatein Assembly

In both demosponges (Murr and Morse 2005, 2007a) and hexactinellids (Wang et al. 2008; Müller et al. 2009c), the self-assembly process of silicatein proceeds via the intermediary formation of fractal structures of silicatein molecules, which finally form long filaments. In the present model based on the results obtained so far, the assembly process consists of three steps (Müller et al. 2007a) (i) initial assembly of four silicatein-α monomers to a tetramer which forms a pentamer by binding a central silicatein-β molecule, (ii) further assembly of the pentamers to fractal-like structures, and (iii) formation of long filaments via the intermediate fractal-like structures formed in step (ii) (Schloßmacher et al. 2011). The morphology of the filaments formed by the recombinant protein strongly resembles that of the natural axial filaments. FT-IR analyses of silicatein protein present in sponge spicules revealed a large percentage of β-sheet conformation (Croce et al. 2004), indicating that the interaction of the silicatein molecule with its silica product might induce a conformational change of the protein.

8.2.3 *Silicatein-Associated Proteins*

The assembly process of silicatein, acting as a template for biosilica deposition, and the structure-directed synthesis of the spicules are controlled by specific silicatein-interacting proteins. Two silicatein interactors have been isolated so far: silintaphin-1 (Müller et al. 2009c) and silintaphin-2 (Wiens et al. 2011; Wang et al. 2012a). The discovery of these silicatein-interacting proteins significantly contributed to the understanding of formation of 3D silica structures (Müller et al. 2009c). The *S. domuncula* proteins are required for the proper assembly, activity, and spatial arrangement of silicatein molecules, as well as for the hardening of the biosilica deposits (Wang et al. 2013c). In addition, galectin is expressed, which forms together with silicatein the "organic cylinders" (Schröder et al. 2006, 2007b) that surround the growing spicule and into which the biosilica is deposited (Fig. 8.7d).

8.2.3.1 Silintaphin-1

The silicatein interactor silintaphin-1 has been identified by application of the yeast two-hybrid system (Wiens et al. 2009). This sponge-specific protein is located within the axial filament, as well as in the organic cylinder surrounding the growing spicules. It comprises an amino acid stretch that shows a significant similarity to the pleckstrin homology (PH) domain (Wiens et al. 2009), a common protein interaction domain (Lemmon and Ferguson 2000). Silintaphin-1 significantly enhances the biosilica-forming activity of silicatein; at a molar ratio of 4:1 [silicatein-α/silintaphin-1], the maximum increase of the biosilica-forming activity was measured (Schloßmacher et al. 2011). In addition, this protein directs the assembly of silicatein filaments. Self-assembly experiments with the recombinant proteins, silicatein-α and silintaphin-1 (molecular ratio 4:1), revealed the formation of fractal-like intermediates, not found in the absence of silintaphin-1, which increase in size and compactness and finally form filaments with lengths of 5 μm and more (Schloßmacher et al. 2011). If these filaments are incubated with orthosilicate

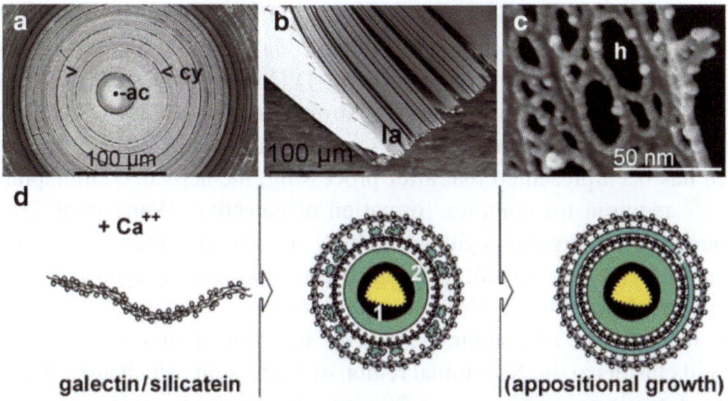

Fig. 8.7 Formation of organic cylinders and appositional growth of biosilica spicules. (**a**) Cross section through the giant basal spicule of the glass sponge *M. chuni* showing the solid axial cylinder (> < cy) surrounded by multiple biosilica lamellae; ac, axial canal (SEM). (**b**) Opening of a stacked layer of lamellae (la). (**c**) Protein scaffold obtained by etching of a lamella with HF vapor, leaving open holes (h). (**d**) Schematic representation of the appositional growth of spicules in the extracellular space. *Left*: Formation of strings/nets by galectin molecules in the presence of Ca^{2+}, to which silicatein molecules bind. *Middle*: Concentric arrangement of the silicatein-galectin strings under formation of organic cylinders around the axial filament and deposition of biosilica between these strings and from the spicule surface. *Right*: A third silica lamella (*3*) is layered around the first two lamellae (*1* and *2*) [Modified after Schröder et al. (2006)]

(prehydrolyzed TEOS), the silicatein fibers organize with the 30-nm-thick silica particles (nanospheres) under formation of densely packed rods (Schloßmacher et al. 2011). These results show that silintaphin-1 not only enhances the enzymatic activity of silicatein but is also crucial for the assembly of silicatein to organized filaments. High-resolution TEM and SEM studies revealed that in vivo during the initial phase of spicule synthesis, within the silicasomes, nanofibrils with a diameter of around 10 nm are formed that comprise bundles of between 10 and 20 nanofibrils (Müller et al. 2013a); see Fig. 8.3. These silicatein bundles condense and progressively form the axial filament. Silintaphin-1 has also been shown to assemble silicatein-coated nanoparticles to 3D structures with a distinct morphology (Müller et al. 2009c).

8.2.3.2 Silintaphin-2

The second silicatein-α interactor, silintaphin-2, has been identified by a solid-phase pull-down assay and yeast two-hybrid library screening (Wiens et al. 2011; Wang et al. 2012a). Silintaphin-2 contains four Ca^{2+}-binding sites and is co-localized with silicatein within the axial filament and the organic cylinder that surrounds the axis of the growing spicules. This organic cylinder functions as a scaffold for the radial apposition of new silica layers during the growth of the spicules in the extracellular space (for a schematic representation, see Fig. 8.7d).

Silintaphin-2 is processed from a longer-sized 15-kDa precursor to a truncated, shorter-sized 13-kDa calcium-binding protein via proteolytic cleavage, mediated by the bone morphogenetic protein-1 (BMP-1) (Müller et al. 2011a). The expression of the gene encoding this protease is strongly upregulated by retinoic acid that regulates the formation of the organic cylinder (Müller et al. 2011a). Experimental evidence has been presented that after processing, the activated silintaphin-2 provides Ca^{2+} required for complex formation of galectins (Wang et al. 2010), one component of the organic cylinder, which in turn polymerize and allow the silicatein molecules to interact, leading to the appositional layering of silica and radial growth of the spicules (Müller et al. 2011a).

More recently, the silintaphin-2 molecule has been shown to comprise anionic amino acid clusters in the N-terminal region of the protein, which are rich in aspartic acid and glutamic acid (Wang et al. 2013c). The 22-mer peptide corresponding to this region (containing eight Asp and four Glu residues) was found to significantly enhance the silicatein-driven bio-silicification process (Wang et al. 2013c). Analyses using inductively coupled plasma-atomic emission spectroscopy (ICP-AES) and energy-dispersive X-ray spectroscopy (EDX) revealed that newly formed spicules contain relatively high concentrations of sodium and potassium (~1 w/w% and 0.3 % w/w%, resp.; Wang et al. 2013c). About 80 % of these alkali metals are removed during maturation of the spicules. This process is accompanied by the removal of water. The resulting progress in the silicatein-driven polycondensation process (decrease in the $Q^2/Q^3/Q^4$ ratios, describing an increase in siloxane bond formation) has been explained by the exchange of the alkali ions, Na^+ and K^+, from the enzymatic biosilica product to the negatively charged amino acids, Asp and Glu, of the silintaphin-2 molecule (Wang et al. 2013c).

Further results revealed that besides these proteins also cellular processes are involved in the axial growth of spicules (Wang et al. 2011c). Using the primmorph system (sponge 3D cell cultures), it has been demonstrated that evaginations of the spicule-forming cells (sclerocytes) into the axial canal of the growing and elongating spicules exist (Wang et al. 2011c). The experiments showed that, around a cell extension protruding into the axial canal, silicatein molecules are released from storage vesicles (silicasomes; Schröder et al. 2007b) into the space between the cell membrane and the inner surface of the silica mantel that surrounds the axial canal and catalyze biosilica deposition at the inner surface (Wang et al. 2011c). It has been concluded that spicule formation comprises two processes (1) biosilica deposition mediated by silicatein within the axial canal and (2) appositional growth through layer-by-layer deposition of new silica lamellae from the outer surface of the growing spicule.

8.2.3.3 DUF Proteins

A further scaffolding protein, termed DUF, has been identified in the proteinaceous matrix ("spongin"; Garrone 1978) that surrounds the siliceous spicules

(Wang et al. 2010). Analyses of the amino acid sequence of protein fragments isolated from this matrix from the demosponge *L. baicalensis* by MALDI-TOF-MS allowed the design of degenerate primers that had been used for screening an *L. baicalensis* cDNA library (Müller et al. 2006b). A series of highly related sequences [DUF, "domains of unknown function"; Novatchkova et al. (2006)] have been isolated, with three pronounced hydrophobic domains, each spanning around 35 amino acids (Wang et al. 2010), indicating that spongin is not related to collagen (Gross et al. 1956; Aouacheria et al. 2006). The recombinant DUF proteins might provide a further building block that can be used, in addition to the silintaphins, for the design of novel nanostructured biosilica materials.

8.3 Biosilica Maturation

8.3.1 Hardening

The initially formed biosilica product of the enzymatic silicatein-catalyzed reaction is a soft, gel-like material that must undergo a hardening process during spicule formation by removal of water (syneresis), resulting in shrinkage of the silica network (Wang et al. 2012c); for a schematic outline of the principle, see Fig. 8.8. During this process the biosilica becomes solid and gains the characteristic spicule morphology. The silicatein and some other associated proteins remain embedded in the biosilica, forming a bioinorganic hybrid material (Müller et al. 2008f, 2010b; Wang et al. 2010). In experiments using *S. domuncula* primmorphs, it has been shown that the process of syneresis, the hardening of the biosilica, involves the removal of water that is released during the polycondensation reaction, via cell-membrane-associated aquaporin channels (Müller et al. 2011c). This process can be inhibited in the presence of Mn sulfate. This metal salt causes a downregulation of expression of the aquaporin-8 gene and thereby strongly affects the morphology and the hardness of the spicules (Wang et al. 2011a; Müller et al. 2011c).

This process can be mimicked in vitro by the addition of poly(ethylene glycol) (PEG) during the silicatein reaction. Addition of PEG not only increases the product formation by the recombinant protein but also solidifies the biosilica. Nanoindentation analyses revealed that the elastic modulus of the biosilica product formed during silicatein reaction increases after the addition of silintaphin-1/PEG from 17 MPa (silicatein alone) via 61 MPa (silicatein/silintaphin-1 ratio of 4:1) to 101 MPa (presence of silicatein/silintaphin-1 and PEG) (Schröder et al. 2012a; Wang et al. 2012c). More recently, the accelerating effect of PEG on the silicatein-mediated polycondensation reaction has been demonstrated by applying a two-phase system (Wang et al. 2013a). Moreover, a spicule-binding nidogen-like protein has been identified in *S. domuncula* that causes an increased sol-gel transition of biosilica (Wang et al. 2012c, 2013a) and might be of interest for the

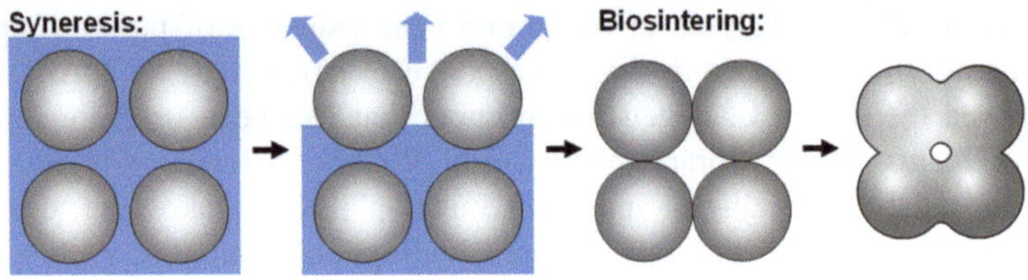

Fig. 8.8 Schematic presentation of the two major steps during maturation of the biosilica spicules, the processes of syneresis (removal of water; *blue*) and biosintering. The process of biosintering results in a reduction of the pore size and a shrinkage of the biosilica material

fabrication of biomimetic silica structures. The recombinant protein was found to induce gelation of biosilica through a polymerization-induced phase separation process.

8.3.2 Precision Biosilica Molding

The fabrication of many devices in nano-optics and microelectronics, especially optical devices, requires precision silica molding processes, but the technical procedures applied usually need acidic pH and high-temperature conditions. There is a need for methods for the production and processing of formable "soft" silica, allowing a molding at ambient conditions, not possible by current techniques. Nature can provide a solution for fabrication of such semisolid silica. We have shown that (1) siliceous sponges are capable of synthesizing biosilica at low precursor (silicic acid) concentrations and ambient conditions enzymatically under formation of a water-rich product; (2) the enzymatically formed biosilica is initially soft and subsequently undergoes a progressive aging/hardening process by removal of water (syneresis) mediated through cellular aquaporin channels (Wang et al. 2011a; Müller et al. 2011c); and (3) the spicules obtain their final morphology by "precision biosilica molding," where the fluid-elastic silica is pressed into a preformed organic casting mold consisting of collagen/galectin (Schröder et al. 2006) or DUF (Wang et al. 2010); during this step biosilica undergoes a hardening process. The "precision biosilica molding" is illustrated in Fig. 8.9a. The formation of precise and tight-fitting connections between individual spicules as a result of "precision biosilica molding" can be seen, in particular, in the demospongian genus Discodermia, e.g., in *Discodermia polydiscus*; the individual spicules, with their axial filaments, remain always separated (Fig. 8.9b). Also the knobs at the blunt ends of the uniradiate pointed spicules of *S. domuncula* are most likely formed by "precision biosilica molding" (Fig. 8.9c). "Precision biosilica molding" is a process only found in the siliceous sponges; all other metazoans form their skeletons differently, e.g., mammalian bone by deposition of hydroxyapatite crystals along an organic template.

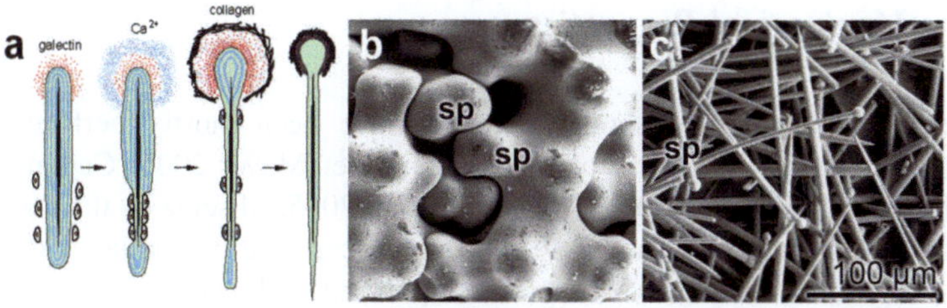

Fig. 8.9 Precision biosilica molding. (**a**) Schematic presentation. The initial biosilica product formed by silicatein is a soft matter material. This material is pressed into a mold which is formed (1) by galectin molecules which are linked together through Ca^{2+} bridges and (2) by collagen fibrils which stabilize the galectin mold. The biosilica material pressed into this mold becomes then hardened by removal of water (syneresis). (**b**) This process is impressively seen in the spicular skeleton of the demosponge *Discodermia polydiscus*. The spicules (sp) closely fit to each other but remain individually separated (SEM). (**c**) Tylostyle spicules (sp) from *S. domuncula* with one pointed end and a knob at the other end, most likely formed by precision molding [Modified after Wang et al. (2012c)]

These principles found in nature, allowing precision biosilica molding at a nano- and micrometer scale under ambient conditions, might be exploited in future bio-inspired approaches, by applying (1) the recombinant silicatein to obtain a water-rich fluid-elastic biosilica product; (2) a synthetic polymer, PEG, mimicking aquaporin function for hardening of the "soft" matter biosilica; and (3) microfluidics techniques substituting the organic mold formed in the living model by collagen or DUF (block-polypeptide) filament networks.

8.3.3 Biosintering

The fusion of the silica nanospheres within the lamellae of the hexactinellid spicules occurs by a sintering-like process ["biosintering"; Müller et al. (2009b, 2010b)]; for a schematic representation, see Fig. 8.8. Biosintering is even more evident during maturation of demosponge spicules which fuse completely (Müller et al. 2009b), in contrast to hexactinellid spicules (Fig. 8.7a, b); individual lamellae are only seen at an early stage of spicule formation, which involves the formation of concentrically arranged string-/netlike silicatein-galectin assemblies around the axial filament (Schröder et al. 2006); see Fig. 8.7d. The spicular biosilica consists of nanospheres with a diameter of 2.8 nm (Woesz et al. 2006), which are most likely the product of the silicatein reaction (Morse 1999) and obviously remain in close contact with those proteins during formation of larger nanospheres, which are then assembled and fused to the appositionally layered lamellae (Wang et al. 2010). In vitro experiments with isolated components from *M. chuni* revealed that biosintering is most likely driven by silicatein which remains within the biosilica matrix after silica formation (Müller et al. 2009b).

8.3.4 Mechanical Properties

Biomechanical studies of sponge spicules have been mostly performed with hexactinellid spicules because of their larger size (Mayer 2005; Cattaneo-Vietti et al. 1996; Müller et al. 2006c; Aizenberg et al. 2005; Miserez et al. 2008; Wang et al. 2009a). These spicules consist of multiple concentric lamellae, surrounding the axial cylinder/axial canal, which are not fused with each other, whereas the spicules from demosponges are rodlike structures consisting of multiple fused lamellae. Therefore, different mechanical as well as optical properties in hexactinellid spicules and demosponge spicules can be expected.

Nanoindentation experiments revealed that the lamellar region of a giant basal spicule from *M. chuni* has an averaged hardness of 2.7 GPa and an averaged modulus of 38.6 GPa. In 3-point bending tests, cracks do not propagate straight through the spicule, but rather traverse several lamellae, along the interface between them; a strength of 800 MPa and an elastic modulus of 25 GPa have been determined for the *M. chuni* spicules (Müller et al. 2013a).

In addition to the lamellar structure, the presence of organic components (proteins) in the biosilica matrix (Fig. 8.7c) is expected to significantly contribute to the mechanical properties of the spicule/biosilica materials. Nevertheless, the inorganic component of the biosilica present in the spicules is almost as pure as quartz glass (Müller et al. 2008c). Studies on spicules of *M. chuni* using laser ablation ICP-MS revealed that trace elements contribute to the total inorganic components of the spicules to less than 0.005-fold with respect to Si (Müller et al. 2008c). This quartz glass-like purity of the sponge silica is most likely the result of two selectivity barriers provided by (1) the energy-dependent Na^+/HCO_3^- ($Si(OH)_4$) cotransporter which pumps $Si(OH)_4$ into the sponge cells (Schröder et al. 2004), a process that occurs under consumption of ATP generated by the arginine kinase (Perović-Ottstadt et al. 2005) to maintain the sodium gradient and (2) the substrate specificity of the silicatein. It has been shown that the content of trace elements changes during aging of the silica material (see above). These trace elements/dopants (mainly sodium and potassium) are expected to have a significant influence on the optical properties (refractive index) of the spicules, in addition to the water content.

8.4 Biomedical Effects of Biosilica

Silica is a morphogenetically active inorganic polymer. Therefore, biosilica offers a number of applications in biomedicine. Biosilica-based strategies for prophylaxis and treatment of osteoporosis have been reviewed in a previous volume of this series (Schröder et al. 2011).

8.4.1 Bone Tissue Engineering and Repair

There is an increasing need for (better) bone substitution materials, especially for bone losses around endoprostheses, vertebral compression fractures of osteoporotic patients, complicated fractures, or the jaw around lost or extracted teeth [reviewed in Epple (2007)]. Inorganic polymers that cause morphogenetic effects on bone cells have not yet introduced into clinics. Scaffolds that are based on bioactive glass or β-tricalcium phosphate usually require supplementation with BMP-2 to induce osteoblastogenesis and mineralization (Waselau et al. 2012). Recent results revealed that the morphogenetically active inorganic polymer biosilica is a suitable material for bone tissue engineering that can even be included in the rapid prototyping chain for the production (3D printing) of customized bone replacement materials (Wang et al. 2013b).

In in vitro experiments, we could demonstrate that biosilica, synthesized by silicatein, induces mineralization (hydroxyapatite formation) in human osteoblast-like SaOS-2 cells (Schröder et al. 2005a; Wiens et al. 2010b). It has been experimentally demonstrated that the growth of these cells on a silicatein/biosilica matrix upregulates the expression of bone morphogenetic protein-2 (BMP-2), an inducer of osteoblast differentiation (Nickel et al. 2001), while the expression of tartrate-resistant acid phosphatase (TRAP), a modulator of bone resorption, remains unaffected (Wiens et al. 2010b). In particular, biosilica is capable of modulating the ratio of expression of the genes encoding osteoprotegerin (OPG) and the receptor activator of NF-kB ligand (RANKL) in human osteoblast-like cells, two proteins which are crucial in pathogenesis of osteoporosis [Wiens et al. (2010c); reviewed in Wang et al. (2012b)]. OPG, a decoy receptor for RANKL, inhibits the function of RANKL to stimulate osteoclast differentiation (Nelson et al. 2012). This cytokine scavenges the soluble RANKL and thereby inhibits the RANKL-induced differentiation of osteoclast precursor cells to mature osteoclasts which are functionally active in bone dissolution (Boyce and Xing 2007). Quantitative real-time RT-PCR analyses revealed that biosilica causes an increased transcription of the OPG gene in SaOS-2 cells, while the steady-state expression of RANKL is not affected, resulting in an enhanced OPG/RANKL ratio (Wiens et al. 2010c). These findings which have been confirmed by other groups (Han et al. 2013) suggest that biosilica displays osteogenic activity on osteoblasts (Wang et al. 2012b). In addition, biosilica has been found to induce the release of factor(s) from bone-forming cells that inhibit the differentiation of osteoclasts (Schröder et al. 2012b). First animal experiments revealed that biosilica is biocompatible (Wiens et al. 2010a). A novel Glu-tag has been developed by bioengineering silicatein-α for a targeted application of the silica-polymerizing enzyme to hydroxyapatite nanofibrils (Natalio et al. 2010a).

In first experiments towards the development of biosilica as a bone substitution material suitable for 3D printing, we used this biopolymer to impregnate 3D printed scaffolds composed of Ca sulfate (Link et al. 2013; see also Chap. 9 of this volume). The impregnated biosilica-containing scaffolds were found to induce the expression of OPG and BMP-2 in SaOS-2 cells, as well as the mineralization of these cells. Based on our result that biosilica can be hardened in the presence of the synthetic polymer PEG (Schröder et al. 2012a; Wang et al. 2013a), further studies are ongoing to demonstrate if this morphogenetically active biopolymer can be used as a suitable matrix or component of a matrix for 3D printing.

More recently, we could show that biosilica also displays morphogenetic activity on cells after embedding in a Na-alginate-based hydrogel (Müller et al. 2013c; Schloßmacher et al. 2013). SaOS-2 cells embedded into the hydrogel showed an increased growth and an increased formation of hydroxyapatite nodules after exposure to a mineralization activation cocktail (containing β-glycerophosphate, ascorbic acid, and dexamethasone) if silica was present in the hydrogel. Moreover, in the silica-containing hydrogels, an enhanced expression of the gene encoding BMP-2, as well as a higher expression rate of the genes encoding collagen 1 [COLI; major fibrillar matrix protein; Viguet-Carrin et al. (2006)] and carbonic anhydrase, an enzyme involved in bone formation/dissolution, was found (Müller et al. 2013c). Based on these findings silica-containing alginate hydrogels have been proposed to be suitable as a morphogenetically active matrix for 3D cell printing. This assumption is supported by previous results showing that Na-alginate hydrogels that are formed in aqueous solution after the addition of $CaCl_2$ are both printable and biocompatible (Wüst et al. 2011; Silva-Correia et al. 2012).

In a further study (Müller et al. 2013e), the effect of the inorganic silica supplement added to the polymeric Na-alginate matrix on the expression on a series of further marker genes of bone formation was studied, in addition to BMP-2 and COLI: osteocalcin (OC), type V collagen (COLV), osteopontin (OPN), osteonectin (ON), bone sialoprotein II (BSP), and alkaline phosphatase (ALP).

OC is a peptide hormone formed by osteoblasts (Karsenty and Ferron 2012) that inhibits bone formation (Ducy et al. 1996). COLI, COLV, ALP, OPN, ON, and BSP are functional and structural proteins required for bone formation that are expressed during differentiation of osteoprogenitor cells to osteoblasts. COLI and COLV are the dominant structural proteins in bone (Bernhardt et al. 2010). The head-to-tail linear polymers formed by COLV molecules are laterally linked to COLI fibrils (Niyibizi and Eyre 1994). OPN, also termed bone sialoprotein I (BSP-1), is a soluble glycoprotein that is formed by pre-osteoblasts, osteoblasts, osteocytes, as well as fibroblasts (Sodek et al. 2000) and can bind to hydroxyapatite through its polyaspartic acid residues. ON is a glycoprotein that is synthesized by osteoblasts and binds to hydroxyapatite and collagen, as well as to sodium ions (Termine et al. 1981). BSP, a major non-collagenous protein in bone tissue, has been shown to stimulate the migration of osteoprogenitor cells, under formation of a triple complex with matrix metallopeptidase 2 (MMP-2) and integrin (Karadag and Fisher 2006). Finally, ALP is an enzyme that activates calcification and is found in those bone regions showing highest ossification (Whyte 2001; Price et al. 2009; Müller et al. 2011b).

In parallel to the transcript levels of these marker proteins, the steady-state expression level of Runt-related transcription factor 2 (RUNX2) was determined. RUNX2 is expressed in cells belonging to the osteoblastic lineage and is a master regulatory transcription factor associated with osteoblast differentiation (Fujita et al. 2004) which is in turn under the control of fibroblast growth factor 2 and noggin (Kalajzic et al. 2003) that interacts with BMPs in an antagonistic manner during the initial differentiation of the osteoprogenitor cells (Gazzerro et al. 1998).

The results revealed that silica present in the hydrogel does not affect the expression of RUNX2 but causes an upregulation of BMP-2 expression (Müller et al. 2013e). In addition, the expression studies revealed a significant increase in the steady-state transcript levels of the genes encoding the collagenous scaffold proteins, COLI and to a lower extent COLV, as well as the OPN and ON genes. OPN may be involved in the mineralization process through its property to bind calcium and collagen (Chen et al. 1992) and ON through binding to collagen (Termine et al. 1981; Ishizeki et al. 2009). On the other hand, silica does not change the transcript levels of OC and BSP. The latter result is in line with the finding that the expression of the *OC* gene (Zhou et al. 2006) and likewise the expression of the *BSP* gene are primarily under the control of RUNX2 (Takagi et al. 2004). It has been concluded that silica causes its morphogenetic effect with respect to some bone-specific genes (*COLI*, *COLV*, *OPN*, and *ON*) via BMP-2 rather than the BMP-2-independent RUNX2 pathway (Müller et al. 2013e).

8.4.2 Dentistry

Human teeth are composed of four tissues: enamel, dentin, dental pulp, and cementum. Enamel mainly consists of carbonated hydroxyapatite and is the hardest mineralized tissue in the body. The enamel crystals are formed on an organic matrix, which is produced by the ameloblasts. Crystal formation is mainly controlled by the enamel matrix proteins amelogenin, ameloblastin, and enamelin (Bartlett et al. 2006). The amelogenins are involved in the initial mineralization process. They are secreted from the ameloblasts and undergo posttranslational modification under formation of aggregates (Fincham et al. 1995). The 15-nm large nanospheres formed by assembly of 100 to 200 amelogenin molecules (Wen et al. 2001) guide the spacing and the orientation of enamel crystallites (Aoba 1996; Fincham et al. 1999; Fig. 8.10). Among the non-amelogenin proteins, ameloblastin controls the elongation of the enamel crystals and serves as a sheet protein to stabilize the Tomes processes to the enamel matrix (Uchida et al. 1998; Fukumoto et al. 2004; Bartlett et al. 2006). Enamelin binds to enamel crystallites and is required for correct mineralization of the enamel matrix (Masuya et al. 2005). The enamel crystals are tightly packed to rodlike structures. The underlying dentin is less mineralized. Dentin contains canals, so-called dentinal tubules, with diameters in the μm range. These tubules radiate outward to the dentin/enamel border and may cause, by exposure of nerves located at the pulpal aspect, dentin hypersensitivity.

The high content of carbonated hydroxyapatite renders teeth susceptible to demineralization and is responsible for the development of dental caries, one of

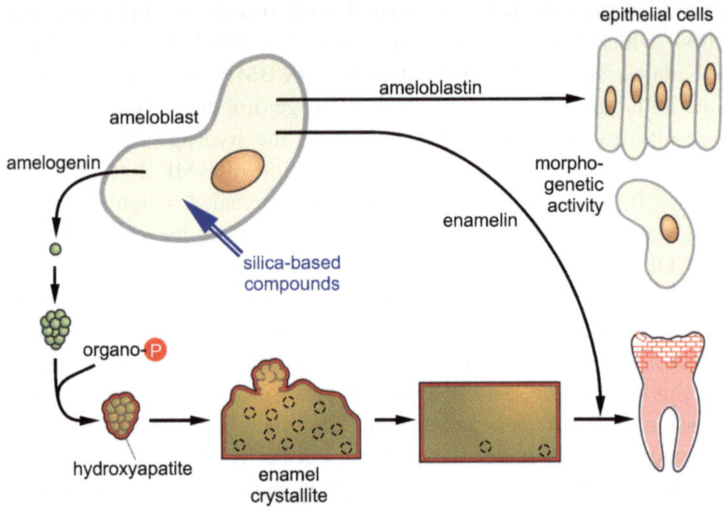

Fig. 8.10 Schematic outline of enamel formation and the effect of silica-based components. Exposure to silica-based components causes a strong upregulation of *amelogenin* gene expression. The amelogenin molecules, after posttranscriptional modification, form nanospheres around which, in the presence of a phosphate source, hydroxyapatite is deposited. In the presence of enamelin which is likewise upregulated by the silica-based components, the hydroxyapatite crystallites aggregate to larger entities. Ameloblastin shows morphogenetic activity and controls the differentiation state of ameloblasts. The expression of this protein formed in ameloblasts, is not changed in the presence of the silica-based components [Modified after Müller et al. (2007c)]

the most common diseases worldwide. Caries is caused by acid-producing bacteria, which metabolize carbohydrates. Two groups of bacteria are responsible for initiating caries, *S. mutans* and *Lactobacillus*. Pits, fissures, and grooves in the enamel layer facilitate caries formation.

Enzymatically formed silica (biosilica) has been proposed to be of interest for dental applications, for example, as a protective dental layer to prevent caries formation (Natalio et al. 2010a). Thereby binding of silica to the hydroxyapatite surface could be achieved through the protein (silicatein) component. The silica layer formed by the application of silicatein, which binds to the dental surface, could be used for sealing of dental pits, fissures, and dentinal tubules, to prevent caries formation, and to reduce dental hypersensitivity.

The effect of biosilica and other silica-based components on the expression level of the genes encoding amelogenin, ameloblastin, and enamelin has been determined in SaOS-2 cells (Müller et al. 2007c). The results revealed that these silica-based materials cause a strong upregulation of the expression level of *amelogenin* and *enamelin*, while the expression of *ameloblastin* remains unchained. In contrast to amelogenin and enamelin, ameloblastin is a cell adhesion molecule that regulates the differentiation state of the ameloblasts (Fukumoto et al. 2005). In parallel to the increase in *amelogenin* and *enamelin* gene expression, an increase in ß-glycerophosphate-induced mineralization is observed (Müller et al. 2007c).

8.4.3 Cell Encapsulation

The development of methods for bioencapsulation of enzymes or even whole cells has gained increasing interest in biotechnology and biomedicine. Bioencapsulated molecules or cells can be used, for example, as bioreactors, biosensors, delivery systems, or for tissue engineering. So far, the sol-gel technique has mainly been used to immobilize cells or living bacteria in silica. Two methods are used: (1) hydrolysis and condensation of metal alkoxides and (2) use of the ability of aqueous metal salts for gel formation (Brinker and Scherer 1990; Coradin and Livage 2007). The discovery of the silicateins now makes it possible to produce silica ("biosilica") biocatalytically under physiological (mild) conditions. It would therefore be an attractive method to apply the technique of enzyme-catalyzed (silicatein-mediated) biosilica synthesis for the production of silica-encapsulated bacterial biosensors and bioreactors. Such novel bacterial bioreactors might, for example, be used for the production of industrially relevant recombinant proteins. It has been shown that the heterologous expression of silicatein on the surface of bacteria transformed with a silicatein cDNA can be applied for the synthesis of porous biocompatible silica shells surrounding the bacteria (Müller et al. 2008b). It is expected that the advantage of such a biosilica encapsulation is an easier handling of the bacteria and an increase in their mechanical and chemical stability. This will facilitate their industrial applicability as bacterial bioreactors or biosensors (see Sect. 8.5.2). In addition, the encapsulation of bacterial cells in biosilica has the advantage that the enzymatically formed silica is flexible and can be partly degraded and resynthesized, enabling a remodeling of such shells, because of the fact that the embedded silicatein molecules show both silica polymerase and silica esterase activity. Therefore, the encapsulated bacteria do not show any impairment of their growth kinetics (Müller et al. 2008b).

8.4.4 Antifouling

Biofouling has an enormous impact in the biomedical field and may concern medical implants like vascular stents, catheters, prosthetic joints, or heart valve implants. The rate of biomaterial-associated infections for initially inserted implants may range up to 30 % and the mortality risk up to 25 % (Darouiche 2001). Thrombosis is one of the fatal events that as a consequence of biofouling of medical implants can occur. Biofilms formed by microbial cells are often resistant against antibiotic treatment. They are the cause of many persistent infections, such infections of the urinary tract, dental plaques, or respiratory diseases (Costerton et al. 1999; Parsek and Singh 2003). Several strategies have been developed for the design of surfaces that are able to prevent fouling [for a review, see Banerjee et al. (2011)]. Biofouling is also of great concern in many other areas, e.g., in food industry (Kumar and Anand 1998), in water systems (Flemming 2002), and in particular in the marine environment (fouling of ship hulls) (Schultz 2007).

Biofouling is a multistep process that initially involves the formation of a conditioning film by adsorption of biomacromolecules, such as proteins, carbohydrates, and glycoproteins. This conditioning film then facilitates the attachment of bacteria and other microorganisms, leading to biofilm formation (Banerjee et al. 2011). The bacteria communicate with each other through a "quorum-sensing" network. In the marine milieu, at a later step, macrofouling organisms, e.g., barnacles, can settle on this surface (Müller et al. 2013g).

In the biofilm layer, the microorganisms are embedded in a gel-like matrix, consisting of extracellular polymeric substances, secreted by them, which provide them with an increased resistance to biocides.

Major antifouling strategies are based on either the prevention of the attachment of bacteria to surfaces or killing them. Prevention of adhesion can be achieved, for example, by functionalization of surfaces with PEG, one of the most used polymers in nonadhesive coatings. Killing of biofouling organisms can be achieved by antifouling coatings that release biocides, such as antibiotics, silver, or quaternary ammonium salts. Due to the development of pathogenic strains showing resistances to antibiotics or silver, and the requirement that, in particular for medical applications, antifoulants should be nontoxic, the development of alternative strategies is urgently needed.

In nature, the siliceous sponges might provide such a solution. These animals have evolved environmentally friendly antifouling strategies that can be used as a model for novel bio-inspired approaches to prevent fouling processes.

Sponges protect themselves from biofouling by a dynamic surface coating consisting of a dynamic biosilica shield (Müller et al. 2013g). This biosilica layer that covers the sponge surface is formed despite the relatively low concentration of orthosilicate in seawater (3–180 μM). The explanation is that this biosilica layer is formed by an enzymatic process, mediated by silicatein, and this low concentration is high enough for this enzyme to perform the enzymatic reaction (Müller et al. 2008d). The biocatalyst, silicatein, catalyzing this reaction has been shown to exist not only in the spicules but also in the extra-spicular space, including the subsurface tissue layer of sponges (Müller et al. 2005).

This biosilica coating has the property of adaptation to changing environmental conditions. The morphology of the sponge surface biosilica layer and its adaptive nature makes this surface highly repellant to physical adsorption of organic particles and hence prevents conditioning film formation. It is assumed that in the marine environment, this dynamic biosilica layer can prevent, for example, the formation of adhesive bonds by mussel or barnacle cement that occurs via adhesive L-3,4-DOPA-containing proteins.

Recent results obtained in in vitro studies using *S. domuncula* revealed that during biosilica aging the processing of the soft matter biosilica to the gelatinous semihard material is driven by a nidogen-like/mucin-like protein (Wang et al. 2012c, 2013a). It was found that the recombinant sponge protein promotes phase separation and in turn hardening of biosilica (Wang et al. 2013a). In addition, it has been demonstrated that biosilica, especially if supplemented with PEG to induce phase separation, causes cell death of bacteria (Müller et al. 2013g). Based on these data it has been concluded that biosilica does not only act as a protective coating of the sponge surface but also inhibits the growth of biofilm-producing bacteria (Müller et al. 2013g).

Moreover, the biosilica coating together with the hedgehog-like spination by outward-radiating spicules generates a suitable surface roughness and spike spacing formation that do not allow cell surface receptors, e.g., integrins, to attach (Dalsin and Messersmith 2005). Hence, sponges might provide an excellent model for the design of microtextured antifouling surfaces.

8.5 Nanotechnological Application

Silicatein is capable of catalyzing not only the synthesis of silica but also the synthesis of other metal oxides, including titania (TiO_2) and zirconia (ZrO_2), in the presence of appropriate water-stable precursors (Tahir et al. 2005). Such metal oxides are of great interest because of their semiconductor, piezoelectric, dielectric, and/or electrooptic properties. It has even been demonstrated that titania formed from titanium bis(ammonium-lactato) dihydroxide can be incorporated into growing spicules in the 3D sponge tissue culture system (*S. domuncula* primmorphs), as revealed by STEM-EDX analysis (Natalio et al. 2010b).

Mimicking the biological model (formation of silica lamellae), a series of biomimetic studies have been performed, based on the application of silicatein. Silicatein can be used for the preparation of silica/metal oxide coatings on various surfaces. For example, this enzyme can be provided with an affinity tag, e.g., a His-tag; more recently, also other tags have been used, e.g., a Glu-tag, allowing silicatein binding to calcium phosphate surfaces (Wiens et al. 2010b). The His-tag enables the silicatein molecule to bind to the surface via Ni^{2+} complexation with appropriate nitrilotriacetic acid (NTA)-containing linkers. Following this approach, immobilization of His-tagged silicatein on gold surfaces has been achieved by applying NTA alkanethiol (Tahir et al. 2004) or an ω-terminated NTA amine which binds at one end to silicatein and at the other end to a reactive ester polymer which specifically reacts with primary amines, e.g., on amino-functionalized gold or other surfaces (Tahir et al. 2005). Such reactive ester polymers have also been used to immobilize silicatein onto silica surfaces, e.g., spicules, after surface functionalization with 3-aminopropyltriethoxysilane (Schröder et al. 2007b). The immobilized silicatein remains enzymatically active (Tahir et al. 2004, 2005). Silicatein immobilized onto the inner walls of nanopores via Ni^{2+} complexation, using NTA bound onto the pore surface via carbodiimide coupling, has recently been used for measuring pH-dependent current-voltage curves to determine the isoelectric point of the protein (Tahir et al. 2013).

Surface functionalization of metal oxides (TiO_2 and Fe_2O_3) can also be achieved by applying a polymeric ligand carrying dopamine (catechol) anchor groups to attach the polymer onto metal oxide surfaces and NTA groups to immobilize His-tagged silicatein (Tahir et al. 2006b). This method has been used for silicatein immobilization onto TiO_2 nanowires (Tahir et al. 2006b) and γ-Fe_2O_3 nanoparticles (Shukoor et al. 2007a, b, 2008a, b, c).

Nanotubes and nanowires are important building blocks used in nanotechnology. The preparation of Au-decorated TiO_2 nanowires functionalized with His-tagged silicatein has been reported (Tahir et al. 2006b). Immobilized silicatein has also been used to generate TiO_2 coatings on functionalized WS_2 chalcogenide nanotubes (Tahir et al. 2009).

Using His-tagged silicatein-α immobilized on glass, dense tin dioxide (SnO_2) films have been generated biocatalytically in the presence of a water-stable tin precursor (Na_2SnF_6) (André et al. 2011b, 2013). Such SnO_2-coated surfaces are applied for transparent semiconductors. Binding of silicatein-α to the glass surface has been achieved through Ni^{2+} complexation, after treatment of the surface with an epoxide-terminated silicane and, further, amine-terminated NTA.

The discovery of the scaffolding proteins, e.g., silintaphin-1, substantially extends the applicability of silicatein in biomimetic approaches. The combined action of silicatein and silintaphin-1 is expected to allow the generation of a great variety of hierarchically structured materials including optical fibers with advantageous mechanical/optical property combinations (Müller et al. 2009c).

Silicatein-α has also been recently used as a template for the formation of synthetic non-siliceous spicules, calcitic spicules, mediated by the self-assembly properties of the protein and its specific adhesion to $CaCO_3$ nanocrystals. The generated spicules displayed waveguiding properties, as demonstrated by using free space coupling of a monochromatic laser light (Natalio et al. 2013).

8.5.1 Core-Shell Particles for Human Therapy

Core-shell materials have gained increasing interest because of their importance for many applications in nanotechnology, in particular in nanomedicine. They can be applied, among others, for cell targeting (Liong et al. 2008), as drug delivery agents, e.g., in cancer therapy (Park et al. 2008), for (biomagnetic) protein separation (Shukoor et al. 2007b, 2008b), for pathogen detection (Gu et al. 2006), and in fluorescence or magnetic resonance imaging (MRI) (Schladt et al. 2010a; Jun et al. 2008) or used in photothermal therapy (Kim et al. 2006). The generation of such nanoparticles by applying a biochemical, enzymatic mechanism has become possible only recently [for a review, see Müller et al. (2009c)].

Besides magnetic nanoparticles, consisting of magnetite (Fe_3O_4) and maghemite (γ-Fe_2O_3), manganese oxide (MnO) nanoparticles have gained much clinical interest, in particular as contrast agents (Na et al. 2007).

In the last years, much progress has been achieved in the development of efficient strategies to improve the biocompatibility, solubility, and stability in biological fluids of these nanoparticles by appropriate surface modifications, a prerequisite for their applicability in patients (Tahir et al. 2006a; Frey et al. 2009). For example, an increased half-life of blood circulation of the nanoparticles can be achieved by using PEG-grafted polymers for surface functionalization (Gref et al. 1994; Moghimi et al. 2001; Torchilin 2006). The hydrophilic PEG layer around the nanoparticles protects them against opsonization, i.e., this layer prevents the absorption of opsonin proteins present in the blood serum and the nonspecific uptake of the nanoparticles by macrophages, impairing the therapeutic (drug delivery) efficiency (Owens and Peppas 2006; Prencipe et al. 2009).

Much progress has also been achieved in the development of strategies to enhance the efficiency of nanoparticles for drug delivery to cancer cells, e.g., chemotherapeutic agents bound to injected magnetic nanoparticles can be concentrated at their desired site of action, in the tumor area, with the help of an external magnetic field (Alexiou et al. 2000). Magnetic nanoparticles can also be applied in clinical diagnostics for the detection of pathogens by ligand-receptor interactions (Gu et al. 2003; Sun et al. 2006). Recently, the synthesis of water-soluble MnO nanoparticles functionalized with a dopamine-PEG-protoporphyrin IX ligand has been described (Schladt et al. 2010a). Protoporphyrin IX acts as a photosensitizer. It has been demonstrated that these protoporphyrin IX-tagged MnO nanoparticles can be used as photodynamic therapeutic agents to induce, after internalization, localized apoptosis in human kidney cancer cells (Caki-1) cells (Schladt et al. 2010a). Such nanoparticles could be used to simultaneously detect and treat tumor tissues.

In the recent years, various metal oxide nanoparticles have been synthesized, following a bio-inspired route. André et al. (2011a) succeeded to generate core-shell $TiO_2@SiO_2$ and $TiO_2@ZrO_2$ nanofibers by functionalization of a TiO_2 nanowire with silicatein-α, followed by a co-assembly of silintaphin-1 through the specific interaction of both proteins. Functionalization of the metal oxide has been achieved using Glu-tagged silicatein molecule. Silintaphin-1 promoted the silicatein-mediated formation of a SiO_2 or ZrO_2 layer on the TiO_2 nanowire backbone. Even more complex bio-inspired core-shell materials have been designed, such as polymer-coated Au@MnO nanoflowers (Schladt et al. 2010b) which had been separately functionalized by selective attachment of catechol anchor groups of a multifunctional, fluorescent dye-tagged polymeric ligand to the metal oxide and a thiol-modified and Texas Red-tagged oligonucleotide on the gold core. Several methods for functionalization of the metal oxide nanoparticles with biocatalytically active silicatein have been applied. In a simple one-step procedure, a Glu-tagged silicatein-α molecule, containing eight glutamate residues, has been demonstrated to efficiently bind to metal oxide surfaces via the carboxylic groups (André et al. 2012).

8.5.2 Nanosensors

There is an increasing interest in biosensor techniques that allow a sensitive and fast detection of molecular and cellular interactions in real time. The optical waveguide lightmode spectroscopy (OWLS) sensing technique has been used to demonstrate the silicatein-mediated formation of biosilica on the surface of SiO_2 sensors (Adányi et al. 2013a). This technique allows in situ and label-free studies of surface processes at molecular level. It is based on the measurement of the resonance angle of polarized laser light which is diffracted by a grating and incoupled into a thin waveguide layer (Ramsden 1999). The mass of the bound molecules or cells can be calculated from the change of the resonance angle. The OWLS technique can also be applied for the detection and discrimination of living and damaged bacterial cells (Adányi et al. 2006). The formation of a biosilica shell around *E. coli* transformed with the silicatein gene (Müller et al. 2008b) has been used to develop a bacterial biosensor by immobilizing the bacterial cells at mild conditions in the presence of TEOS on the surface of the SiO_2-containing sensor chips (Adányi et al. 2013b). Applying this mild immobilization technique, a stable *E. coli* layer could be created on the sensor surface. This bacterial biosensor has been successfully used as an effective tool for the detection of various stressors (e.g., hydrogen peroxide), environmental pollutants (pesticides, e.g., carbofuran), and antibiotics (e.g., chloramphenicol) in real-time measurement (Adányi et al. 2013b).

8.5.3 Microelectronics and Nano-optics

The glass-fiber-like spicules can act as optical fibers that transmit light with high efficiency both in vitro (Fig. 8.11a; Cattaneo-Vietti et al. 1996; Aizenberg et al. 2004; Müller et al. 2006c, 2007d; Wang et al. 2009b) and in vivo (Wiens et al. 2010d). They show a striking similarity to technical optical fibers used in telecommunication, impressively demonstrated for the up to 3-m-long giant spicules from deep-sea glass sponges (Fig. 8.11b, c; Wang et al. 2011b, 2012a). Technical optical fibers consist of a high-refractive-index core and a lower-refractive-index cladding. The light is transmitted in the core by total internal reflection. Based on their composite structure and lamellar architecture, sponge spicules have advantages over manufactured optical fibers, including enhanced fracture toughness, low-temperature synthesis, and amenability to doping with diverse organic and inorganic molecules.

It has been proposed that the waveguiding spicular network of sponges might act as a "nervous"-like signal transmission system comprising a luciferase (light-generating enzyme) (Müller et al. 2009a), the spicule (phototransduction system), and a cryptochrome-associated photoreception system (Müller et al. 2010a). The bioluminescence emission spectrum of the luciferase [470–640 nm; Wang et al. (2010)] overlaps with the light transmission spectrum of the spicules (620–1,250 nm); this would allow a sensitive regulation of the coupling of light emission and absorption in sponges (Müller et al. 2010a). Moreover, we have demonstrated that *S. domuncula* is capable of flashing in vivo, i.e., to endogenously produce light (Wiens et al. 2010d).

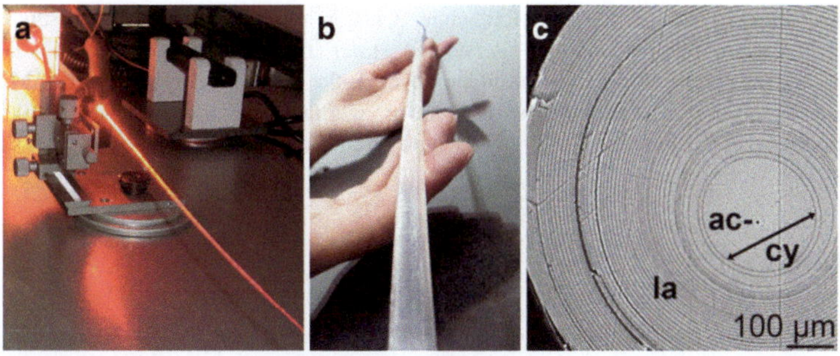

Fig. 8.11 Sponge spicules as light transmitters. (**a**) Light transmission through a giant basal spicule of the hexactinellid *M. chuni*. (**b**) Giant basal spicule of *M. chuni*. (**c**) Cross section through a giant basal spicule of the hexactinellid *H. sieboldi* showing the axial cylinder (cy) and the concentric lamellae (la) that surround the axial canal (ac) [Modified after Wang et al. (2010)]

Recently, we found that *S. domuncula* contains *nocturnin*, a light-/dark-controlled gene that controls the half-life of mRNAs (Müller et al. 2012). Nocturnin is a poly(A)-specific 3′ exoribonuclease, an enzyme previously discovered by us (Schröder et al. 1980). The sponge enzyme specifically degrades the poly(A) tract in glycogenin mRNA, a crucial enzyme in glycogen metabolism. qPCR analyses revealed that in sponges the expression of *nocturnin* is upregulated in the dark and undergoes decay during light exposure. The expression of *glycogenin* is inversely correlated to *nocturnin*. It is concluded that sponges are provided with the molecular circadian clock mechanism (Müller et al. 2012).

First steps have been undertaken to use the recombinant biosilica-forming enzymes as a tool to fabricate optical components under physiological processing conditions. By application of the silicatein-mediated bio-polycondensation technology, a controlled fabrication of silica layers with electrically insulating properties has already been achieved (Polini et al. 2011; Pagliara et al. 2012). Another example is the successful fabrication of glass fibers/components suitable for confinement of light within waveguides (Polini et al. 2012).

The advantage of the thereby applied bio-inspired technique compared to current techniques is the fact that biosilica molding can be performed at ambient temperatures, e.g., at room temperature or even at temperatures down to zero degrees, neutral pH, and silicate concentrations below 1 mM, while industrially applied processes require temperatures above the glass transition temperature of 350 °C. In addition, it is expected that this technique might allow the formation of miniaturized optical structures with superior surface smoothness and high-volume accuracy that cannot be fabricated industrially.

Acknowledgments W.E.G.M. is a holder of an ERC Advanced Investigator grant (no. 268476 BIOSILICA) as well as of an ERC Proof-of-Concept grant (no. 324564; silica-based nanobiomedical approaches for treatment of bone diseases). This work was supported by grants from the European Commission (large-scale integrating project no. 311848, BlueGenics; project no. FP7-KBBE-2010-4-266033, SPECIAL; project no. PIRSES-GA-2009-246987,

European-Chinese Research Staff Exchange Cluster MarBioTec*EU-CN*), the German Bundesministerium für Bildung und Forschung (International Bureau) (no. CHN 09/1AP – German-Chinese Joint Lab on Bio-Nano-Composites), the Public Welfare Project of Ministry of Land and Resources of the People's Republic of China (grant no. 201011005-06), and the International S&T Cooperation Program of China (grant no. 2008DFA00980).

References

Adányi N, Németh E, Halász A, Szendro I, Váradi M (2006) Application of electrochemical optical waveguide lightmode spectroscopy for studying the effect of different stress factors on lactic acid bacteria. Anal Chim Acta 573–574:41–47

Adányi N, Bori Z, Szendrő I, Erdélyi K, Wang XH, Schröder HC, Müller WEG (2013a) Biosilica-based immobilization strategy for label-free OWLS sensors. Sens Actuators B Chem 177:1–7

Adányi N, Bori Z, Szendrő I, Erdélyi K, Wang X, Schröder HC, Müller WEG (2013b) Bacterial sensors based on biosilica immobilization for label-free OWLS detection. N Biotechnol 30: 493–499

Aizenberg J, Sundar V, Yablon AD, Weaver JC, Chen G (2004) Biological glass fibers: correlation between optical and structural properties. Proc Natl Acad Sci USA 101:3358–3363

Aizenberg J, Weaver JC, Thanawala MS, Sundar VC, Morse DE, Fratzl P (2005) Skeleton of *Euplectella* sp: structural hierarchy from the nanoscale to the macroscale. Science 309: 275–278

Alexiou C, Arnold W, Klein RJ, Parak FG, Hulin P, Bergemann C, Erhardt W, Wagenpfeil S, Lübbe AS (2000) Locoregional cancer treatment with magnetic drug targeting. Cancer Res 60:6641–6648

André R, Tahir MN, Link T, Jochum FD, Kolb U, Theato P, Berger R, Wiens M, Schröder HC, Müller WEG, Tremel W (2011a) Growth of SiO_2 and ZrO_2 particles on TiO_2 nanowires controlled by surface-bound silicatein and silintaphin-1. Chemical mimicry: hierarchical 1D TiO_2@ZrO_2 core-shell structures reminiscent of sponge spicules by the synergistic effect of silicatein-α and silintaphin-1. Langmuir 27:5464–5471

André R, Tahir MN, Schröder HC, Müller WEG, Tremel W (2011b) Enzymatic synthesis and surface deposition of tin dioxide using silicatein-α. Chem Mater 23:5358–5365

André R, Tahir MN, Natalio F, Tremel W (2012) Bioinspired synthesis of multifunctional inorganic and bio-organic hybrid materials. FEBS J 279:1737–1749

André R, Natalio F, Tahir MN, Tremel W (2013) Self-cleaning antimicrobial surfaces by bio-enabled growth of SnO_2 coatings on glass. Nanoscale 5:3447–3456

Aoba T (1996) Recent observations on enamel crystal formation during mammalian amelogenesis. Anat Rec 245:208–218

Aouacheria A, Geourjon C, Aghajari N, Navratil V, Deléage G, Lethias C, Exposito JY (2006) Insights into early extracellular matrix evolution: spongin short chain collagen-related proteins are homologous to basement membrane type IV collagens and form a novel family widely distributed in invertebrates. Mol Biol Evol 23:2288–2302

Armirotti A, Damonte G, Pozzolini M, Mussino F, Cerrano C, Salis A, Benatti U, Giovine M (2009) Primary structure and post-translational modifications of silicatein beta from the marine sponge *Petrosia ficiformis* (Poiret, 1789). J Proteome Res 8:3995–4004

Banerjee I, Pangule RC, Kane RS (2011) Antifouling coatings: recent developments in the design of surfaces that prevent fouling by proteins, bacteria, and marine organisms. Adv Mater 23: 690–718

Bartlett JD, Ganss B, Goldberg M, Moradian-Oldak J, Paine ML, Snead ML, Wen X, White SN, Zhou YL (2006) Protein–protein interactions of the developing enamel matrix. Curr Top Dev Biol 74:57–115

Belikov SI, Kaluzhnaya OV, Schröder HC, Krasko A, Müller IM, Müller WEG (2005) Expression of silicatein in spicules from the Baikalian sponge *Lubomirskia baicalensis*. Cell Biol Int 29: 943–951

Bernhardt A, Thieme S, Domaschke H, Springer A, Rösen-Wolff A, Gelinsky M (2010) Crosstalk of osteoblast and osteoclast precursors on mineralized collagen—towards an *in vitro* model for bone remodelling. J Biomed Mater Res A 95:848–856

Boyce BF, Xing L (2007) Biology of RANK, RANKL, and osteoprotegerin. Arthritis Res Ther 9 (Suppl 1):S1

Brinker CJ, Scherer GW (eds) (1990) Sol-gel science, the physics and chemistry of sol-gel processing. Academic, Boston, MA, pp 1–908

Cattaneo-Vietti R, Bavestrello G, Cerrano C, Sara A, Benatti U, Giovine M, Gaino E (1996) Optical fibres in an Antarctic sponge. Nature 383:397–398

Cha JN, Shimizu K, Zhou Y, Christianssen SC, Chmelka BF, Stucky GD, Morse DE (1999) Silicatein filaments and subunits from a marine sponge direct the polymerization of silica and silicones *in vitro*. Proc Natl Acad Sci USA 96:361–365

Chen Y, Bal BS, Gorski JP (1992) Calcium and collagen binding properties of osteopontin, bone sialoprotein, and bone acidic glycoprotein-75 from bone. J Biol Chem 267:24871–24878

Coradin T, Livage J (2007) Aqueous silicates in biological sol-gel applications: new perspectives for old precursors. Acc Chem Res 40:819–826

Costerton JW, Stewart PS, Greenberg EP (1999) Bacterial biofilms: a common cause of persistent infections. Science 284:1318–1322

Croce G, Frache A, Milanesio M, Marchese L, Causà M, Viterbo D, Barbaglia A, Bolis V, Bavestrello G, Cerrano C, Benatti U, Pozzolini M, Giovine M, Amenitsch H (2004) Structural characterization of siliceous spicules from marine sponges. Biophys J 86:526–534

Dalsin JL, Messersmith PB (2005) Bioinspired antifouling polymers. Mater Today 8:38–46

Darouiche RO (2001) Device-associated infections: a macroproblem that starts with microadherence. Clin Infect Dis 33:1567–1572

Ducy P, Desbois C, Boyce B, Pinero G, Story B, Dunstan C, Smith E, Bonadio J, Goldstein S, Gundberg C, Bradley A, Karsenty G (1996) Increased bone formation in osteocalcin-deficient mice. Nature 382:448–452

Epple M (2007) Biomimetic bone substitution materials. In: Epple M, Baeuerlein E (eds) Biomineralisation: medical and clinical aspects. Wiley-VCH, Weinheim, pp 81–95

Fairhead M, Johnson KA, Kowatz T, Mcmahon SA, Carter LG, Oke M, Liu H, Naismith JH, Van Der Walle CF (2008) Crystal structure and silica condensing activities of silicatein alpha-cathepsin L chimeras. Chem Commun: 1765–1767

Fincham AG, Moradian-Oldak J, Diekwisch TGH, Lyaruu DM, Wright JT, Bringas P, Slavkin HC (1995) Evidence for amelogenin "nanospheres" as functional components of secretory-stage enamel matrix. J Struct Biol 115:50–59

Fincham AG, Moradian-Oldak J, Simmer JP (1999) The structural biology of the developing dental enamel matrix. J Struct Biol 126:270–299

Flemming HC (2002) Biofouling in water systems—cases, causes and countermeasures. Appl Microbiol Biotechnol 59:629–640

Frey NA, Peng S, Cheng K, Sun S (2009) Magnetic nanoparticles: synthesis, functionalization, and applications in bioimaging and magnetic energy storage. Chem Soc Rev 38:2532–2542

Fujita T, Azuma Y, Fukuyama R, Hattori Y, Yoshida C, Koida M, Ogita K, Komori T (2004) Runx2 induces osteoblast and chondrocyte differentiation and enhances their migration by coupling with PI3K-Akt signalling. J Cell Biol 166:85–95

Fukumoto S, Kiba T, Hall B, Iehara N, Nakamura T, Longenecker G, Krebsbach PH, Nanci A, Kulkarni AB, Yamada Y (2004) Ameloblastin is a cell adhesion molecule required for maintaining the differentiation state of ameloblasts. J Cell Biol 167:973–983

Fukumoto S, Yamada A, Nonaka K, Yamada Y (2005) Essential roles of ameloblastin in maintaining ameloblast differentiation and enamel formation. Cells Tissues Organs 181: 189–195

Garrone R (1978) Phylogenesis of connective tissue. Morphological aspects and biosynthesis of sponge intercellular matrix. S Karger, Basel

Gazzerro E, Gangji V, Canalis E (1998) Bone morphogenetic proteins induce the expression of *noggin*, which limits their activity in cultured rat osteoblasts. J Clin Invest 102:2106–2114

Gref R, Minamitake Y, Peracchia MT, Trubetskoy V, Torchilin V, Langer R (1994) Biodegradable long-circulating polymeric nanospheres. Science 263:1600–1603

Gross J, Sokal Z, Rougvie M (1956) Structural and chemical studies on the connective tissue of marine sponges. J Histochem Cytochem 4:227–246

Gu H, Ho PL, Tsang KW, Wang L, Xu B (2003) Using biofunctional magnetic nanoparticles to capture vancomycin-resistant enterococci and other gram-positive bacteria at ultralow concentration. J Am Chem Soc 125:15702–15703

Gu H, Xu K, Xu C, Xu B (2006) Biofunctional magnetic nanoparticles for protein separation and pathogen detection. Chem Commun 2006:941–949

Han P, Wu C, Xiao Y (2013) The effect of silicate ions on proliferation, osteogenic differentiation and cell signalling pathways (WNT and SHH) of bone marrow stromal cells. Biomater Sci 1: 379–392

Ishizeki K, Kagiya T, Fujiwara N, Otsu K, Harada H (2009) Expression of osteogenic proteins during the intrasplenic transplantation of Meckel's chondrocytes: a histochemical and immunohistochemical study. Arch Histol Cytol 72:1–12

Jun YW, Lee JH, Cheon J (2008) Chemical design of nanoparticle probes for high-performance magnetic resonance imaging. Angew Chem Int Ed 47:5122–5135

Kalajzic I, Kalajzic Z, Hurley MM, Lichtler AC, Rowe DW (2003) Stage specific inhibition of osteoblast lineage differentiation by FGF2 and noggin. J Cell Biochem 88:1168–1176

Kaluzhnaya OV, Belikov SI, Schröder HC, Wiens M, Giovine M, Krasko A, Müller IM, Müller WEG (2005) Dynamics of skeleton formation in the Lake Baikal sponge *Lubomirskia baicalensis*. Part II. Molecular biological studies. Naturwissenschaften 92:134–138

Karadag A, Fisher LW (2006) Bone sialoprotein enhances migration of bone marrow stromal cells through matrices by bridging MMP-2 to $\alpha_v\beta_3$-integrin. J Bone Miner Res 21:1627–1636

Karsenty G, Ferron M (2012) The contribution of bone to whole-organism physiology. Nature 481:314–320

Kim J, Park S, Lee JE, Jin SM, Lee JH, Lee IS, Yang I, Kim JS, Kim SK, Cho MH, Hyeon T (2006) Designed fabrication of multifunctional magnetic gold nanoshells and their application to magnetic resonance imaging and photothermal therapy. Angew Chem Int Ed 45:7754–7758

Krasko A, Gamulin V, Seack J, Steffen R, Schröder HC, Müller WEG (1997) Cathepsin, a major protease of the marine sponge *Geodia cydonium*: purification of the enzyme and molecular cloning of cDNA. Mol Mar Biol Biotechnol 6:296–307

Krasko A, Lorenz B, Batel R, Schröder HC, Müller IM, Müller WEG (2000) Expression of silicatein and collagen genes in the marine sponge *Suberites domuncula* is controlled by silicate and myotrophin. Eur J Biochem 267:4878–4887

Kruse M, Müller IM, Müller WEG (1997) Early evolution of metazoan serine/threonine and tyrosine kinases: identification of selected kinases in marine sponges. Mol Biol Evol 14: 1326–1334

Kruse M, Leys SP, Müller IM, Müller WEG (1998) Phylogenetic position of the hexactinellida within the phylum Porifera based on the amino acid sequence of the protein kinase C from *Rhabdocalyptus dawsoni*. J Mol Evol 46:721–728

Kumar CG, Anand SK (1998) Significance of microbial biofilms in food industry: a review. Int J Food Microbiol 42:9–27

Lemmon MA, Ferguson KM (2000) Signal-dependent membrane targeting by pleckstrin homology (PH) domains. Biochem J 350:1–18

Link T, Wang XH, Schloßmacher U, Feng QL, Schröder HC, Müller WEG (2013) An approach to a biomimetic bone scaffold: increased expression of BMP-2 and of osteoprotegerin in SaOS-2 cells grown onto silica-biologized 3D printed scaffolds. RSC Adv 3:11140–11147

Liong M, Lu J, Kovochich M, Xia T, Ruehm SG, Nel AE, Tamanoi F, Zink JI (2008) Multifunctional inorganic nanoparticles for imaging, targeting, and drug delivery. ACS Nano 2:889–896

Masuya H, Shimizu K, Sezutsu H, Sakuraba Y, Nagano J, Shimizu A, Fujimoto N, Kawai A, Miura I, Kaneda H, Kobayashi K, Ishijima J, Maeda T, Gondo Y, Noda T, Wakana S, Shiroishi T (2005) Enamelin (Enam) is essential for amelogenesis: ENU-induced mouse mutants as models for different clinical subtypes of human amelogenesis imperfecta (AI). Hum Mol Genet 14: 575–583

Mayer G (2005) Rigid biological systems as models for synthetic composites. Science 310: 1144–1147

Miserez A, Weaver JC, Thurner PJ, Aizenberg J, Dauphin Y, Fratzl P, Morse DE, Zok FW (2008) Effects of laminate architecture on fracture resistance of sponge biosilica: lessons from nature. Adv Funct Mater 18:1241–1248

Moghimi SM, Hunter AC, Murray JC (2001) Long-circulating and target-specific nanoparticles: theory to practice. Pharmacol Rev 53:283–318

Morse DE (1999) Silicon biotechnology: harnessing biological silica production to construct new materials. Trends Biotechnol 17:230–232

Müller WEG, Krasko A, Le Pennec G, Schröder HC (2003) Biochemistry and cell biology of silica formation in sponges. Microsc Res Tech 62:368–377

Müller WEG, Rothenberger M, Boreiko A, Tremel W, Reiber A, Schröder HC (2005) Formation of siliceous spicules in the marine demosponge *Suberites domuncula*. Cell Tissue Res 321: 285–297

Müller WEG, Belikov SI, Tremel W, Perry CC, Gieskes WWC, Boreiko A, Schröder HC (2006a) Siliceous spicules in marine demosponges (example *Suberites domuncula*). Micron 37: 107–120

Müller WEG, Kaluzhnaya OV, Belikov SI, Rothenberger M, Schröder HC, Reiber A, Kaandorp JA, Manz B, Mietchen D, Volke F (2006b) Magnetic resonance imaging of the siliceous skeleton of the demosponge *Lubomirskia baicalensis*. J Struct Biol 153:31–41

Müller WEG, Wendt K, Geppert C, Wiens M, Reiber A, Schröder HC (2006c) Novel photoreception system in sponges? unique transmission properties of the stalk spicules from the hexactinellid *Hyalonema sieboldi*. Biosens Bioelectron 21:1149–1155

Müller WEG, Boreiko A, Schloßmacher U, Wang XH, Tahir MN, Tremel W, Brandt D, Kaandorp JA, Schröder HC (2007a) Fractal-related assembly of the axial filament in the demosponge *Suberites domuncula*. Biomaterials 28:4501–4511

Müller WEG, Boreiko A, Wang X, Belikov SI, Wiens M, Grebenjuk VA, Schloßmacher U, Schröder HC (2007b) Silicateins, the major biosilica forming enzymes present in demosponges: protein analysis and phylogenetic relationship. Gene 395:62–71

Müller WEG, Boreiko A, Wang X, Krasko A, Geurtsen W, Custódio MR, Winkler T, Lukić-Bilela L, Link T, Schröder HC (2007c) Morphogenetic activity of silica and bio-silica on the expression of genes controlling biomineralization using SaOS-2 cells. Calcif Tissue Int 81:382–393

Müller WEG, Eckert C, Kropf K, Wang X, Schloßmacher U, Seckert C, Wolf SE, Tremel W, Schröder HC (2007d) Formation of giant spicules in the deep-sea hexactinellid *Monorhaphis chuni* (Schulze 1904): electron-microscopic and biochemical studies. Cell Tissue Res 329: 363–378

Müller WEG, Li J, Schröder HC, Qiao L, Wang X (2007e) The unique skeleton of siliceous sponges (Porifera; Hexactinellida and Demospongiae) that evolved first from the Urmetazoa during the Proterozoic: a review. Biogeosciences 4:219–232

Müller WEG, Schloßmacher U, Eckert C, Krasko A, Boreiko A, Ushijima H, Wolf SE, Tremel W, Müller IM, Schröder HC (2007f) Analysis of the axial filament in spicules of the demosponge *Geodia cydonium*: Different silicatein composition in microscleres (asters) and megascleres (oxeas and triaenes). Eur J Cell Biol 86:473–487

Müller WEG, Wang X, Belikov SI, Tremel W, Schloßmacher U, Natoli A, Brandt D, Boreiko A, Tahir MN, Müller IM, Schröder HC (2007g) Formation of siliceous spicules in demosponges: example *Suberites domuncula*. In: Bäuerlein E (ed) Handbook of Biomineralization, vol 1, Biological aspects and structure formation. Wiley-VCH, Weinheim, pp 59–82

Müller WEG, Boreiko A, Schloßmacher U, Wang X, Eckert C, Kropf K, Li J, Schröder HC (2008a) Identification of a silicatein(-related) protease in the giant spicules of the deep sea hexactinellid *Monorhaphis chuni*. J Exp Biol 211:300–309

Müller WEG, Engel S, Wang X, Wolf SE, Tremel W, Thakur NL, Krasko A, Divekar M, Schröder HC (2008b) Bioencapsulation of living bacteria (*Escherichia coli*) with poly(silicate) after transformation with silicatein-α gene. Biomaterials 29:771–779

Müller WEG, Jochum K, Stoll B, Wang XH (2008c) Formation of giant spicule from quartz glass by the deep sea sponge *Monorhaphis*. Chem Mater 20:4703–4711

Müller WEG, Schloßmacher U, Wang XH, Boreiko A, Brandt D, Wolf SE, Tremel W, Schröder HC (2008d) Poly(silicate)-metabolizing silicatein in siliceous spicules and silicasomes of demosponges comprises dual enzymatic activities (silica-polymerase and silica-esterase). FEBS J 275:362–370

Müller WEG, Wang XH, Kropf K, Boreiko A, Schloßmacher U, Brandt D, Schröder HC, Wiens M (2008e) Silicatein expression in the hexactinellid *Crateromorpha meyeri*: the lead marker gene restricted to siliceous sponges. Cell Tissue Res 333:339–351

Müller WEG, Wang XH, Kropf K, Ushijima H, Geurtsen W, Eckert C, Tahir MN, Tremel W, Boreiko A, Schloßmacher U, Li J, Schröder HC (2008f) Bioorganic/inorganic hybrid composition of sponge spicules. J Struct Biol 161:188–203

Müller WEG, Kasueske M, Wang XH, Schröder HC, Wang Y, Pisignano D, Wiens M (2009a) Luciferase a light source for the silica-based optical waveguides (spicules) in the demosponge *Suberites domuncula*. Cell Mol Life Sci 66:537–552

Müller WEG, Wang X, Burghard Z, Bill J, Krasko A, Boreiko A, Schloßmacher U, Schröder HC, Wiens M (2009b) Bio-sintering processes in hexactinellid sponges: fusion of biosilica in giant basal spicules from *Monorhaphis chuni*. J Struct Biol 168:548–561

Müller WEG, Wang X, Cui FZ, Jochum KP, Tremel W, Bill J, Schröder HC, Natalio F, Schloßmacher U, Wiens M (2009c) Sponge spicules as blueprints for the biofabrication of inorganic-organic composites and biomaterials. Appl Microbiol Biotechnol 83:397–413

Müller WEG, Wang XH, Schröder HC, Korzhev M, Grebenyuk V, Markl J, Jochum KP, Pisignano D, Wiens M (2010a) A cryptochrome-based photosensory system in the siliceous sponge *Suberites domuncula* (Demospongiae). FEBS J 277:1182–1201

Müller WEG, Wang X, Sinha B, Wiens M, Schröder HC, Jochum KP (2010b) NanoSIMS: insights into the organization of the proteinaceous scaffold within hexactinellid sponge spicules. Chembiochem 11:1077–1082

Müller WEG, Binder M, von Lintig J, Guo YW, Wang XH, Kaandorp JA, Wiens M, Schröder HC (2011a) Interaction of the retinoic acid signaling pathway with spicule formation in the marine sponge *Suberites domuncula* through activation of bone morphogenetic protein-1. Biochim Biophys Acta 1810:1178–1194

Müller WEG, Wang XH, Diehl-Seifert B, Kropf K, Schloßmacher U, Lieberwirth I, Glasser G, Wiens M, Schröder HC (2011b) Inorganic polymeric phosphate/polyphosphate as an inducer of alkaline phosphatase and a modulator of intracellular Ca^{2+} level in osteoblasts (SaOS-2 cells) *in vitro*. Acta Biomater 7:2661–2671

Müller WEG, Wang XH, Wiens M, Schloßmacher U, Jochum KP, Schröder HC (2011c) Hardening of biosilica in sponge spicules involves an aging process after its enzymatic polycondensation: evidence for an aquaporin-mediated water absorption. Biochim Biophys Acta 1810:713–726

Müller WEG, Wang XH, Grebenjuk VA, Korzhev M, Wiens M, Schloßmacher U, Schröder HC (2012) Nocturnin in the demosponge *Suberites domuncula*: a potential circadian clock protein controlling glycogenin synthesis in sponges. Biochem J 448:233–242

Müller WEG, Mugnaioli E, Schröder HC, Schloßmacher U, Giovine M, Kolb U, Wang XH (2013a) Hierarchical composition of the axial filament from spicules of the siliceous sponge *Suberites domuncula*: from biosilica-synthesizing nanofibrils to structure- and morphology-guiding triangular stems. Cell Tissue Res 351:49–58

Müller WEG, Schröder HC, Burghard Z, Pisignano D, Wang XH (2013b) Silicateins: a paradigm shift in bioinorganic chemistry. enzymatic synthesis of inorganic polymeric silica. Chem Eur J 19:5790–5804

Müller WEG, Schröder HC, Feng QL, Schlossmacher U, Link T, Wang XH (2013c) Development of a morphogenetically active scaffold for three-dimensional growth of bone cells: biosilica-alginate hydrogel for SaOS-2 cell cultivation. J Tissue Eng Regen Med, doi:10.1002/term.1745

Müller WEG, Schröder HC, Muth S, Gietzen S, Korzhev M, Grebenjuk V, Wiens M, Schloßmacher U, Wang XH (2013d) The silicatein-propeptide acts as inhibitor/modulator of self-organization during spicule axial filament formation. FEBS J 280:1693–1708

Müller WEG, Wang XH, Grebenjuk V, Diehl-Seifert B, Steffen R, Schloßmacher U, Trautwein A, Neumann S, Schröder HC (2013e) Silica as a morphogenetically active inorganic polymer. Biomater Sci 1:669–678

Müller WEG, Wang XH, Jochum K, Schröder HC (2013f) Self-healing, an intrinsic property of biomineralization processes. IUBMB Life 65:382–396

Müller WEG, Wang XH, Proksch P, Perry CC, Osinga R, Garderes J, Schröder HC (2013g) Principles of biofouling protection in marine sponges. Design of novel biomimetic and bioinspired coatings in the marine environment? Mar Biotechnol 15:375–398

Murr MM, Morse DE (2005) Fractal intermediates in the self-assembly of silicatein filaments. Proc Natl Acad Sci USA 102:11657–11662

Na HB, Lee JH, An K, Park YI, Park M, Lee IS, Nam DH, Kim ST, Kim SH, Kim SW, Lim KH, Kim KS, Kim SO, Hyeon T (2007) Development of a T1 contrast agent for magnetic resonance imaging using MnO nanoparticles. Angew Chem Int Ed 46:5397–5401

Natalio F, Link T, Müller WEG, Schröder HC, Cui FZ, Wang XH, Wiens M (2010a) Bioengineering of the silica-polymerizing enzyme silicatein-α for a targeted application to hydroxyapatite. Acta Biomater 6:3720–3728

Natalio F, Mugnaioli E, Wiens M, Wang X, Schröder HC, Tahir MN, Tremel W, Kolb U, Müller WEG (2010b) Silicatein-mediated incorporation of titanium into spicules from the demosponge *Suberites domuncula*. Cell Tissue Res 339:429–436

Natalio F, Corrales T, Panthöfer M, Lieberwirth I, Schollmeyer D, Müller WEG, Kappl M, Butt HJ, Tremel W (2013) Flexible minerals: self-assembled calcite spicules with extreme bending strength. Science 339:1298–1302

Nelson CA, Warren JT, Wang MW, Teitelbaum SL, Fremont DH (2012) RANKL employs distinct binding modes to engage RANK and the osteoprotegerin decoy receptor. Structure 20: 1971–1982

Nickel J, Dreyer MK, Kirsch T, Sebald W (2001) The crystal structure of the BMP-2–BMPR-IA complex and the generation of BMP-2 antagonists. J Bone Joint Surg Am 83-A(Suppl 1): S7–S14

Niyibizi C, Eyre DR (1994) Structural characteristics of cross-linking sites in type V collagen of bone. Chain specificities and heterotypic links to type I collagen. Eur J Biochem 224:943–950

Novatchkova M, Schneider G, Fritz R, Eisenhaber F, Schleiffer A (2006) DOUTfinder—identification of distant domain outliers using subsignificant sequence similarity. Nucleic Acids Res 34:W214–W218

Owens DE, Peppas NA (2006) Opsonization, biodistribution, and pharmacokinetics of polymeric nanoparticles. Int J Pharm 307:93–102

Pagliara S, Polini A, Camposeo A, Schröder HC, Müller WEG, Pisignano D (2012) Electrical properties of *in vitro* biomineralized recombinant silicatein deposited by microfluidics. Appl Phys Lett 101:193702

Park H, Yang J, Seo S, Kim K, Suh J, Kim D, Haam S, Yoo KH (2008) Multifunctional nanoparticles for photothermally controlled drug delivery and magnetic resonance imaging enhancement. Small 4:192–196

Parsek MR, Singh PK (2003) Bacterial biofilms: an emerging link to disease pathogenesis. Annu Rev Microbiol 57:677–701

Perović-Ottstadt S, Wiens M, Schröder HC, Batel R, Giovine M, Krasko A, Müller IM, Müller WEG (2005) Arginine kinase in the demosponge *Suberites domuncula*: regulation of its expression and catalytic activity by silicic acid. J Exp Biol 208:637–646

Polini A, Pagliara S, Camposeo A, Biasco A, Schröder HC, Müller WEG, Pisignano D (2011) Biosilica electrically-insulating layers by soft lithography-assisted biomineralisation with recombinant silicatein. Adv Mater 23:4674–4678

Polini A, Pagliara S, Camposeo A, Cingolani R, Wang XH, Schröder HC, Müller WEG, Pisignano D (2012) Optical properties of *in-vitro* biomineralised silica. Sci Rep 2:607

Prencipe G, Tabakman SM, Welsher K, Liu Z, Goodwin AP, Zhang L, Henry J, Dai H (2009) PEG branched polymer for functionalization of nanomaterials with ultralong blood circulation. J Am Chem Soc 131:4783–4787

Price PA, Toroian D, Chan WS (2009) Tissue-nonspecific alkaline phosphatase is required for the calcification of collagen in serum: a possible mechanism for biomineralization. J Biol Chem 284:4594–4604

Ramsden JJ (1999) OWLS: a versatile technique for sensing with bioarrays. Chimia 53:67–71

Schladt TD, Schneider K, Shukoor MI, Natalio F, Tahir MN, Weber S, Schreiber LM, Schröder HC, Müller WEG, Tremel W (2010a) Highly soluble multifunctional MnO nanoparticles for simultaneous optical and MRI imaging and cancer treatment using photodynamic therapy. J Mater Chem 20:8297–8304

Schladt TD, Shukoor MI, Schneider K, Tahir MN, Natalio F, Ament I, Becker J, Jochum FD, Weber S, Köhler O, Theato P, Schreiber LM, Sönnichsen C, Schröder HC, Müller WEG, Tremel W (2010b) Au@MnO nanoflowers: hybrid nanocomposites for selective dual functionalization and imaging. Angew Chem Int Ed 49:3976–3980

Schloßmacher U, Wiens M, Schröder HC, Wang XH, Jochum KP, Müller WEG (2011) Silintaphin-1: interaction with silicatein during structure-guiding biosilica formation. FEBS J 278:1145–1155

Schloßmacher U, Schröder HC, Wang XH, Feng QL, Diehl-Seifert B, Neumann S, Trautwein A, Müller WEG (2013) Alginate/silica composite hydrogel as a potential morphogenetically active scaffold for three-dimensional tissue engineering. RSC Adv 3:11185–11194

Schröder HC, Zahn RK, Dose K, Müller WEG (1980) Purification and characterization of a poly (A)-specific exoribonuclease from calf thymus. J Biol Chem 255:4535–4538

Schröder HC, Perović-Ottstadt S, Rothenberger M, Wiens M, Schwertner H, Batel R, Korzhev M, Müller IM, Müller WEG (2004) Silica transport in the demosponge *Suberites domuncula*: fluorescence emission analysis using the PDMPO probe and cloning of a potential transporter. Biochem J 381:665–673

Schröder HC, Boreiko O, Krasko A, Reiber A, Schwertner H, Müller WEG (2005a) Mineralisation of SaOS-2 cells on enzymatically (silicatein) modified bioactive osteoblast-stimulating surfaces. J Biomed Mater Res Part B Appl Biomater 75B:387–392

Schröder HC, Perović-Ottstadt S, Grebenjuk VA, Engel S, Müller IM, Müller WEG (2005b) Biosilica formation in spicules of the sponge *Suberites domuncula*: synchronous expression of a gene cluster. Genomics 85:666–678

Schröder HC, Boreiko A, Korzhev M, Tahir MN, Tremel W, Eckert C, Ushijima H, Müller IM, Müller WEG (2006) Co-expression and functional interaction of silicatein with galectin: matrix-guided formation of siliceous spicules in *Suberites domuncula*. J Biol Chem 281: 12001–12009

Schröder HC, Brandt D, Schloßmacher U, Wang X, Tahir MN, Tremel W, Belikov SI, Müller WEG (2007a) Enzymatic production of biosilica glass using enzymes from sponges: basic aspects and application in nanobiotechnology (material sciences and medicine). Naturwissenschaften 94:339–359

Schröder HC, Natalio F, Shukoor I, Tremel W, Schloßmacher U, Wang XH, Müller WEG (2007b) Apposition of silica lamellae during growth of spicules in the demosponge *Suberites domuncula*: biological/biochemical studies and chemical/biomimetical confirmation. J Struct Biol 159: 325–334

Schröder HC, Wang XH, Tremel W, Ushijima H, Müller WEG (2008) Biofabrication of biosilica-glass by living organisms. Nat Prod Rep 25:455–474

Schröder HC, Wiens M, Wang X, Schloßmacher U, Müller WEG (2011) Biosilica-based strategies for treatment of osteoporosis and other bone diseases. Prog Mol Subcell Biol 52:283–312

Schröder HC, Wang XH, Manfrin A, Yu SH, Grebenjuk VA, Korzhev M, Wiens M, Schloßmacher U, Müller WEG (2012a) Silicatein: acquisition of structure-guiding and structure-forming properties during maturation from the pro-silicatein to the silicatein form. J Biol Chem 287:22196–22205

Schröder HC, Wang XH, Wiens M, Diehl-Seifert B, Kropf K, Schloßmacher U, Müller WEG (2012b) Silicate modulates the cross-talk between osteoblasts (SaOS-2) and osteoclasts (RAW 264.7 cells): inhibition of osteoclast growth and differentiation. J Cell Biochem 113: 3197–3206

Schröder HC, Wiens M, Schloßmacher U, Brandt D, Müller WEG (2012c) Silicatein-mediated polycondensation of orthosilicic acid: modeling of catalytic mechanism involving ring formation. Silicon 4:33–38

Schultz MP (2007) Effects of coating roughness and biofouling on ship resistance and powering. Biofouling 23:331–341

Shimizu K, Cha J, Stucky GD, Morse DE (1998) Silicatein alpha: cathepsin L-like protein in sponge biosilica. Proc Natl Acad Sci USA 95:6234–6238

Shukoor MI, Natalio F, Ksenofontov V, Tahir MN, Eberhardt M, Theato P, Schröder HC, Müller WEG, Tremel W (2007a) Double-stranded RNA polyinosinic-polycytidylic acid immobilized onto γ-Fe_2O_3 nanoparticles by using a multifunctional polymeric linker. Small 3:1374–1378

Shukoor MI, Natalio F, Tahir MN, Ksenofontov V, Therese HA, Theato P, Schröder HC, Müller WEG, Tremel W (2007b) Superparamagnetic γ-Fe_2O_3 nanoparticles with tailored functionality for protein separation. Chem Commun 2007:4677–4679

Shukoor MI, Natalio F, Metz N, Glube N, Tahir MN, Therese HA, Ksenofontov V, Theato P, Langguth P, Boissel JP, Schröder HC, Müller WEG, Tremel W (2008a) dsRNA functionalized multifunctional γ-Fe_2O_3 nanocrystals: a tool for targeting cell surface receptors. Angew Chem Int Ed 47:4748–4752

Shukoor MI, Natalio F, Tahir MN, Divekar M, Metz N, Therese HA, Theato P, Ksenofontov V, Schröder HC, Müller WEG, Tremel W (2008b) Multifunctional polymer-derivatized γ-Fe_2O_3 nanocrystals as a methodology for the biomagnetic separation of recombinant His-tagged proteins. J Magn Magn Mater 320:2339–2344

Shukoor MI, Natalio F, Therese HA, Tahir MN, Ksenofontov V, Panthöfer M, Eberhardt M, Theato P, Schröder HC, Müller WEG, Tremel W (2008c) Fabrication of a silica coating on magnetic γ-Fe_2O_3 nanoparticles by an immobilized enzyme. Chem Mater 20:3567–3573

Silva-Correia J, Miranda-Gonçalves V, Salgado AJ, Sousa N, Oliveira JM, Reis RM, Reis RL (2012) Angiogenic potential of gellan gum-based hydrogels for application in nucleus pulposus regeneration: *in vivo* study. Tissue Eng Part A 18:1203–1212

Sodek J, Ganss B, McKee MD (2000) Osteopontin. Crit Rev Oral Biol Med 11:279–303

Sun EY, Josephson L, Kelly KA, Weissleder R (2006) Development of nanoparticle libraries for biosensing. Bioconjug Chem 17:109–113

Tahir MN, Théato P, Müller WEG, Schröder HC, Janshoff A, Zhang J, Huth J, Tremel W (2004) Monitoring the formation of biosilica catalysed by histidin-tagged silicatein. Chem Commun 2004:2848–2849

Tahir MN, Théato P, Müller WEG, Schröder HC, Borejko A, Faiß S, Janshoff A, Huth J, Tremel W (2005) Formation of layered titania and zirconia catalysed by surface-bound silicatein. Chem Commun 28:5533–5535

Tahir MN, Eberhardt M, Theato P, Faiss S, Janshoff A, Gorelik T, Kolb U, Tremel W (2006a) Reactive polymers: a versatile toolbox for the immobilization of functional molecules on TiO_2 nanoparticles. Angew Chem Int Ed 45:908–912

Tahir MN, Eberhardt M, Therese HA, Kolb U, Theato P, Müller WEG, Schröder HC, Tremel W (2006b) From single molecules to nanoscopically structured functional materials: Au nanocrystal growth on TiO_2 nanowires controlled by surface bound silicatein. Angew Chem Int Ed 45:4803–4809

Tahir MN, Natalio F, Therese HA, Yella A, Metz N, Shah MR, Mugnaioli E, Berger R, Theato P, Schröder HC, Müller WEG, Tremel W (2009) Enzyme-mediated deposition of a TiO_2 coating onto biofunctionalized WS_2 chalcogenide nanotubes. Adv Funct Mater 19:285–291

Tahir MN, Ali M, Andre R, Müller WEG, Schröder HC, Tremel W, Ensinger W (2013) Silicatein conjugation inside nanoconfined geometries through immobilized NTA–Ni(II) chelates. Chem Commun 49:2210–2212

Takagi M, Kamiya N, Takahashi T, Ito S, Hasegawa M, Suzuki N, Nakanishi K (2004) Effects of bone morphogenetic protein-2 and transforming growth factor beta1 on gene expression of transcription factors, AJ18 and Runx2 in cultured osteoblastic cells. J Mol Histol 35:81–90

Termine JD, Kleinman HK, Whitson SW, Conn KM, McGarvey ML, Martin GR (1981) Osteonectin, a bone-specific protein linking mineral to collagen. Cell 26:99–105

Torchilin VP (2006) Multifunctional nanocarriers. Adv Drug Deliv Rev 58:1532–1555

Uchida T, Murakami C, Wakida K, Dohi N, Iwai Y, Simmer JP, Fukae M, Satado T, Takahashi O (1998) Sheet proteins: synthesis, secretion, degradation and fate in forming enamel. Eur J Oral Sci 106:308–314

Viguet-Carrin S, Garnero P, Delmas PD (2006) The role of collagen in bone strength. Osteoporos Int 17:319–336

Wang XH, Boreiko A, Schloßmacher U, Brandt D, Schröder HC, Li J, Kaandorp JA, Götz H, Duschner H, Müller WEG (2008) Axial growth of hexactinellid spicules: formation of cone-like structural units in the giant basal spicules of the hexactinellid *Monorhaphis*. J Struct Biol 164:270–280

Wang XH, Schröder HC, Müller WEG (2009a) Giant siliceous spicules from the deep-sea glass sponge *Monorhaphis chuni*: morphology, biochemistry and molecular biology. Int Rev Cell Mol Biol 273:69–115

Wang X, Zhang X, Schröder HC, Müller WEG (2009b) Giant basal spicule from the deep-sea glass sponge *Monorhaphis chuni*: synthesis of the largest bio-silica structure on earth by silicatein. Front Mater Sci China 3:226–240

Wang X, Wiens M, Schröder HC, Hu S, Mugnaioli E, Kolb U, Tremel W, Pisignano D, Müller WEG (2010) Morphology of sponge spicules: silicatein a structural protein for bio-silica formation. Adv Eng Mater 12:B422–B437

Wang XH, Schröder HC, Brandt D, Wiens M, Lieberwirth I, Glasser G, Schloßmacher U, Wang S, Müller WEG (2011a) Sponge bio-silica formation involves syneresis following polycondensation *in vivo*. Chembiochem 12:2316–2324

Wang XH, Wiens M, Schröder HC, Jochum KP, Schloßmacher U, Götz H, Duschner H, Müller WEG (2011b) Circumferential spicule growth by pericellular silica deposition in the hexactinellid sponge *Monorhaphis chuni*. J Exp Biol 214:2047–2056

Wang XH, Wiens M, Schröder HC, Schloßmacher U, Pisignano D, Jochum KP, Müller WEG (2011c) Evagination of cells controls bio-silica formation and maturation during spicule formation in sponges. PLoS One 6:e20523

Wang XH, Schloßmacher U, Wiens M, Batel R, Schröder HC, Müller WEG (2012a) Silicateins, silicatein interactors, and cellular interplay in sponge skeletogenesis: formation of the glass fiber-like spicules. FEBS J 279:1721–1736

Wang XH, Schröder HC, Wiens M, Ushijima H, Müller WEG (2012b) Bio-silica and bio-polyphosphate: applications in biomedicine (bone formation). Curr Opin Biotechnol 23: 570–578

Wang XH, Schröder HC, Wang K, Kaandorp JA, Müller WEG (2012c) Genetic, biological and structural hierarchies during sponge spicule formation: from soft sol-gels to solid 3D silica composite structures. Soft Matter 8:9501–9518

Wang XH, Wiens M, Schloßmacher U, Jochum KP, Schröder HC, Müller WEG (2012d) Bio-sintering/bio-fusion of silica in sponge spicules: a review. Adv Eng Mater 14:B4–B12

Wang XH, Schloßmacher U, Schröder HC, Müller WEG (2013a) Biologically-induced transition of bio-silica sol to mesoscopic gelatinous flocs: a biomimetic approach to a controlled fabrication of bio-silica structures. Soft Matter 9:654–664

Wang XH, Schröder HC, Feng QL, Draenert F, Müller WEG (2013b) The deep-sea natural products, biogenic polyphosphate (bio-polyP) and biogenic silica (bio-silica) as biomimetic scaffolds for bone tissue engineering: fabrication of a morphogenetically-active polymer. Mar Drugs 11:718–746

Wang XH, Schröder HC, Schloßmacher U, Jiang L, Korzhev M, Müller WEG (2013c) Biosilica aging: from enzyme-driven gelation via syneresis to chemical/biochemical hardening. Biochim Biophys Acta 1830:3437–3446

Waselau M, Patrikoski M, Juntunen M, Kujala K, Kääriäinen M, Kuokkanen H, Sándor GK, Vapaavuori O, Suuronen R, Mannerström B, von Rechenberg B, Miettinen S (2012) Effects of bioactive glass S53P4 or beta-tricalcium phosphate and bone morphogenetic protein-2 and bone morphogenetic protein-7 on osteogenic differentiation of human adipose stem cells. J Tissue Eng. doi:10.1177/2041731412467789

Wen HB, Fincham AG, Moradian-Oldak J (2001) Progressive accretion of amelogenin molecules during nanospheres assembly revealed by atomic force microscopy. Matrix Biol 20:387–395

Whyte MP (2001) Hypophosphatasia. In: Scriver CR, Beaudet AL, Sly WS, Valle D, Childs B, Kinzler KW, Vogelstein B (eds) The metabolic and molecular bases of inherited diseases. McGraw-Hill Inc., New York, NY, pp 5313–5329

Wiens M, Belikov SI, Kaluzhnaya OV, Krasko A, Schröder HC, Perovic-Ottstadt S, Müller WEG (2006) Molecular control of serial module formation along the apical-basal axis in the sponge *Lubomirskia baicalensis*: silicateins, mannose-binding lectin and mago nashi. Dev Genes Evol 216:229–242

Wiens M, Belikov SI, Kaluzhnaya OV, Adell T, Schröder HC, Perovic-Ottstadt S, Kaandorp JA, Müller WEG (2008) Regional and modular expression of morphogenetic factors in the demosponge *Lubomirskia baicalensis*. Micron 39:447–460

Wiens M, Bausen M, Natalio F, Link T, Schlossmacher U, Müller WEG (2009) The role of the silicatein-α interactor silintaphin-1 in biomimetic biomineralization. Biomaterials 30:1648–1656

Wiens M, Wang X, Natalio F, Schröder HC, Schloßmacher U, Wang S, Korzhev M, Geurtsen W, Müller WEG (2010a) Bioinspired fabrication of bio-silica-based bone-substitution materials. Adv Eng Biomater 12:B438–B450

Wiens M, Wang X, Schloßmacher U, Lieberwirth I, Glasser G, Ushijima H, Schröder HC, Müller WEG (2010b) Osteogenic potential of biosilica on human osteoblast-like (SaOS-2) cells. Calcif Tissue Int 87:513–524

Wiens M, Wang X, Schröder HC, Kolb U, Schloßmacher U, Ushijima H, Müller WEG (2010c) The role of biosilica in the osteoprotegerin/RANKL ratio in human osteoblast-like cells. Biomaterials 31:7716–7725

Wiens M, Wang XH, Unger A, Schröder HC, Grebenjuk V, Pisignano D, Jochum K, Müller WEG (2010d) Flashing light signaling circuit in sponges: endogenous light generation after tissue ablation in *Suberites domuncula*. J Cell Biochem 111:1377–1389

Wiens M, Schröder HC, Wang XH, Link T, Steindorf D, Müller WEG (2011) Isolation of the silicatein-α interactor silintaphin-2 by a novel solid-phase pull-down assay. Biochemistry 50: 1981–1990

Woesz A, Weaver JC, Kazanci M, Dauphin Y, Aizenberg J, Morse DE, Fratzl P (2006) Micromechanical properties of biological silica in skeletons of deep-sea sponges. J Mater Res 21:2068–2078
Wolf SE, Schlossmacher U, Pietuch A, Mathiasch B, Schröder HC, Müller WEG, Tremel W (2010) Formation of silicones mediated by the sponge enzyme silicatein-α. Dalton Trans 39: 9245–9249
Wüst S, Müller R, Hofmann S (2011) Controlled positioning of cells in biomaterials—approaches towards 3D tissue printing. J Funct Biomater 2:119–154
Zhou G, Zheng Q, Engin F, Munivez E, Chen Y, Sebald E, Krakow D, Lee B (2006) Dominance of SOX9 function over RUNX2 during skeletogenesis. Proc Natl Acad Sci USA 103: 19004–19009

Chapter 9
Inorganic Polymers: Morphogenic Inorganic Biopolymers for Rapid Prototyping Chain

Werner E.G. Müller, Heinz C. Schröder, Zhijian Shen, Qingling Feng, and Xiaohong Wang

Contents

W.E.G. Müller (✉) • H.C. Schröder
ERC Advanced Investigator Group, Institute for Physiological Chemistry, University Medical Center of the Johannes Gutenberg University, Duesbergweg 6, 55128 Mainz, Germany

NanotecMARIN GmbH, Duesbergweg 6, 55128 Mainz, Germany
e-mail: wmueller@uni-mainz.de

Z. Shen • Q. Feng
Department of Materials Science and Engineering, Tsinghua University, Beijing 100084, People's Republic of China

X.H. Wang
ERC Advanced Investigator Group, Institute for Physiological Chemistry, University Medical Center of the Johannes Gutenberg University, Duesbergweg 6, 55128 Mainz, Germany

W.E.G. Müller et al. (eds.), *Biomedical Inorganic Polymers*, Progress in Molecular and Subcellular Biology 54, DOI 10.1007/978-3-642-41004-8_9,

Abstract In recent years, considerable progress has been achieved towards the development of customized scaffold materials, in particular for bone tissue engineering and repair, by the introduction of rapid prototyping or solid freeform fabrication techniques. These new fabrication techniques allow to overcome many problems associated with conventional bone implants, such as inadequate external morphology and internal architecture, porosity and interconnectivity, and low reproducibility. However, the applicability of these new techniques is still hampered by the fact that high processing temperature or a postsintering is often required to increase the mechanical stability of the generated scaffold, as well as a post-processing, i.e., surface modification/functionalization to enhance the biocompatibility of the scaffold or to bind some bioactive component. A solution might be provided by the introduction of novel inorganic biopolymers, biosilica and polyphosphate, which resist harsh conditions applied in the RP chain and are morphogenetically active and do not need supplementation by growth factors/cytokines to stimulate the growth and the differentiation of bone-forming cells.

9.1 Introduction

The introduction of rapid prototyping (RP) or solid freeform fabrication (SFF) techniques has opened entirely new and growing possibilities not only in a variety of technical areas, such as automotive industry and tool technology, but also in the biomedical field, in particular in surgery and dentistry (Yeong et al. 2004; Butscher et al. 2011; Silva et al. 2011). These techniques allow the production of customized components following a bottom-up approach that has been termed as additive manufacturing. Such an approach has enormous advantages compared to subtractive manufacturing such as cost effectiveness, saving of resources, and avoidance of waste. In the biomedical field, the fabrication of custom-made scaffolds for bone regeneration and repair has gained the most attention, and various additive manufacturing techniques, including selective laser sintering (SLS), robocasting, and three-dimensional (3D) printing, have been developed that are intended to be used for the generation of scaffolds/implant materials, perfectly fitting into the cavity of the bone defect. However, these developments, even though impressive results already exist (e.g., Rengier et al. 2010; Bagaria et al. 2011), are still in a very preliminary phase, and there are many challenges both in terms of the technologies to be applied and with respect to the material properties and the patient's response that deserve much attention and further research efforts. There is still an urgent need in materials that can be introduced in the RP chain to enable a fast regeneration of bone and its integration in both the hard and soft tissues. Other requirements for such a material are biodegradability and preferably, stimulation of cell growth, and differentiation. Research on the lowest multicellular organism, the sponges, revealed that nature might offer such a material: biosilica, an enzymatically formed inorganic biopolymer consisting of a network of polycondensated silica units with various degrees of Si–O–Si linkages [for a review, see Müller et al. (2009) and Wang et al. (2012a)]. This bioinorganic material, biosilica, turned out to be morphogenetically active, i.e., it acts as an inducer of a number of cytokines and growth factors that are involved in differentiation and control of activity of cells involved in bone anabolism and catabolism (Wang et al. 2012c) and also has self-adapting/self-repairing properties (Müller et al. 2013c). Moreover, this biopolymer can be used as a component of a printable matrix used for cell embedding (Wang et al. 2013b). In addition, another inorganic biopolymer that can be introduced into the RP chain has turned out to be a promising candidate for a morphogenetically active scaffold material, polyphosphate (polyP), that can be produced both enzymatically, as bio-polyP, at ambient conditions by bacterial polyP kinases, and chemically at high temperatures (Wang et al. 2012c). Both inorganic biopolymers can be applied for preparation of customized materials, either alone or in combination, taking advantage of the partly complementary spectrum of biological activities of these polymers (Wang et al. 2013b). Both polymers proved not only to positively and favorably influence osteoblasts growth and function but also to be biocompatible and biodegradable, and finally, they can be totally replaced by the body's own bone material [reviewed in Wang et al. (2012c)].

9.2 Bone Formation

Bone formation is controlled by both anabolic and catabolic processes, involving two classes of cells, the bone-forming osteoblasts and the bone dissolving osteoclasts, that allow a remodeling and adaptation of bone to the changing mechanical requirements during bone growth, repair, and aging. The tuned interaction and the cross talk between these cells is regulated by the specific expression pattern of a series of morphogenetically active growth factors or signaling molecules that direct the maturation/differentiation of the osteoblasts from their mesenchymal progenitor stem cells and of the osteoclasts from their hemopoietic precursor cells (Porter et al. 2009). Therefore, biomimetic scaffold materials for healing bone defects should preferably be not only osteoconductive, i.e., guide cells involved in bone repair to the site of the lesion, but also, if possible, osteoinductive, i.e., stimulate the differentiation of osteoprogenitor stem cells into mature, functionally active osteoblasts that form new bone (Albrektsson and Johansson 2001; Epple 2007). However, osteoinductivity of bone scaffold materials is a challenging task that usually needs the addition of exogenous growth factors/cytokines to the scaffold material. Because the production of scaffolds in rapid prototyping processes requires a sintering or postsintering step, these heat-sensitive biomolecules can only be added to the scaffold after this processing step. Therefore, it would be a great progress to have a scaffold material, like biosilica or polyP, as proposed in this chapter, that is morphogenetically active, mimicking the role of the inductive stimuli of the natural extracellular matrix (ECM) and does not require additional processing by modification/functionalization with some further, biologically active component.

9.3 Bone Defects

Bone defects, in particular critical-sized defects in bone, are a widespread clinical problem (Porter et al. 2009). Common causes of bone defects are trauma, inflammation, and cancer. Bone defects are often complex and difficult to repair, in particular maxillofacial bone defects, such as larger defects of the upper or lower jaw that often cause surgical problems (Dimitriou et al. 2011). Consequently, there is an increasing demand for bone graft substitutes. In particular, injuries of bone caused by traffic accidents are dramatically increasing worldwide, especially in the developing countries. About 40 % of the women over 50 years are expected to suffer from osteoporotic fractures; the number of fractures caused by osteoporosis has doubled in the last decade. At present, about 500,000 total hip replacements are annually performed in Europe (Holzwarth and Cotogno 2012).

There are a number of properties that are expected from an ideal bone substitute material (see also, e.g., Janicki and Schmidmaier 2011), including biodegradability; osteoconductive, osteogenic, and preferably also osteoinductive properties; promotion of fusion with bone but not with soft tissue; and similar or the same complex morphology and mechanical properties as the normal bone tissue to guarantee the physiological function of bone during regeneration and to prevent additional stress caused by imperfect contact of the substitution material with the body's own bone that will result in destruction of normal bone tissue.

9.4 Porous Scaffolds

Scaffold materials for bone tissue engineering should provide a 3D platform that allows the surrounding bone cells to attach, to differentiate, and to mineralize. The current gold standard in the therapy of critical-sized bone defects is the use of autologous bone grafts. These autografts show all properties that are required to guarantee optimal bone growth and regeneration, such as osteoconductivity, osteogenicity, and osteoinductivity, but there are a number of limitations that restrict their use, such as insufficient availability of material and additional risk of infection and pain at the donor site, e.g., iliac crest (Nishida and Shimamura 2008). Allografts, on the other hand, that can be used as an alternative are marred by other uncertainties, such as immunogenicity and disease transmission (Laurencin and El-Amin 2008). Consequently, there is an increasing interest in synthetic bone scaffolds, allowing to avoid the disadvantages of natural bone grafts. Such synthetic bone scaffolds should mimic the nanofibrous architecture of the natural bone tissue and show a sufficient porosity allowing the ingrowth of cells into the implant and an efficient transport of nutrients and oxygen, as well as cytokines and growth factors that are necessary to induce growth and differentiation of cells (osteoblastogenesis, osteoclastogenesis, and angiogenesis) (Holzwarth and Ma 2011). The scaffold material should possess an adequate mechanical stability and should be osteoconductive and preferably also osteoinductive (Albrektsson and Johansson 2001). In addition, this material should allow a sufficient vascularization. Furthermore, the material should be biodegradable and support the formation of a collagen fiber network and finally to be replaced by the body's own bone tissue.

These demands on biocompatibility and biodegradability are best met by organic scaffold materials. These materials comprise both synthetic polymers, e.g., poly (lactide) (Giordano et al. 1996; Park et al. 1998; Gugala and Gogolewski 2005), poly(lactide/glycolide) (Yoon et al. 2005), and polycaprolactone (Li et al. 2008), and natural (processed or unprocessed) polymers, such as collagen (Warren et al. 2001), silk fibroin (Unger et al. 2004), starch (Tuzlakoglu et al. 2010), chitosan (Seol et al. 2004), and poly(3-hydroxybutyrate) (Sombatmankhong et al. 2007). The advantage of these polymers is that they mimic the structure of the extracellular matrix of bone tissue. However, both synthetic polymers and processed natural biomaterials lack the ligands that bind to the bone cell surface receptors. Therefore, supplementation with growth factors is necessary to induce cell-specific differentiation processes, such as bone morphogenetic protein 2 (BMP-2), platelet-derived growth factor (PDGF), insulin-like growth factor 2 (IGF-2), or transforming growth factor β (TGF-β), that have to be added to the scaffolds after fabrication.

Further strategies towards the design of suitable scaffolds for bone tissue engineering, mimicking the complex natural environment that supports bone cells to grow and to differentiate, comprise (1) seeding of the cells onto prefabricated porous scaffolds or (2) encapsulation of the cells during scaffold formation (Nicodemus and Bryant 2008). The latter strategy using hydrogels has been successfully applied for engineering bone tissue; see Sect. 9.10.

Porous scaffolds made of ceramic materials have sufficient stability and minimal infection risks, but these materials are mostly not biodegradable and cannot be fully replaced by the body's own bone tissue. These materials have been mostly used in the fabrication of scaffold materials for bone tissue engineering using RP procedures, allowing a customization of both external geometry and internal architecture of the implant (Yeong et al. 2004; Rengier et al. 2010; Butscher et al. 2011). However, the ceramic material alone is not able to support the growth and differentiation of bone cells. These scaffold materials after fabrication by RP techniques require a further processing step consisting in the addition of exogenous organic factors to induce the growth/differentiation, as well as the ingrowth of cells involved in new bone formation. Such growth factors/cytokines often become inhomogeneously distributed when added to the scaffold or they show unsatisfactory stability or release kinetics. Therefore, it would be of great benefit to have a material that can be included in the RP chain but does not require supplementation by further organic components. Such novel materials would overcome the hurdles associated with materials that are currently in use. In addition, their combinations with bioceramics/bioglasses are expected to fulfill the mechanical requirements for application in surgery of bone defects.

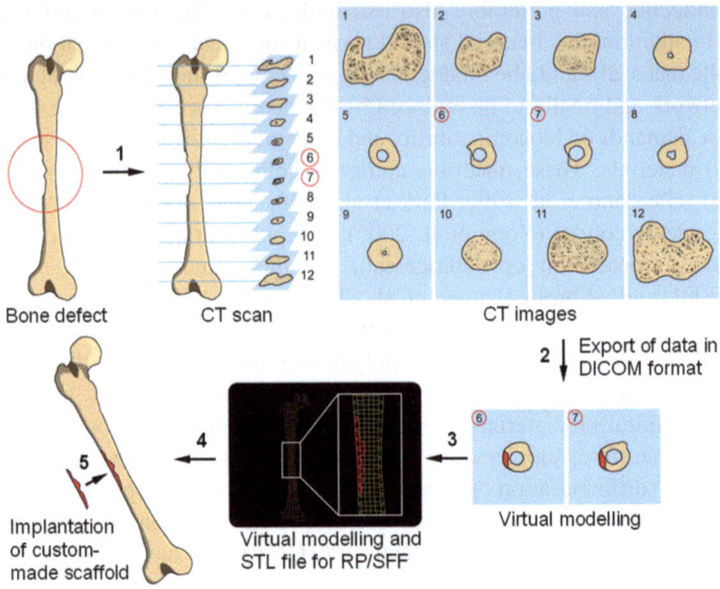

Fig. 9.1 Steps of RP chain: (1) morphological data acquisition by medical imaging (CT or MRI), (2) export of data in DICOM format, (3) virtual modeling and generation of STL file, (4) RP/SFF manufacturing of the customized implant, and (5) surgery using customized scaffold

9.5 Rapid Prototyping Chain

The application of the RP or SFF techniques in the medical field has made possible by the rapid development of advanced medical imaging techniques, in particular computed tomography (CT) and magnetic resonance imaging (MRI). The spatial resolution of these techniques has been dramatically increased and greatly improved in the last few years. The RP process chain, from image acquisition to the fabrication of the 3D RP model, consists of the following three steps: (i) extraction of high-resolution 3D imaging data (from CT and MRI), (ii) computer-aided design (CAD)-/computer-aided manufacturing (CAM)-based RP design, and (iii) manufacturing (3D printing) of the scaffold (Fig. 9.1). In the first and second step, the data from high-resolution CT scans, saved in the common Digital Imaging and Communications in Medicine (DICOM) format, are processed to Surface Tessellation Language (STL) files using CAD software. In the third step, the scaffolds are 3D printed using the STL data sent to the printing machine. As a result of the 3D printing/RP process, a 3D CAD data set and a 3D physical model are generated based on the image data from 2D scans (Rengier et al. 2010; Bagaria et al. 2011). In that way, parts/scaffolds with customized shapes and performances for individual patients can be designed and fabricated.

9.6 Rapid Prototyping Technologies

In principle, customized porous scaffolds can be prepared using two processes: (1) additive manufacturing (a bottom-up approach) and (2) subtractive manufacturing (a top-down approach) (Butscher et al. 2011). Additive manufacturing comprises the techniques of 3D ink-jet printing, robocasting, and selective laser sintering. Subtractive manufacturing techniques such as CNC milling of porous preforms are traditionally used for fabrication of customized dense solid implants in surgery/dentistry; e.g., most implants are currently manufactured by 3D CNC milling. However, subtractive techniques have a series of limitations that are mainly due to the high material consumption associated with them. Accordingly, RP/SFF techniques have attracted much attention for the manufacture of customized scaffolds in surgery/dentistry. These techniques make use of 3D digital data to fabricate physical objects layer by layer. They allow the design of 3D structures with predetermined external dimensions and pore architecture.

The RP/SFF techniques used for biomedical applications can be grouped into (1) printer-based, (2) extrusion-based, and (3) beam-based methods (Billiet et al. 2012). In general, the fabrication of scaffolds using these techniques is based on a 3D design of the scaffold which is transferred into an STL file format. This is then split into a series of consecutive virtual slices to allow the 3D printer/RP machine to produce the scaffold in a layer-by-layer fashion. It should be noted, however, that conventional RP/SFF techniques often require a postsintering step, resulting in a shrinkage of the material that might require an additional treatment by 3D CNC milling to meet the morphological requirements at the site of the implant.

9.6.1 3D Printing

The 3D printing technique is based on the conventional ink-jet technology. A schematic illustration of the printing process is shown in Fig. 9.2. The 3D ink-jet printing technique is a layer-by-layer fabrication technology to manufacture 3D components based on CAD data. Thereby, a liquid “binder” is printed onto a layer of, e.g., a ceramics or composite powder using an ink-jet printing head. The “binder” then bonds together adjacent powder particles present in the powder bed (Seitz et al. 2005). The first step is the distribution of the powder with a roller onto the platform of the fabrication chamber (Fig. 9.2, from left to right). The ink-jet print head then deposits the liquid “binder” in a two-dimensional pattern onto the powder layer (Fig. 9.2, left). Subsequently, the next powder layer is distributed onto the platform, and the printing process starts again (Fig. 9.2, from left to right). These steps are repeated until the part is completed. The excess powder in each layer acts as support for the part and is blown away at the end of the procedure.

The 3D printing process has been successfully applied for the fabrication of scaffold materials for bone tissue engineering, using bioactive glasses and calcium

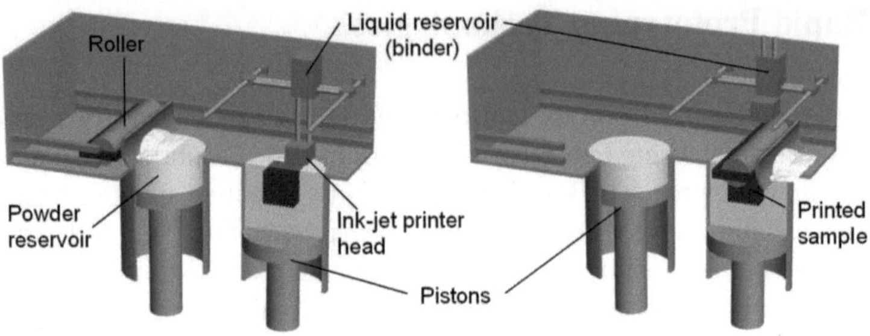

Fig. 9.2 Schematic representation of the 3D printing process using a 3D-powder printing system

phosphate (Bergmann et al. 2010). The 3D printer developed by Zcorp uses plaster of Paris (gypsum plaster) as well as natural polymers in combination with a water-based ink (Utela et al. 2008). We used this printer, starting with plaster of Paris, to generate, for the first time, a bioactive scaffold that supported growth and cytokine production of bone cells by impregnation of the Ca-sulfate grids with silica (Link et al. 2013). 3D printing can also be applied for organic polymers like polylactic acid (PLA), a bioresorbable polymer (Giordano et al. 1996). The fabrication of 3D-printed scaffold materials for bone tissue engineering usually requires a sintering process of the initially formed aggregates of the calcium phosphate and bioactive glass granules, preventing an addition of organic growth factors during the scaffold fabrication (Fielding et al. 2012). Therefore, the is a strong interest in alternative scaffold materials allowing the fabrication of 3D scaffolds by 3D printing without (post)sintering. One candidate might be biosilica that can be hardened by applying a bioinspired method mimicking the hardening of "soft" biosilica initially formed during sponge spicule formation, based on addition of poly(ethylene glycol) (PEG) as "binder" (see below, Schröder et al. 2012a).

9.6.2 Robocasting

Robocasting is an extrusion-based RP technique. In this process, a colloidal suspension/ink is loaded onto a syringe and extruded through a nozzle attached to the syringe to form a continuously growing filament that is organized to a 3D structure in a computer-controlled process (Miranda et al. 2006) (Fig. 9.3). This 3D structure consists of a mesh of interpenetrating rods that is formed layer by layer by extrusion of the ink through the nozzle. This robocasting technique has been successfully applied for the fabrication of porous scaffolds from hydroxyapatite (Michna et al. 2005) and β-tricalcium phosphate (Miranda et al. 2006; Martínez-Vázquez et al. 2010), as well as bioactive glasses (Deliormanli and Rahaman 2012; Eqtesadi et al. 2013). The material properties of the generated scaffolds have been investigated (Miranda et al. 2008).

Fig. 9.3 Schematic presentation of the robocasting process

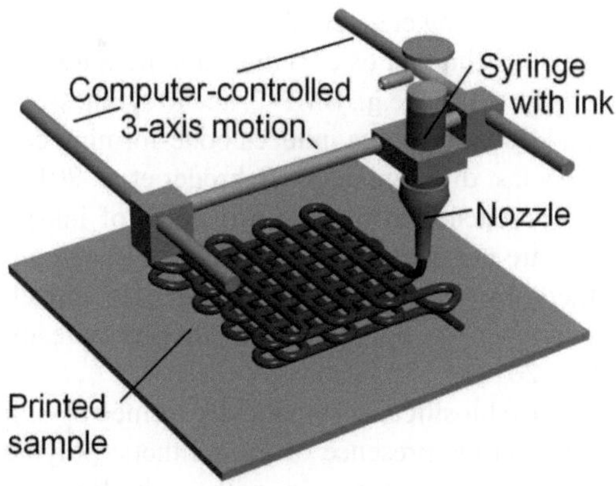

9.6.3 *Selective Laser Sintering*

In selective laser sintering (SLS), 3D objects are generated layer by layer from CAD data files using a laser beam (Stotko 2009; Mazzoli 2013). Exposure of thin layers of powder to a scanning laser beam induces a local melting of the material. The melt droplets then fuse together under formation of 3D objects with the desired morphology after cooling. The SLS technique is not based on a sintering process but on a local melting and solidification. The whole process consists of one single processing step. This technique has a great potential to fabricate hierarchical architectures with complex geometries (Gruner and Shen 2011). The SLS can also be applied for formation of scaffolds by certain polymers which can be melted by laser and mixed with bioceramic powders, e.g., hydroxyapatite (Tan et al. 2003). Thereby, the ceramic components are not melted by the laser, but bonded together by the polymer phase. In the technique of surface-selective laser sintering (SSLS), a laser radiation absorber (e.g., carbon microparticles) is added, allowing even to apply this process for polymers that do not absorb laser radiation (Antonov et al. 2004).

9.7 3D Cell Printing

It is generally agreed that the technique of 3D printing of cells has an enormous potential for the fabrication of complex 3D tissues and organs [reviewed in Wüst et al. (2011)]. However, there is still an unmet demand of a suitable printable scaffold material into which the cells can be embedded. The desired scaffold should not only serve as a mechanical framework but also should be functionally active as a template that is able to elicit the expression of essential morphogens and cytokines by the cells in a spatially and temporally controlled way. It has been proposed that biosilica might provide such a scaffold for 3D cell printing due to its morphogenetic activity (Wang et al. 2013b). This biopolymer has been shown to induce osteoblast-like cells (human osteogenic sarcoma SaOS-2 cells) to express genes involved in bone formation, e.g., BMP-2, and to synthesize hydroxyapatite (Wiens et al. 2010b). In addition, biosilica induces bone-forming cells to release a factor(s) that inhibits osteoclast differentiation (Schröder et al. 2012b).

Approaches to use the principles of ink-jet patterning for 3D bio-printing of cells are not new (Calvert 2007; Grayson et al. 2008), but they are limited, among others, by the property of hydrogels to spread during printing. Efforts have been undertaken to optimize microdroplet spreading and gelation behavior (Pataky et al. 2012).

Since biosilica, enzymatically formed by silicatein, can be hardened to a gel-like matrix in the presence of the synthetic polymer PEG (Schröder et al. 2012a), we used biosilica, together with a hydrogel, prepared of Na-alginate (Müller et al. 2013b), as a matrix for embedding the SaOS-2 cells, and applied this matrix for 3D printing. Our results (see Sect. 9.10) revealed that biosilica-alginate hydrogels provide a promising basis for 3D printing of bone-forming cells to repair bone defects. Embedding of cells in such a biosilica-alginate hydrogel matrix has been found not only to maintain cell viability and to support cell growth but also to cause the expression of genes that facilitate the release of the entrapped cells and of osteogenetically active silicate (see Fig. 9.7 in Sect. 9.10; Müller et al. 2013b).

9.8 Bioinorganic Polymers for Scaffolds

In the last few years, the naturally occurring polymers, biosilica and polyP, have been identified to cause morphogenetic activity on both osteoblasts and osteoclasts. These bioinorganic polymers are strongly osteogenic, biocompatible, and biodegradable [reviewed in Wang et al. (2012c)]. Both polymers can be blended with synthetic polymers used in bone substitution materials and have been successfully applied in first animal experiments (e.g., Wiens et al. 2010a). These promising, morphogenetically active polymers which can be combined with rigid bioceramics and bioactive glasses to enhance their mechanical performance have recently been applied in the RP chain.

9.8.1 Biosilica

The importance of silicon as an essential factor for human bone formation has been recognized already more than 40 years ago (Carlisle 1972), but only in recent years, we began to understand the metabolismus and mode of action of this element to some of extent. This became possible because the simplest animals, the sponges, do not possess a calcium-based (calcium phosphate or calcium carbonate) skeleton such as humans do, but a skeleton made of silica glass, most likely a consequence of

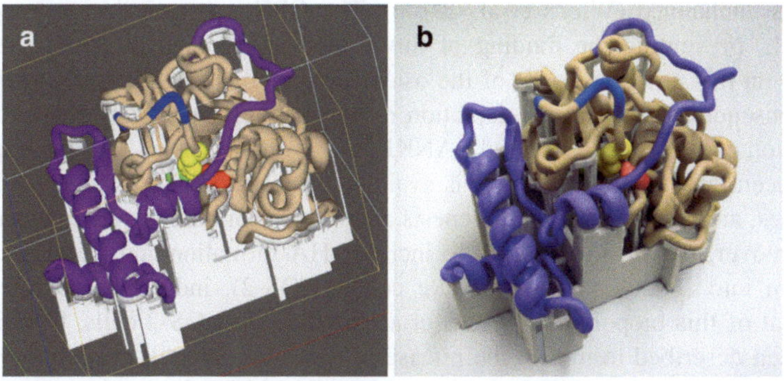

Fig. 9.4 Silicatein-α. (**a**) Computer-generated virtual 3D model of silicatein-α for 3D printing. (**b**) Photo of the printed molecule

the much higher concentration of this mineral in the ancient oceans than today. As a result of the great research efforts on animal molecular evolution, with the (siliceous) sponges as the oldest, still extant multicellular organisms that are closest to the Urmetazoa, and the strong interest in the mechanism of formation of the intriguing, hierarchically ordered biosilica skeleton of these animals, intense knowledge exists about the molecular basis of biosilica formation—from the genes via the biosilica-forming enzymes/proteins (silicateins) to the inorganic–organic biosilica hybrid structures and the complex skeletal architectures formed by them (Wang et al. 2012b).

The key enzyme of the siliceous sponges involved in biosilica formation is silicatein (Fig. 9.4)—an enzyme that is unique in its ability to synthesize the inorganic polymer, silica, previously thought to be available only in a chemical way, either at high temperatures or by sol–gel chemistry [reviewed in Schröder et al. (2008) and Wang et al. (2012a)]. All siliceous sponges investigated so far express this enzyme, from freshwater sponges living in surface waters of rivers and lakes to the deep-sea glass sponges (hexactinellids), with their up to 3 m long biosilica spicules (Müller et al. 2008b), an enzyme that is biocatalytically active at ambient conditions at temperatures ranging from above 30 °C (tropical sponges, e. g., the "tree sponge" *Drulia browni* from Rio Negro area, Brazil) to nearly zero degrees (Lake Baikal sponges, e.g., *Lubomirskia baicalensis*) and at concentrations of silica precursors, far below those which are required for the chemical condensation reaction [reviewed in Wang et al. (2012a, b) and Müller et al. (2013a)].

Biosilica, produced by silicatein, turned out to be morphogenetically active: it induces hydroxyapatite (HA) formation in osteoblast-like SaOS-2 cells (Schröder et al. 2005). In addition, it upregulates the expression of BMP-2 (Wiens et al. 2010b), an inducer of bone formation and bone regeneration (see Nickel et al. 2001). Moreover, biosilica differentially affects the gene expression of the two cytokines which are crucially involved in the development of osteoporosis, osteoprotegerin (OPG), and receptor activator of NF-kB ligand (RANKL) (Simonet et al. 1997; Wittrant et al. 2004; Baud'huin et al. 2007). Exposure of SaOS-2 cells to biosilica strongly enhances the expression of OPG, while the expression of RANKL remains unchanged (Wiens et al. 2010c). The OPG acts as a decoy receptor of RANKL, preventing the binding of this osteoclastogenic ligand to its receptor RANK on the surface of cells of the osteoclast cell lineage. In osteoporosis, e.g., as a consequence of estrogen depletion in postmenopausal osteoporosis, the concentration ratio between OPG and RANKL is lower than normal. As a consequence, the concentration of RANKL that is available to interact with its receptor RANK increases, and hence, osteoclastogenesis and bone resorption become enhanced.

Moreover, biosilica turned out to increase HA formation and (^{3}H)dThd incorporation into DNA in osteoblast-like cells (SaOS-2), indicating an osteogenic potential of this biopolymer (Wiens et al. 2010b). In SaOS-2 cells, biosilica has also been described to induce the release of a soluble osteoclast inhibitory factor that prevents growth and differentiation of RAW 264.7 cells (osteoclast-like cells derived from a mouse monocyte/macrophage precursor cells; Vincent et al. 2009) into osteoclasts (Schröder et al. 2012b).

The osteogenic activity of biosilica has also been demonstrated in first animal experiments. A two-component moldable and biodegradable implant material has been formulated by encapsulation of silicatein together with its substrate, orthosilicate, in poly(D,L-lactide)-/poly(vinyl pyrrolidone)-based microspheres and embedding these microspheres in a poly(vinyl pyrrolidone)-/starch-based matrix (Wiens et al. 2010a). This implant material hardens at a rate suitable for clinical applications. It has been found to be tightly integrated in artificial defects of rabbit femurs (Wiens et al. 2010a). The material turned out to be biocompatible; no toxic reactions has been observed so far both in vitro and in vivo.

It should be noted that the hardness of the biosilica material produced by silicatein can be significantly increased if supplemented with its natural interactor, silintaphin-1 (Wiens et al. 2009; Schloßmacher et al. 2011), or the synthetic polymer PEG (Schröder et al. 2012a; Wang et al. 2012b). Therefore, silicatein-produced, morphogenetically active biosilica is suitable not only for impregnation/coating of other scaffold materials showing an appropriate hardness but might also be applied as the basic component of newly designed scaffold structures.

Biosilica, produced by silicatein, is a self-adapting and -repairing material [recent review: Wang et al. (2012a)]. Moreover, it is a biogenic material, in contrast to "bioglass," e.g., 45S5 Bioglass® and 13–93, currently used in bone/tissue engineering. Biosilica differs from bioglass by the fact that it consists of pure silica (with an integrated silicatein-protein network) and therefore shows extreme mechanical stability and low brittleness, compared to bioglass. In addition, the bioactivity of bioglass is lost if the network connectivity N_c is above 2.4; usually bioglass has a N_c of 1.9 (Martin et al. 2012). The network connectivity N_c in bioactive glasses is the average number of bridging oxygen atoms per silicon, which can range between 0 (Q^0) and 4 (Q^4) (Hill 1996). At low N_c values, the glass rapidly dissolves. In contrast, biosilica has a high percentage of Q^4 and Q^3 and a low percentage Q^2 and Q^1 and hence a high N_c value (high stability but easy biodegradability because of the protein component) (Martin et al. 2012). Even under these conditions, this material is strongly bioactive and does not need sintering that leads to crystallization and reduction of bioactivity of bioglass.

The excellent mechanical properties of biosilica are also a result of its mechanism of formation, catalyzed by silicatein, an auto-chaperone protein; self-cleavage of this auto-chaperone results in (1) formation of enzymatically active silicatein (biocatalytic function) and (2) self-assembly of the silicatein molecules to ordered structures used as a matrix of silica deposition (Schröder et al. 2012a). In that way the protein molecules become integrated in the enzymatically synthesized silica product under formation of a bioactive and extremely stable but well degradable inorganic–organic hybrid material ("biosilica").

The composite material formed behaves like a single-phase hybrid material: the self-assembled silicatein filaments interlock between the enzymatically formed silica nanoparticles (Müller et al. 2010) and are not encased in a silica "block" and hence are degraded at a similar rate. This reduces the possibility of immune reactions. Such a material cannot be realized by applying conventional rapid prototyping procedures using other materials.

It should be noted that the biosilica nanoparticles formed by silicatein have the suitable size (50 nm) to be bioactive. Mesoporous silica particles with diameters of around 100 nm have been shown not to show any bioactivity (Huang et al. 2005), as opposed to silica nanoparticles with a diameter of 50 nm, which act as potent stimulators of osteoblast differentiation and mineralization (Beck et al. 2012).

The application of the recombinant silicatein and biosilica in bone tissue engineering has been patented (Müller et al. 2003, 2012). Due to the fact that biosilica particles are synthesized by a natural, enzymatic mechanism, used by sponges since millions of years, it can be expected that their acceptance by patients will be higher than that of synthetic silica nanoparticles (e.g., Beck et al. 2012) because of their nanotechnology origin.

9.8.2 Bio-polyP

Biogenic polyphosphates are naturally occurring linear polymers which can consist of several hundreds of phosphate (P_i) residues (Kulaev et al. 2004; Omelon and Grynpas 2008; Rao et al. 2009). Their synthesis is catalyzed by specific polyphosphate kinases that polymerize polyP by transfer of the terminal P_i from ATP molecules to the growing polyP chain and have been identified in bacteria [reviewed in Schröder et al. (1999)]. The recombinant enzymes can be used for the preparation of long polyP chains under physiological, ambient conditions (Ahn and Kornberg 1990). Besides the enzymatically produced biopolymer (bio-polyP), polyP can be synthesized chemically, but the chemical synthesis requires high temperatures, not needed during formation of the biogenically formed bio-polyP. PolyP can be metabolized (degraded) by specific exopolyphosphatases and endopolyphosphatases that have been isolated from both prokaryotic and eukaryotic organisms (e.g., Lorenz et al 1994; Lorenz and Schröder 2001).

We discovered, already 15 years ago, that polyP is present in bone tissue (Leyhausen et al. 1998; Schröder et al. 2000) and is a substrate for the principle enzyme involved in bone formation, the bone specific alkaline phosphatase (Lorenz and Schröder 2001); this enzyme is assumed to provide the inorganic phosphate required for HA synthesis. These results that had been subsequently confirmed by other groups [e.g., St-Pierre et al. (2010), recent review: Kulakovskaya et al. (2012)] provided the basis for the development of scaffold materials or coatings for bone replacement/dentistry. However, only recently we succeeded to find out how these polymers become bioactive: only if applied as a complex with Ca^{2+} ions, these polymers are able to induce hydroxyapatite formation and expression of growth factors, e.g., BMP-2 (Müller et al. 2011; Wang et al. 2013a, b). The Ca^{2+} ions that are released from the polyP Ca^{2+} complex in the extracellular space as a result of its enzymatic hydrolysis have been proposed to be utilized during hydroxyapatite formation (Omelon and Grynpas 2008); they might also be responsible for the increase in intracellular calcium concentration, as seen in SaOS-2 cells that had been exposed to the polyP Ca^{2+} complex (Müller et al. 2011).

The morphogenetic activity of polyP used as a Ca^{2+} complex has been demonstrated in osteoblast-like SaOS-2 cells (Müller et al. 2011). The polyP Ca^{2+} complex (polyP chain length: 45 phosphate residues) was found not only to induce hydroxyapatite formation but also enhances the expression of the gene encoding BMP-2 in SaOS-2 cells (Wang et al. 2013a). In addition, this complex induces the alkaline phosphatase, both the activity of this enzyme and the expression of the gene encoding the bone alkaline phosphatase, the tissue nonspecific alkaline phosphatase. On the other hand, the polyP Ca^{2+} complex inhibits the progression of RAW 264.7 cells into osteoclasts (Müller et al. 2011), most likely by impairment of the phosphorylation of IκBα by the IκBα kinase, a process that inhibits the differentiation of pre-osteoclasts into mature osteoclasts (Wang et al. 2013a).

Like biosilica, polyP is a promising candidate for the fabrication of novel bone biomimetic scaffold materials that show growth-promoting and differentiation-inducing activity, especially if combined with biosilica. The property of the polymer to be soluble as a Na^+ or K^+ salt at physiological pH and to becoming insoluble as a Ca^{2+} complex at higher concentrations and with increasing chain length of the polymer allows to develop strategies for hardening of the material required for its application in the RP process.

Further, it should be mentioned that both silicate bioceramics (Zhai et al. 2012) and calcium polyphosphate ceramics, e.g., strontium-doped calcium polyphosphate (Liu et al. 2011), have been reported to induce angiogenesis during bone regeneration.

It is expected that this novel approach, if introduced in the RP chain, has the potential to overcome current limitations in the use of this technique and to replace hitherto used procedures in surgery based on autologous or heterologous bone materials which are prone to infection or immunogenicity or, after sterilization, lack essential growth factors needed for tissue regeneration.

9.8.3 Modification of Scaffolds with Biosilica/Bio-polyP

Essential requirements for a successful scaffold material to be used in bone tissue engineering are a sufficient mechanical stability and a controlled macropore architecture that mimics the architecture of the natural extracellular matrix and allows cell adhesion molecules to adsorb to their receptors in an appropriate spatial arrangement (Pedersen and Swartz 2005). Based on their morphogenetic activity, biosilica and polyP have been proposed to be promising molecules for the modification of bioceramic scaffolds or scaffolds consisting of bioactive glasses, fabricated by SFF techniques (Wang et al. 2013b).

The modification/impregnation of the rigid sintered scaffolds fabricated from bioceramic or bioactive glass powders by biosilica or polyP can be performed by dip coating. In situ polymerization of the bioinorganic polymers might be an alternative procedure. This could be achieved by impregnation of nanoceramic powders with Glu-tagged silicatein followed by incubation with substrate (orthosilicate). Hardening could be achieved by printing of PEG (Schröder et al. 2012a) which acts as a "binder". This approach exploits the unique feature of the biopolymer, biosilica, to undergo a hardening process at ambient temperature without heating ("biosintering").

In initial experiments, to elucidate if biosilica or silica retains its morphogenetic activity for SaOS-2 cells if included during the 3D printing process, calcium sulfate scaffolds have been printed by 3D ink-jet printing (Link et al. 2013). In Fig. 9.5a, b, the 3D fabrication of a layered rectangular block using a ZPrinter 450 (Z Corporation) and plaster of Paris powder (Ca-sulfate hemihydrates) is shown. The computer-designed model and the microscopic images of one grid layer with regular quadratic channel structures are shown in Fig. 9.5c and Fig. 9.5d, e, resp. The powder was solidified by ink spread onto the Ca-sulfate powder to create a layer-by-layer model. For cell culture experiments, Ca-sulfate grid layers were hardened by impregnation with Mg-sulfate and then processed ("biologized") with Ca-phosphate or silica. The results revealed that the impregnated grids are not toxic. The SaOS-2 cells cultivated on these grids showed an enhanced growth and mineralization (Link et al. 2013). In addition, the silica impregnation was found to cause a strong increase in BMP-2 and OPG expression in the cells (Link et al. 2013).

These results show that the 3D-printed biosilica-modified material retains the osteogenic properties of biosilica. They contribute towards a solution of a fundamental problem in scaffold fabrication using RP/SFF techniques: the need of a post-processing/postsintering of the 3D scaffold that excludes any addition/presence of heat-sensitive growth factors/cytokines during the printing process. The subsequent addition of these factors usually results in nonoptimal distribution of those molecules within the scaffold and hence an insufficient functioning of the implant.

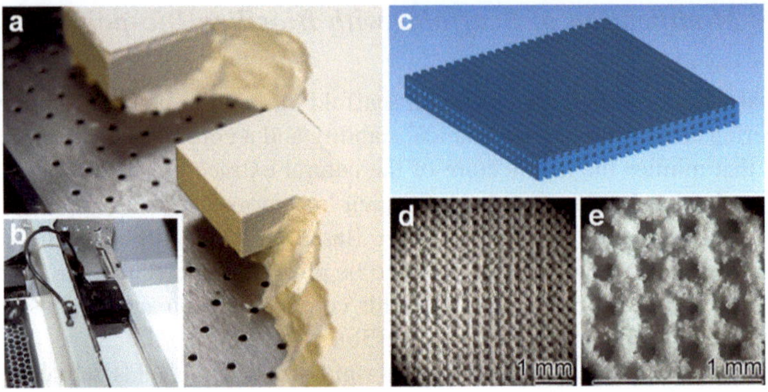

Fig. 9.5 3D ink-jet printing. (**a**) Printed regular block, made of Ca-sulfate. (**b**) Ink-jet printer head of ZPrinter 450 (Z Corporation) during printing. (**c**) 3D model for a porous scaffold designed for 3D printing. (**d**) Individual grid layer prepared from a regular block. (**e**) Higher magnification of the grid layer [modified after Link et al. (2013)]

9.9 Morphogenetically Active Scaffold: Mesenchymal Stem Cells

In bone tissue engineering, the mesenchymal stems cells (MSCs), taken from the donor site, e.g., the iliac crest, and seeded onto an inert scaffold, require the presence the natural growth factors and cytokines for their differentiation to osteoblasts that are needed for new bone (hydroxyapatite) formation. Therefore, the cultures must be supplemented with these growth factors/cytokines, such as PDGF, BMP-2, IGF, and TGF-β (Fig. 9.6, left panel), and these molecules must be present phase-specifically depending on the respective differentiation stage of the cells. It would be a great success if it would be possible to get the differentiating cells to produce these factors in a spatially and temporally correct way by themselves. The morphogenetically active inorganic biopolymers, biosilica and bio-polyP, offer a new strategy: these polymers, either constituting the scaffold themselves or incorporated into a ceramic structure, are able to elicit the expression and release of the morphogen/ligand molecules, such as BMP-2 and RANKL, and thereby promote the cell differentiation and hydroxyapatite deposition (Fig. 9.6, right panel).

9.10 Biosilica-Alginate Hydrogel

As a suitable matrix for 3D cell printing that is morphogenetically active and biocompatible, a matrix consisting of a biosilica and alginate hydrogel has been developed (Müller et al. 2013b). Alginate is a polysaccharide from algae and

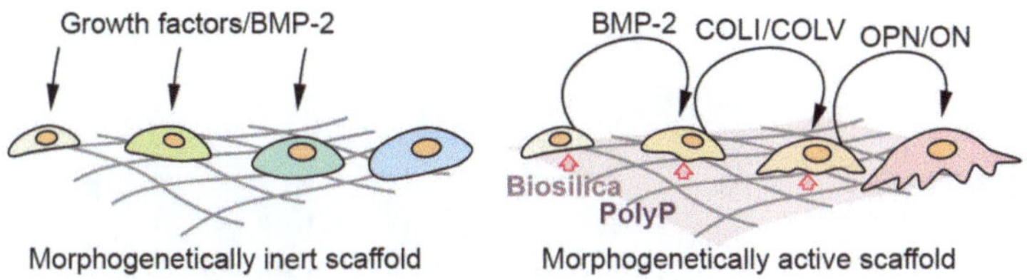

Fig. 9.6 Strategy to fabricate morphogenetically active scaffolds for bone tissue engineering compared to conventional strategy. *Left*: morphogenetically inert scaffold (conventional strategy). *Right*: morphogenetically active scaffold, prepared from biosilica or polyP (new strategy). Conventional scaffolds require the addition of growth factors/cytokines to support the growth/differentiation of precursor cells that can be taken, e.g., from iliac crest and cultivated ex vivo, while morphogenetically active scaffolds are able to induce the expression/release of these factors by the cells themselves in a regulated way

bacteria, consisting of β-(1–4)-linked D-mannuronic acid and α-(1–4)-linked L-guluronic acid residues (Smidsrød and Skåk-Bræk 1990). This matrix has been developed based on the finding that encasement of living cells in a biosilica shell does not affect the viability of the cells and their ability to divide (Müller et al. 2008a). Hydrogels formed after addition of $CaCl_2$ to aqueous solutions of Na-alginate have been shown to be biocompatible and suitable for 3D cell printing (Wüst et al. 2011).

The results revealed an increased growth and an increased mineralization of SaOS-2 cells if biosilica, enzymatically prepared by recombinant silicatein (Krasko et al. 2000), is present in the hydrogel (Müller et al. 2013b; Wang et al. 2013a). Staining of the hydroxyapatite clusters formed by SaOS-2 cells using fluorescence OsteoImage mineralization assay shows that mineralization of the cells in hydrogels with orthosilicate and recombinant silicatein (catalyzing biosilica synthesis, Fig. 9.7g, right panel) is highest compared to hydrogels with orthosilicate alone (no enzyme, Fig. 9.7f) or with orthosilicate and bovine serum albumin (control protein, Fig. 9.7h), while in hydrogels without silica supplement, only a very small mineral deposition is detected (Fig. 9.7e). No mineralization is observed in the absence of β-glycerophosphate-containing mineralization cocktail (Fig. 9.7a–d). In addition, the cells present in silica-containing hydrogels show an enhanced gene expression of BMP-2, collagen 1, and carbonic anhydrase (Müller et al. 2013b); a schematic outline is given in Fig. 9.7, left panel. The latter finding indicates that biosilica retains its morphogenetic/osteogenic potential even after entrapment into the Na-alginate hydrogel. It is assumed that the carbonic anhydrase which is able to degrade biosilica (Müller et al. 2007) might cause a controlled release of the biosilica-embedded cells as well as of silicate. Therefore, it can be expected that the biosilica entrapped in the hydrogel matrix will be morphogenetically active not only to the embedded cells but also to other cells ingrowing into the porous 3D scaffold. Further, it is known that the encasement of cells in a biosilica shell makes the cells more resistant against mechanical stress. These properties

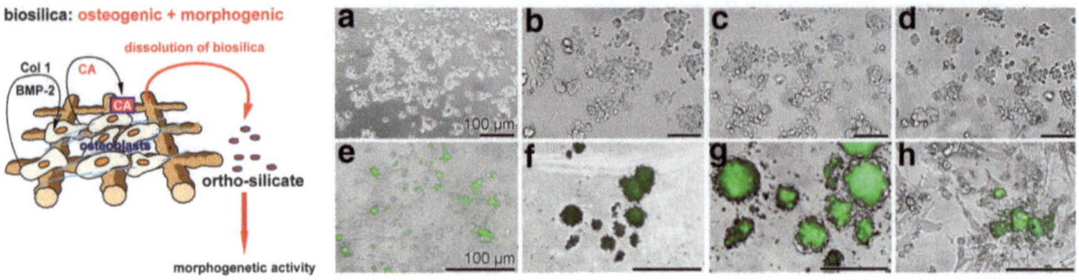

Fig. 9.7 3D cell printing on a morphogenetically active scaffold. *Left*: schematic outline: morphogenetically active biosilica-alginate scaffold. Osteoblasts embedded in the scaffold or attached to the scaffold start to synthesize the BMP-2 and collagen 1 (Col 1), as well as the silica-degrading enzyme carbonic anhydrase (CA). The released ortho- and oligo-silicate products are morphogenetically active. *Right*: mineralization of SaOS-2 cells in hydrogels. The cells were embedded into Na-alginate hydrogel either alone (**a**, **e**) or with silica (**b**, **f**), silica and silicatein (**c**, **g**), or silica and bovine serum albumin (**d**, **h**). The cultures were incubated in the absence (**a–d**) or presence of the mineralization cocktail (**e–h**). In the presence of biosilica, an increased formation of hydroxyapatite (*stained green*) is observed [modified after Wang et al. (2013a)]

might qualify biosilica hydrogels as a suitable matrix for 3D cell printing process (Müller et al. 2013b).

In a further step towards the development of a morphogenetically active matrix for 3D cell printing, alginate/silica composite hydrogels into which cells are embedded have been developed (Schloßmacher et al. 2013). Both osteoblast-related SaOS-2 cells and osteoclast-like RAW 264.7 cells have been separately embedded into beads formed of a Na-alginate-based or a silica-containing Na-alginate-based hydrogel matrix (Schloßmacher et al. 2013). The beads have been prepared by suspending the cells into the hydrogel matrix and dropping the resulting cell-hydrogel suspensions through a 0.45 mm wide needle, attached at a syringe, into a $CaCl_2$ solution. Thereby, the alginate beads become hardened. The beads (diameter about 1 mm) containing encapsulated SaOS-2 and RAW 264.7 have then been incubated either separately or together (co-incubation of SaOS-2- and RAW 264.7-containing beads), in the absence or presence of silica. For analysis, the cells can be released from the hydrogel matrix by chelating out Ca^{2+} from the gel using Na-citrate (Shoichet et al. 1996; Yamaoka et al. 2006). The alginate and alginate/silica hydrogel matrices have been found not to impair the viability of the encapsulated cells. Further, it has been found that the SaOS-2 cells entrapped in these matrices retain their capacity to synthesize hydroxyapatite (Schloßmacher et al. 2013). In SaOS-2 cells, the gene expression ratio of *OPG*: *RANKL* has been found to increase in cocultivation experiments with RAW 264.7 cell beads if silica had been included in the hydrogel matrix (Schloßmacher et al. 2013), indicating a decreased osteoclastogenic activity of osteoblasts (SaOS-2) and an impaired differentiation of osteoclasts (RAW 264.7) (Thomas et al. 2001; Wang et al. 2012c). This conclusion is supported by the finding that the expression of the gene encoding tartrate-resistant acid phosphatase (TRAP), a marker for osteoclast differentiation (Ballanti et al. 1997; Walsh et al. 2003), is strongly reduced in RAW 264.7 if co-incubated with SaOS-2 embedded into silica-supplemented hydrogel beads (Schloßmacher et al. 2013). The results are in line with previous findings suggesting that co-incubation of SaOS-2 cells and RAW 264.7 cells results in the release of an osteoclast inhibitory factor (Schröder et al. 2012b).

It has been concluded that the bead encapsulation of bone cells is a useful technique to produce bioactive programmable hydrogels (Schloßmacher et al. 2013). Both components are degradable and undergo dissolution, the alginate hydrogel by removal of divalent and trivalent cations (Jejurikar et al. 2012) and the silica, if not aged/hardened, at concentrations below the solubility product. Like silica, the product of its hydrolysis, orthosilicate, has been reported to show morphogenetic activity (Carlisle 1972, 1986; Schwarz and Milne 1972; Wang et al. 2012c).

9.11 Outlook

The development of advanced RP/SFF techniques has started to revolutionize surgery, orthopaedics, and dentistry. These techniques allow the fabrication of customized scaffolds and implants which are able to meet the demands of individual patients. One important task for future research is to develop scaffold materials that obviate the need of addition of exogenous growth factors/cytokines in order to achieve a sufficient response of the cells involved in bone formation. Another task will be the fabrication of scaffold materials using RP techniques that are customized not only to the geometry of the bone defect but also take into account the topologically distinct mechanical and metabolic requirements for the implant or the specific disease conditions of the patient, e.g., in the case of osteoporosis. The natural bioinorganic polymer, biosilica, is extremely suitable for that purpose because its hardness, elasticity, bioactivity, and biodegradation rate can be adapted to the desired conditions, e.g., the hardness of this material can be modulated by supplementation with various amounts of PEG. In addition, biosilica can restore the balance between bone formation by osteoblasts and bone dissolution by osteoclasts and, hence, might be a preferential scaffold material for treatment of osteoporotic patients. This disease is characterized by a rarefication of the trabecular network in the cancellous bone (Sambrook and Cooper 2006; Lane and Yao 2009) caused by a reduced level of OPG compared to RANKL, resulting in an increased bone resorption (Jabbar et al. 2011). Moreover, it is conceivable that the porosity and interconnectivity of the enzymatically (via silicatein) produced material might be adjusted at the micropore level, e.g., by combining biosilica with high- and low-pore additives, while the macroporosity is controllable through the RP process. Hence, the specific architectural design and function of cortical (compact) bone and cancellous (trabecular) bone can be realized in the same implant.

The two inorganic biopolymer materials, biosilica and polyP, which are the only inorganic biopolymers that are morphogenetically active in bone tissue engineering, open new avenues for 3D printing in bone repair.

Acknowledgments W.E.G.M. is a holder of an ERC Advanced Investigator Grant (no. 268476 BIOSILICA) as well as of an ERC proof-of-concept grant (no. 324564; Silica-based nanobiomedical approaches for treatment of bone diseases). This work was supported by grants from the European Commission (FP7-NMP-2013-EU-CHINA project no. 604036—Bio-Scaffolds; large-scale integrating project, project no. 311848—BlueGenics; project no. FP7-KBBE-2010-4-266033—SPECIAL; project no. PIRSES-GA-2009-246987—European-Chinese Research Staff Exchange Cluster MarBioTec*EU-CN*), the German Bundesministerium für Bildung und Forschung—International Bureau (no. CHN 09/1AP—German-Chinese Joint Lab on Bio-Nano-Composites), the Public Welfare Project of Ministry of Land and Resources of the People's Republic of China (grant no. 201011005–06), and the International S & T Cooperation Program of China (grant no. 2008DFA00980).

References

Ahn K, Kornberg A (1990) Polyphosphate kinase from *Escherichia coli*. Purification and demonstration of a phosphoenzyme intermediate. J Biol Chem 265:11734–11739

Albrektsson T, Johansson C (2001) Osteoinduction, osteoconduction and osseointegration. Eur Spine J 10:S96–S101

Antonov EN, Bagratashvili VN, Whitaker MJ, Barry JJ, Shakesheff KM, Konovalov AN, Popov VK, Howdle SM (2004) Three-dimensional bioactive and biodegradable scaffolds fabricated by surface-selective laser sintering. Adv Mater 17:327–330

Bagaria V, Deshpande S, Rasalkar DD, Kuthe A, Paunipagar BK (2011) Use of rapid prototyping and three-dimensional reconstruction modeling in the management of complex fractures. Eur J Radiol 80:814–820

Ballanti P, Minisola S, Pacitti MT, Scarnecchia L, Rosso R, Mazzuoli GF, Bonucci E (1997) Tartrate-resistant acid phosphate activity as osteoclastic marker: sensitivity of cytochemical assessment and serum assay in comparison with standardized osteoclast histomorphometry. Osteoporos Int 7:39–43

Baud'huin M, Lamoureux F, Duplomb L, Rédini F, Heymann D (2007) RANKL, RANK, osteoprotegerin: key partners of osteoimmunology and vascular diseases. Cell Mol Life Sci 64:2334–2350

Beck GR Jr, Ha SW, Camalier CE, Yamaguchi M, Li Y, Lee JK, Weitzmann MN (2012) Bioactive silica-based nanoparticles stimulate bone-forming osteoblasts, suppress bone-resorbing osteoclasts, and enhance bone mineral density *in vivo*. Nanomedicine 8:793–803

Bergmann C, Lindner M, Zhang W, Koczur K, Kirsten A, Telle R, Fischer H (2010) 3D printing of bone substitute implants using calcium phosphate and bioactive glasses. J Eur Ceram Soc 30: 2563–2567

Billiet T, Vandenhaute M, Schelfhout J, Van Vlierberghe S, Dubruel P (2012) A review of trends and limitations in hydrogel-rapid prototyping for tissue engineering. Biomaterials 33: 6020–6041

Butscher A, Bohner M, Hofmann S, Gauckler L, Müller R (2011) Structural and material approaches to bone tissue engineering in powder-based three-dimensional printing. Acta Biomater 7:907–920

Calvert P (2007) Printing cells. Science 318:208–209

Carlisle EM (1972) Silicon: an essential element for the chick. Science 178:619–621

Carlisle EM (1986) Silicon as an essential trace element in animal nutrition. In: Evered D, O'Connor M (eds) Silicon biochemistry, vol 121, Ciba Foundation symposium. Wiley, Chichester, pp 123–136

Deliormanli AM, Rahaman MN (2012) Direct-write assembly of silicate and borate bioactive glass scaffolds for bone repair. J Eur Ceram Soc 32:3637–3646

Dimitriou R, Jones E, McGonagle D, Giannoudis PV (2011) Bone regeneration: current concepts and future directions. BMC Med 9:66

Epple M (2007) Biomimetic bone substitution materials. In: Epple M, Baeuerlein E (eds) Biomineralisation: medical and clinical aspects. Wiley-VCH, Weinheim, pp 81–95

Eqtesadi S, Motealleh A, Miranda P, Lemos A, Rebelo A, Ferreira JMF (2013) A simple recipe for direct writing complex 45S5 Bioglass. sup. [R] 3D scaffolds. Mater Lett 93:68–71

Fielding GA, Bandyopadhyay A, Bose S (2012) Effects of silica and zinc oxide doping on mechanical and biological properties of 3D printed tricalcium phosphate tissue engineering scaffolds. Dent Mater 28:113–122

Giordano RA, Wu BM, Borland SW, Cima LG, Sachs EM, Cima MJ (1996) Mechanical properties of dense polylactic acid structures fabricated by three dimensional printing. J Biomater Sci Polym Ed 8:63–75

Grayson WL, Chao PH, Marolt D, Kaplan DL, Vunjak-Novakovic G (2008) Engineering custom-designed osteochondral tissue grafts. Trends Biotechnol 26:181–189

Gruner D, Shen ZJ (2011) Ordered coalescence of nano-crystals during rapid solidification of ceramic melts. CrystEngComm 13:5303–5305

Gugala Z, Gogolewski S (2005) The in vitro growth and activity of sheep osteoblasts on three-dimensional scaffolds from poly(L/DL-lactide) 80/20%. J Biomed Mater Res A 75A:702–709

Hill R (1996) An alternative view of the degradation of bioglass. J Mater Sci Lett 15:1122–1125

Holzwarth U, Cotogno G (2012) Total hip arthroplasty – state of the art, challenges and prospects. Joint Research Centre of the European Commission, Publications Office of the European Union, Luxembourg

Holzwarth JM, Ma PX (2011) Biomimetic nanofibrous scaffolds for bone tissue engineering. Biomaterials 32:9622–9629

Huang DM, Hung Y, Ko BS, Hsu SC, Chen WH, Chien CL, Tsai CP, Kuo CT, Kang JC, Yang CS, Mou CY, Chen YC (2005) Highly efficient cellular labeling of mesoporous nanoparticles in human mesenchymal stem cells: implication for stem cell tracking. FASEB J 19:2014–2016

Jabbar S, Drury J, Fordham JN, Datta HK, Francis RM, Tuck SP (2011) Osteoprotegerin, RANKL and bone turnover in postmenopausal osteoporosis. J Clin Pathol 64:354–357

Janicki P, Schmidmaier G (2011) What should be the characteristics of the ideal bone graft substitute? Combining scaffolds with growth factors and/or stem cells. Injury 42(Suppl 2): S77–S81

Jejurikar A, Seow XT, Lawrie G, Martin D, Jayakrishnan A, Grøndahl L (2012) Degradable alginate hydrogels crosslinked by the macromolecular crosslinker alginate dialdehyde. J Mater Chem 22:9751–9758

Krasko A, Lorenz B, Batel R, Schröder HC, Müller IM, Müller WEG (2000) Expression of silicatein and collagen genes in the marine sponge *Suberites domuncula* is controlled by silicate and myotrophin. Eur J Biochem 267:4878–4887

Kulaev IS, Vagabov V, Kulakovskaya T (2004) The biochemistry of inorganic polyphosphates. Wiley, New York, NY

Kulakovskaya TV, Vagabov VM, Kulaev IS (2012) Inorganic polyphosphate in industry, agriculture and medicine: modern state and outlook. Process Biochem 47:1–10

Lane NE, Yao W (2009) Developments in the scientific understanding of osteoporosis. Arthritis Res Ther 11:228

Laurencin CT, El-Amin SF (2008) Xenotransplantation in orthopedic surgery. J Am Acad Orthop Surg 16:4–8

Leyhausen G, Lorenz B, Zhu H, Geurtsen W, Bohnensack R, Müller WEG, Schröder HC (1998) Inorganic polyphosphate in human osteoblast-like cells. J Bone Miner Res 13:803–812

Li T, Yu YT, Wang J, Tang TS (2008) 1,25-Dihydroxyvitamin D(3) stimulates bone neo-vascularization by enhancing the interactions of osteoblasts-like cells and endothelial cells. J Biomed Mater Res A 86:583–588

Link T, Wang XH, Schloßmacher U, Feng QL, Schröder HC, Müller WEG (2013) An approach to a biomimetic bone scaffold: increased expression of BMP-2 and of osteoprotegerin in SaOS-2 cells grown onto silica-biologized 3D printed scaffolds. RSC Adv 3:11140–11147

Liu F, Zhang X, Yu X, Xu Y, Feng T, Ren D (2011) In vitro study in stimulating the secretion of angiogenic growth factors of strontium-doped calcium polyphosphate for bone tissue engineering. J Mater Sci Mater Med 22:683–692

Lorenz B, Schröder HC (2001) Mammalian intestinal alkaline phosphatase acts as highly active exopolyphosphatase. Biochim Biophys Acta 1547:254–261

Lorenz B, Müller WEG, Kulaev IS, Schröder HC (1994) Purification and characterization of an exopolyphosphatase activity from *Saccharomyces cerevisiae*. J Biol Chem 269:22198–22204

Martin RA, Yue S, Hanna JV, Lee PD, Newport RJ, Smith ME, Jones JR (2012) Characterizing the hierarchical structures of bioactive sol–gel silicate glass and hybrid scaffolds for bone regeneration. Philos Trans A Math Phys Eng Sci 370:1422–1443

Martínez-Vázquez FJ, Perera FH, Miranda P, Pajares A, Guiberteau F (2010) Improving the compressive strength of bioceramic robocast scaffolds by polymer infiltration. Acta Biomater 6:4361–4368

Mazzoli A (2013) Selective laser sintering in biomedical engineering. Med Biol Eng Comput 51:245–256

Michna S, Wu W, Lewis JA (2005) Concentrated hydroxyapatite inks for direct-write assembly of 3-D periodic scaffolds. Biomaterials 26:5632–5639

Miranda P, Saiz E, Gryn K, Tomsia AP (2006) Sintering and robocasting of beta-tricalcium phosphate scaffolds for orthopaedic applications. Acta Biomater 2:457–466

Miranda P, Pajares A, Guiberteau F (2008) Finite element modeling as a tool for predicting the fracture behavior of robocast scaffolds. Acta Biomater 4:1715–1724

Müller WEG, Lorenz A, Krasko A, Schröder HC (2003) Silicatein-mediated synthesis of amorphous silicates and siloxanes and their uses. European Patent 1320624, US Patent 7169589

Müller WEG, Krasko A, Schröder HC (2007) Decomposition and modification of silicate and silicone by silicase and use of the reversible enzyme. European Patent EP1546319

Müller WEG, Engel S, Wang X, Wolf SE, Tremel W, Thakur NL, Krasko A, Divekar M, Schröder HC (2008a) Bioencapsulation of living bacteria (*Escherichia coli*) with poly(silicate) after transformation with silicatein-α gene. Biomaterials 29:771–779

Müller WEG, Wang X, Kropf K, Boreiko A, Schloßmacher U, Brandt D, Schröder HC, Wiens M (2008b) Silicatein expression in the hexactinellid *Crateromorpha meyeri*: the lead marker gene restricted to siliceous sponges. Cell Tissue Res 333:339–351

Müller WEG, Wang XH, Cui FZ, Jochum KP, Tremel W, Bill J, Schröder HC, Natalio F, Schloßmacher U, Wiens M (2009) Sponge spicules as blueprints for the biofabrication of inorganic–organic composites and biomaterials. Appl Microbiol Biotechnol 83:397–413

Müller WEG, Wang X, Sinha B, Wiens M, Schröder HC, Jochum KP (2010) NanoSIMS: insights into the organization of the proteinaceous scaffold within hexactinellid sponge spicules. Chembiochem 11:1077–1082

Müller WEG, Wang XH, Diehl-Seifert B, Kropf K, Schloßmacher U, Lieberwirth I, Glasser G, Wiens M, Schröder HC (2011) Inorganic polymeric phosphate/polyphosphate is an inducer of alkaline phosphatase and a modulator of intracellular Ca^{2+} level in osteoblasts (SaOS-2 cells) *in vitro*. Acta Biomater 7:2661–2671

Müller WEG, Schwertner H, Schröder HC (2012) Enzymatic method for producing bioactive, osteoblast-stimulating surfaces and use thereof. European Patent 1740706

Müller WEG, Schröder HC, Burghard Z, Pisignano D, Wang XH (2013a) Silicateins: a paradigm shift in bioinorganic chemistry. Enzymatic synthesis of inorganic polymeric silica. Chem Eur J 19:5790–5804

Müller WEG, Schröder HC, Feng QL, Schlossmacher U, Link T, Wang XH (2013b) Development of a morphogenetically active scaffold for three-dimensional growth of bone cells: biosilica-alginate hydrogel for SaOS-2 cell cultivation. J Tissue Eng Regen Med. doi:10.1002/term.1745

Müller WEG, Wang XH, Jochum K, Schröder HC (2013c) Self-healing, an intrinsic property of biomineralization processes. IUBMB Life 65:382–396

Nickel J, Dreyer MK, Kirsch T, Sebald W (2001) The crystal structure of the BMP-2: BMPR-IA complex and the generation of BMP-2 antagonists. J Bone Joint Surg 83:S7–S14

Nicodemus GD, Bryant SJ (2008) Cell encapsulation in biodegradable hydrogels for tissue engineering applications. Tissue Eng Part B Rev 14:149–165

Nishida J, Shimamura T (2008) Methods of reconstruction for bone defect after tumor excision: a review of alternatives. Med Sci Monit 14:RA107–RA113

Omelon SJ, Grynpas MD (2008) Relationships between polyphosphate chemistry, biochemistry and apatite biomineralization. Chem Rev 108:4694–4715

Park A, Wu BM, Cima MJ (1998) Griffith LG (1998) Integration of surface modification and 3D fabrication techniques to prepare patterned poly(L-lactide) scaffolds allowing regionally selective hepatocyte and fibroblast adhesion. J Biomater Sci Polym Ed 9:89–110

Pataky K, Braschler T, Negro A, Renaud P, Lutolf MP, Brugger J (2012) Microdrop printing of hydrogel bioinks into 3D tissue-like geometries. Adv Mater 24:391–396

Pedersen JA, Swartz MA (2005) Mechanobiology in the third dimension. Ann Biomed Eng 33: 1469–1490

Porter JR, Ruckh TT, Popat KC (2009) Bone tissue engineering: a review in bone biomimetics and drug delivery strategies. Biotechnol Prog 25:1539–1560

Rao NN, Gómez-Garcʑa MR, Kornberg A (2009) Inorganic polyphosphate: essential for growth and survival. Annu Rev Biochem 78:605–647

Rengier F, Mehndiratta A, von Tengg-Kobligk H, Zechmann CM, Unterhinninghofen R, Kauczor HU, Giesel FL (2010) 3D printing based on imaging data: review of medical applications. Int J Comput Assist Radiol Surg 5:335–341

Sambrook P, Cooper C (2006) Osteoporosis. Lancet 367:2010–2018

Schloßmacher U, Wiens M, Schröder HC, Wang XH, Jochum KP, Müller WEG (2011) Silintaphin-1: interaction with silicatein during structure guiding biosilica formation. FEBS J 278:1145–1155

Schloßmacher U, Schröder HC, Wang XH, Feng QL, Diehl-Seifert B, Neumann S, Trautwein A, Müller WEG (2013) Alginate/silica composite hydrogel as a potential morphogenetically active scaffold for three-dimensional tissue engineering. RSC Adv 3:11185–11194

Schröder HC, Lorenz B, Kurz L, Müller WEG (1999) Inorganic polyP in eukaryotes: enzymes, metabolism and function. In: Schröder HC, Müller WEG (eds) Inorganic polyphosphates—biochemistry, biology, biotechnology. Prog Mol Subcell Biol 23:45–81

Schröder HC, Kurz L, Müller WEG, Lorenz B (2000) Polyphosphate in bone. Biochemistry (Mosc) 65:296–303

Schröder HC, Borejko A, Krasko A, Reiber A, Schwertner H, Müller WEG (2005) Mineralization of SaOS-2 cells on enzymatically (Silicatein) modified bioactive osteoblast-stimulating surfaces. J Biomed Mater Res B Appl Biomater 75B:387–392

Schröder HC, Wang X, Tremel W, Ushijima H, Müller WEG (2008) Biofabrication of biosilica-glass by living organisms. Nat Prod Rep 25:455–474

Schröder HC, Wang XH, Manfrin A, Yu SH, Grebenjuk VA, Korzhev M, Wiens M, Schloßmacher U, Müller WEG (2012a) Silicatein: acquisition of structure-guiding and structure-forming properties during maturation from the pro-silicatein to the silicatein form. J Biol Chem 287:22196–22205

Schröder HC, Wang XH, Wiens M, Diehl-Seifert B, Kropf K, Schloßmacher U, Müller WEG (2012b) Silicate modulates the cross-talk between osteoblasts (SaOS-2) and osteoclasts (RAW 264.7 cells): inhibition of osteoclast growth and differentiation. J Cell Biochem 113:3197–3206

Schwarz K, Milne DB (1972) Growth-promoting effects of silicon in rats. Nature 239:333–334

Seitz H, Rieder W, Irsen S, Leukers B, Tille C (2005) Three-dimensional printing of porous ceramic scaffolds for bone tissue engineering. J Biomed Mater Res B Appl Biomater 74B:782–788

Seol YJ, Lee JY, Park YJ, Lee YM, Young-Ku RIC, Lee SJ, Han SB, Chung CP (2004) Chitosan sponges as tissue engineering scaffolds for bone formation. Biotechnol Lett 26:1037–1041

Shoichet MS, Li RH, White ML, Winn SR (1996) Stability of hydrogels used in cell encapsulation: an in vitro comparison of alginate and agarose. Biotechnol Bioeng 50:374–381

Silva NR, Witek L, Coelho PG, Thompson VP, Rekow ED, Smay J (2011) Additive CAD/CAM process for dental prostheses. J Prosthodont 20:93–96

Simonet WS, Lacey DL, Dunstan CR, Kelley M, Chang MS, Lüthy R, Nguyen HQ, Wooden S, Bennett L, Boone T, Shimamoto G, DeRose M, Elliott R, Colombero A, Tan HL, Trail G, Sullivan J, Davy E, Bucay N, Renshaw-Gegg L, Hughes TM, Hill D, Pattison W, Campbell P, Sander S, Van G, Tarpley J, Derby P, Lee R, Boyle WJ (1997) Osteoprotegerin: a novel secreted protein involved in the regulation of bone density. Cell 89:309–319

Smidsrød O, Skåk-Bræk G (1990) Alginates as immobilization matrix for cells. Trends Biotechnol 8:71–78

Sombatmankhong K, Sanchavanakit N, Pavasant P, Supaphol P (2007) Bone scaffolds from electrospun fiber mats of poly (3-hydroxybutyrate), poly(3-hydroxybutyrate-co-3-hydroxyvalerate) and their blend. Polymer 48:1419–1427

Stotko CM (2009) Laser sintering: layer by layer. Nat Photonics 3:265–266

St-Pierre JP, Pilliar RM, Grynpas MD, Kandel RA (2010) Calcification of cartilage formed *in vitro* on calcium polyphosphate bone substitutes is regulated by inorganic polyphosphate. Acta Biomater 6:3302–3309

Tan KH, Chua CK, Leong KF, Cheah CM, Cheang P, Abu Bakar MS, Cha SW (2003) Scaffold development using selective laser sintering of polyetheretherketone-hydroxyapatite biocomposite blends. Biomaterials 24:3115–3123

Thomas GP, Baker SU, Eisman JA, Gardiner EM (2001) Changing RANKL/OPG mRNA expression in differentiating murine primary osteoblasts. J Endocrinol 170:451–460

Tuzlakoglu K, Pashkuleva I, Rodrigues MR, Gomes VL, Müller R, Reis RL (2010) A new route to produce starch-based fiber mesh scaffolds by wet spinning and the improvement in cell attachment and proliferation by tailoring their surface properties. J Biomed Mater Res A 92: 369–377

Unger RE, Wolf M, Peters K, Motta A, Migliaresi C, Kirkpatrick JC (2004) Growth of human cells on a nonwoven silk fibroin net: a potential for use in tissue engineering. Biomaterials 25: 1069–1075

Utela B, Storti D, Anderson R, Ganter M (2008) A review of process development steps for new material systems in three dimensional printing (3DP). J Manuf Process 10:96–104

Vincent C, Kogawa M, Findlay DM, Atkins GJ (2009) The generation of osteoclasts from RAW 264.7 precursors in defined, serum-free conditions. J Bone Miner Metab 27:114–119

Walsh NC, Cahill M, Carninci P, Kawai J, Okazaki Y, Hayashizaki Y, Hume DA, Cassady AI (2003) Multiple tissue-specific promoters control expression of the murine tartrate-resistant acid phosphatase gene. Gene 307:111–123

Wang XH, Schloßmacher U, Wiens M, Batel R, Schröder HC, Müller WEG (2012a) Silicateins, silicatein interactors, and cellular interplay in sponge skeletogenesis: formation of the glass fiber-like spicules. FEBS J 279:1721–1736

Wang XH, Schröder HC, Wang K, Kaandorp JA, Müller WEG (2012b) Genetic, biological and structural hierarchies during sponge spicule formation: from soft sol-gels to solid 3D silica composite structures. Soft Matter 8:9501–9518

Wang XH, Schröder HC, Wiens M, Ushijima H, Müller WEG (2012c) Biosilica and bio-polyphosphate: applications in biomedicine (bone formation). Curr Opin Biotechnol 23: 570–578

Wang XH, Schröder HC, Diehl-Seifert B, Kropf K, Schloßmacher U, Wiens M, Müller WEG (2013a) Dual effect of inorganic polymeric phosphate/polyphosphate on osteoblasts and osteoclasts *in vitro*. J Tissue Eng Regen Med 7:767–776

Wang XH, Schröder HC, Feng QL, Draenert F, Müller WEG (2013b) The deep-sea natural products, biogenic polyphosphate (bio-polyP) and biogenic silica (biosilica) as biomimetic scaffolds for bone tissue engineering: fabrication of a morphogenetically-active polymer. Mar Drugs 11: 718–746

Warren SM, Steinbrech DS, Mehrara BJ, Saadeh PB, Greenwald JA, Spector JA, Bouletreau PJ, Longaker MT (2001) Hypoxia regulates osteoblast gene expression. J Surg Res 99:147–155

Wiens M, Bausen M, Natalio F, Link T, Schlossmacher U, Müller WEG (2009) The role of the silicatein-α interactor silintaphin-1 in biomimetic biomineralization. Biomaterials 30: 1648–1656

Wiens M, Wang X, Natalio F, Schröder HC, Schloßmacher U, Wang S, Korzhev M, Geurtsen W, Müller WEG (2010a) Bioinspired fabrication of biosilica-based bone-substitution materials. Adv Eng Mater 12:B438–B450

Wiens M, Wang X, Schloßmacher U, Lieberwirth I, Glasser G, Ushijima H, Schröder HC, Müller WEG (2010b) Osteogenic potential of biosilica on human osteoblast-like (SaOS-2) cells. Calcif Tissue Int 87:513–524

Wiens M, Wang X, Schröder HC, Kolb U, Schloßmacher U, Ushijima H, Müller WEG (2010c) The role of biosilica in the osteoprotegerin/RANKL ratio in human osteoblast-like cells. Biomaterials 31:7716–7725

Wittrant Y, Theoleyre S, Chipoy C, Padrines M, Blanchard F, Heymann D, Redini F (2004) RANKL/RANK/OPG: new therapeutic targets in bone tumours and associated osteolysis. Biochim Biophys Acta 1704:49–57

Wüst S, Müller R, Hofmann S (2011) Controlled positioning of cells in biomaterials – approaches towards 3D tissue printing. J Funct Biomater 2:119–154

Yamaoka H, Asato H, Ogasawara T, Nishizawa S, Takahashi T, Nakatsuka T, Koshima I, Nakamura K, Kawaguchi H, Chung UI, Takato T, Hoshi K (2006) Cartilage tissue engineering using human auricular chondrocytes embedded in different hydrogel materials. J Biomed Mater Res A 78:1–11

Yeong WY, Chua CK, Leong KF, Chandrasekaran M (2004) Rapid prototyping in tissue engineering: challenges and potential. Trends Biotechnol 22:643–652

Yoon CH, Hur J, Park KW, Kim JH, Lee CS, Oh IY, Kim TY, Cho HJ, Kang HJ, Chae IH, Yang HK, Oh BH, Park YB, Kim HS (2005) Synergistic neovascularization by mixed transplantation of early endothelial progenitor cells and late outgrowth endothelial cells: the role of angiogenic cytokines and matrix metalloproteinases. Circulation 112:1618–1627

Zhai W, Lu H, Chen L, Lin X, Huang Y, Dai K, Naoki K, Chen G, Chang J (2012) Silicate bioceramics induce angiogenesis during bone regeneration. Acta Biomater 8:341–349

Chapter 10
Inorganic Polyphosphates: Biologically Active Biopolymers for Biomedical Applications

Xiaohong Wang, Heinz C. Schröder, Ute Schloßmacher, and Werner E.G. Müller

Contents

X.H. Wang
ERC Advanced Investigator Group, Institute for Physiological Chemistry, University Medical Center of the Johannes Gutenberg University, Duesbergweg 6, 55128 Mainz, Germany
e-mail: wang013@uni-mainz.de

H.C. Schröder • W.E.G. Müller (✉)
ERC Advanced Investigator Group, Institute for Physiological Chemistry, University Medical Center of the Johannes Gutenberg University, Duesbergweg 6, 55128 Mainz, Germany

NanotecMARIN GmbH, Duesbergweg 6, 55128 Mainz, Germany
e-mail: wmueller@uni-mainz.de

U. Schloßmacher
ERC Advanced Investigator Group, Institute for Physiological Chemistry, University Medical Center of the Johannes Gutenberg University, Duesbergweg 6, 55128 Mainz, Germany

W.E.G. Müller et al. (eds.), *Biomedical Inorganic Polymers*, Progress in Molecular and Subcellular Biology 54, DOI 10.1007/978-3-642-41004-8_10,

Abstract Inorganic polyphosphate (polyP) is a widely occurring but only rarely investigated biopolymer which exists in both prokaryotic and eukaryotic organisms. Only in the last few years, this polymer has been identified to cause morphogenetic activity on cells involved in human bone formation. The calcium complex of polyP was found to display a dual effect on bone-forming osteoblasts and bone-resorbing osteoclasts. Exposure of these cells to polyP (Ca^{2+} complex) elicits the expression of cytokines that promote the mineralization process by osteoblasts and suppress the differentiation of osteoclast precursor cells to the functionally active mature osteoclasts dissolving bone minerals. The effect of polyP on bone formation is associated with an increased release of the bone morphogenetic protein 2 (BMP-2), a key mediator that activates the anabolic processes leading to bone formation. In addition, polyP has been shown to act as a hemostatic regulator that displays various effects on blood coagulation and fibrinolysis and might play an important role in platelet-dependent proinflammatory and procoagulant disorders.

10.1 Introduction

Inorganic polyphosphate (polyP) is a multifunctional molecule existing in a wide range of organisms including bacteria, fungi, algae, plants, and animals [for a review, see Schröder et al. (1999), Kornberg (1999), Kulaev et al. (2004), Rao et al. (2009)]. It consists of linear polymers of tens to hundreds of phosphate units which are linked together via high-energy phosphoanhydride bonds (Fig. 10.1). Living organisms can produce this polymer metabolically at ambient temperatures, e.g., bacteria via polyphosphate kinases [see Rao et al. (2009)], while the chemical synthesis of polyP requires oven temperatures of several hundred degrees (Griffith 1995). PolyP either prepared chemically or by recombinant polyphosphate kinase has turned out to be a promising material for the treatment of human diseases, in particular bone disorders and dysfunctions like osteoporosis (Wang et al. 2012b). At chain lengths below 100 phosphate units, polyP is readily soluble in water (Katchman and Smith 1958; Van Wazer 1958). In aqueous neutral solution, it is relatively stable despite the presence of phosphoanhydride bonds (Kulaev et al. 2004; Rao et al. 2009).

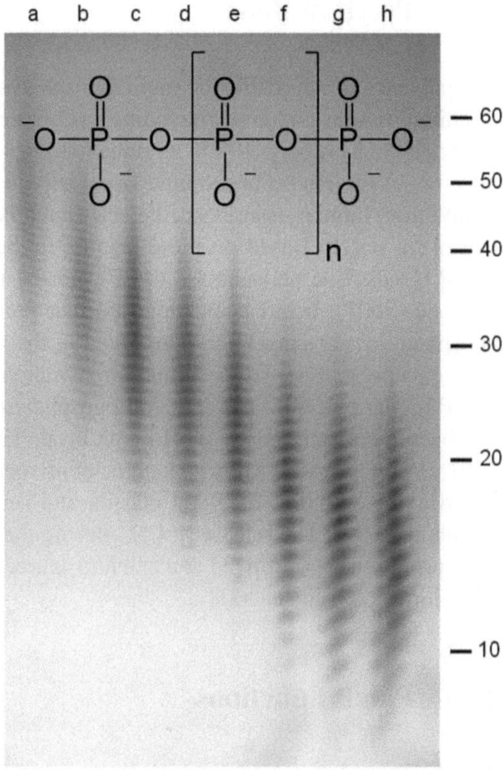

Fig. 10.1 Inorganic polyphosphate. In the background, the separation of various polyP fractions with different ranges of chain lengths on a high-percentage urea polyacrylamide gel is shown

PolyP is a nontoxic polymer and is used as a food additive (Omoto et al. 1997). The nutritional benefit of polyP has been confirmed in animal experiments (Lee et al. 2008, 2009). It also has antioxidant activity (Shahidi et al. 1986). PolyP can bind to or act as a chelator of several divalent cations such as Ca, Mg, Fe, Mn, and Co [see Rao et al. (2009)].

The enzymatic synthesis of polyP, mediated by polyP kinases (polyphosphate: ADP phosphotransferases) [Lorenz et al. (1997b); reviewed in Kulaev et al. (2004), Rao et al. (2009)], involves the transfer of the γ-phosphate of ATP as a precursor. This polyP anabolic enzyme has been identified and characterized in bacteria and yeast (Ahn and Kornberg 1990). The polyP catabolic enzyme, the exopolyphosphatase, is also present in eukaryotes (Lorenz et al. 1994b, 1995; Lorenz and Schröder 2001). Human metastasis regulator protein H-prune has been reported to act as a (short-chain) exopolyphosphatase (Tammenkoski et al. 2008). These enzymes hydrolyze the energy-rich phosphoanhydride bonds in polyP under release of monomeric phosphate (Lorenz et al. 1994b).

Several methods for the analysis of polyP, including quantification and determination of chain length, and of the polyP-metabolizing enzymes have been developed that facilitate the experimental investigation of this polymer and its (enzymatic) formation/degradation [Lorenz et al. (1997c); for a review, see Lorenz and Schröder (1999)].

10.2 PolyP in Bone

The presence of inorganic polymeric phosphates has been demonstrated in a variety of animal tissues and cells, including vertebrates (Lorenz et al. 1997b; Leyhausen et al. 1998; Kornberg 1999). Previous studies revealed that polyP molecules of different chain lengths accumulate especially in bone cells (Leyhausen et al. 1998). In addition, human osteoblast-like cells have been found to contain enzymes that hydrolyze polyP. Based on these results, further studies focused on the effect of polyP on alkaline phosphatase (ALP) present in bone-forming cells (Lorenz and Schröder 2001). It has been proposed that polyP, after hydrolysis to monomeric phosphate by phosphatases, is involved in the control of mineralization processes during vertebrate skeleton formation (Omelon et al. 2009). The degradation of polyP by various exo- and endopolyphosphatases as well as alkaline phosphatases has been demonstrated by us (Lorenz et al. 1994b, 1997b; Lorenz and Schröder 2001). PolyP is a well-known chelator of divalent cations, including Ca^{2+}. It is also found extracellularly, in blood plasma and serum (Kumble and Kornberg 1995; Kornberg 1999). The release of Ca^{2+} during the enzymatic hydrolysis of polyP has been proposed to be utilized as a calcium source for hydroxyapatite (HA) formation (Omelon and Grynpas 2008).

10.3 PolyP Functions

The biological function of polyP in microorganisms has been well studied [reviewed in Kulaev (1979), Wood and Clark (1988)], but further insights into the function of these polymers in animals, in particular humans, have been received only more recently [reviewed in Schröder and Müller (1999), Rao et al. (2009)]. Various functions of polyP have been proposed, and many functions have been demonstrated experimentally. It has been proposed that polyP, among others, acts as a storage substance of energy, as chelator for metal cations, as donor for sugar and adenylate kinase, and as an inducer of apoptosis [Schröder and Müller (1999); reviewed in Kulaev et al. (2004)]. A function of polyP in the mineralization process of bone tissue has already been proposed by us more than 15 year ago [Leyhausen et al. (1998), Schröder et al. (2000); reviewed in Kulaev et al. (2004)]. Since then, a large number of data have been provided that confirmed this function and will be reviewed in this chapter. Later results revealed that polyP, released from platelets (Ruiz et al. 2004), has also an important function in blood coagulation and fibrinolysis (Smith et al. 2006; Müller et al. 2009b). Many data suggest that polyP might act as a modulator of gene expression in bone but also in other tissues and cells. For example, polyP has been shown to cause an increased expression of genes encoding a number of proteins that are crucial for bone formation, like osteocalcin, osterix, bone sialoprotein (Sinha et al. 2010; Sun et al. 2005), and tissue-nonspecific alkaline phosphatase, as determined in the osteoblast-like cell line, MC3T3-E1 (Usui et al. 2010). In addition, polyP has been shown to enhance the mitogenic activity of basic fibroblast growth factor (bFGF) in MC3T3-E1 cells (Yuan et al. 2009). On the other hand, polyP has been reported to inhibit bFGF-induced proliferation and ERK/p38 mitogen-activated protein kinase (MAPK) activation of human endothelial cells, as well as bFGF-induced angiogenesis in vitro and in vivo (Han et al. 2007). PolyP has been shown to accumulate at high levels in the nucleolus of myeloma cells and modulates nucleolar transcription (Jimenez-Nunez et al. 2012). Moreover, polyP causes a specific induction of apoptosis in human plasma cells, myeloma cells, and B lymphoid cell lines, resulting in an inhibition of immunoglobulin secretion (Hernandez-Ruiz et al. 2006).

PolyP has also been found to form complexes and to modulate the activities of many proteins including ion channels. This biopolymer turned out to be a component and a potent activator of the mitochondrial permeability transition pore in the inner mitochondrial membrane of cardiac myocytes, most likely by complex formation with Ca^{2+} (Seidlmayer et al. 2012). This mitochondrial pore has been reported to be associated with mitochondrial dysfunction, resulting in cardiac cell and tissue damage due to an excessive Ca^{2+} accumulation during myocardial infarction and ischemia-reperfusion injury. PolyP is also associated with the transient receptor potential melastatin 8 (TRPM8) channel protein and found to be essential for the function of this channel (Zakharian et al. 2009). In addition, polyP has been proposed to be involved in the energy metabolism of mammalian cells; the production of polyP can be suppressed by the mitochondrial ATP synthase inhibitor oligomycin (Pavlov et al. 2010).

10.4 Bone Formation

Bone formation depends on the functional interaction of HA-forming osteoblasts and HA-dissolving osteoclasts [for a review, see Robins et al. (2006)]. The cross talk between osteoblasts and osteoclasts is controlled by cytokines, such as the macrophage colony-stimulating factor (M-CSF), osteoprotegerin (OPG), and the receptor activator of the NF-kB ligand (RANKL) (Morgan et al. 2008). RANKL stimulates osteoclast precursor cells to differentiate into mature osteoclasts (Zhou et al. 2008), while the maturation of osteoblasts from their precursor cells is promoted by bone morphogenetic protein 2 (BMP-2) (Chen et al. 2004). The stimulatory effect of RANKL is abolished by OPG.

There is only little knowledge about the processes that result in the formation and transport of the primordial calcium phosphate crystals to the extracellular matrix. It is assumed that the early mineralization process starts intracellularly, in matrix vesicles. The crystallites then migrate to the extracellular space (Wuthier et al. 1977; Anderson 1969, 1995). More recently, it has been reported (Rohde and Mayer 2007) that the first product of mineralization of osteoblasts is an amorphous calcium phosphate material that is secreted through an exocytotic mechanism from vacuoles of the osteoblast under formation of mature HA crystallites which are then deposited onto collagen fibrils in the extracellular space. The collagen scaffold, onto which the HA crystallites are deposited, mainly consists of collagen type I (COL-I) (Landis et al. 1996; Robey 2002), as well as non-collagenous proteins that control the differentiation and function of the osteoblasts and osteoclasts. These proteins are synthesized and secreted by osteoblasts [e.g., bone sialoprotein (BSP) (Oldberg et al. 1988), osteopontin (OP) (Oldberg et al. 1986; Sodek et al. 2000), osteocalcin (OC) (Boskey et al. 1998), and osteonectin (Bolander et al. 1988)] or by osteoclasts (e.g., the bone morphogenetic proteins) (Garimella et al. 2008).

The bone-specific ALP (b-ALP) plays a prominent role in bone phosphate metabolism (Lorch 1949). This enzyme (1) generates the inorganic phosphate (P_i) which is required for HA formation (Millan 2006) and (2) hydrolyzes inorganic pyrophosphate (PP_i) which acts as a mineralization inhibitor (Rezende et al. 1998; Hessle et al. 2002). The bone ALP is one potential target of bisphosphonates which are used in treatment of osteoporosis and act as inhibitors of that enzyme (Vaisman et al. 2005).

10.4.1 Osteoblasts

The osteoblasts derive from mesenchymal stem cells (Bruedigam et al. 2010), and their differentiation ends with the osteocytes (Fig. 10.2). The major transcription factor involved in the differentiation of osteoprogenitor cells is the Runt-related transcription factor 2 (RUNX2) which is under the control of BMP-2 (Lee et al. 2010). Runx2 is expressed in the mesenchymal stem cells and along the different stages of the osteoblast lineage (Hu et al. 2011) and causes a stage-dependent increase in gene expression of a series of structural and functional proteins, e.g., b-ALP, COL-I, OP, asialoprotein (ASP), RANKL, BSP, and OC (Fig. 10.2). The HA-producing osteoblasts finally differentiate to osteocytes which become either embedded into the HA deposits or the osteoblasts undergo apoptosis (Bellido and Plotkin 2011). The osteocytes express and release sclerostin, an inhibitable PTH hormone glycoprotein, which functions as inhibitor of BMP-2. Sclerostin, a potent inhibitor of osteoblastogenesis, is downregulated by the parathyroid hormone (PTH) (Agholme et al. 2010). Osteoblasts have been shown to contain relatively large amounts of polyP (400 μM) (Kawazoe et al. 2004).

10.4.2 Osteoclasts

The osteoclasts originate from hematopoietic stem cells (Hayase et al. 1997). Their main differentiation stages including the formation of pre-osteoclasts and finally functionally active, bone-resorbing osteoclasts are given in Fig. 10.2. The differentiation process starts through activation of the transcription factor PU.1 and inflammatory signals. After entering the circulating system, the $CD34^+$ osteoclast precursor cells become recruited onto the bone surface, in the presence of M-CSF

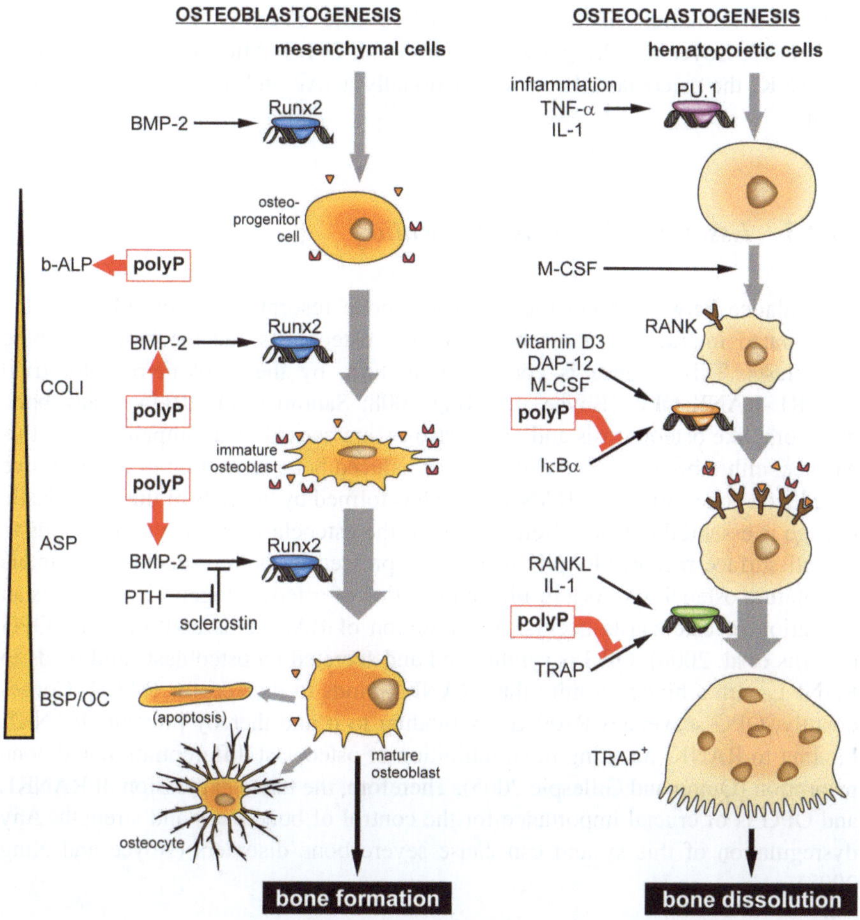

Fig. 10.2 Schematic presentation of the effect of polyP (Ca^{2+} salt) on the differentiation steps of the osteoblast and osteoclast progenitor cells to the functionally active, bone-forming osteoblasts (osteoblastogenesis) and bone-resorbing osteoclasts (osteoclastogenesis). *Left*: Osteoblast differentiation. *Right*: Osteoclast differentiation. The sites at which polyP activates or inhibits the differentiation pathways are highlighted in *red*

and 1,25-dihydroxyvitamin D3 [$1,25(OH)_2D_3$]. Osteoclasts are multinucleated cells that undergo multinucleation after the stimulation of the DAP12 adapter protein/receptor, in contrast to osteoblasts (Boyle et al. 2003). Markers for activated osteoclasts are the tartrate-resistant acid phosphatase (TRAP), the calcitonin receptor (CTR) binding protein, and integrin a_vb_3 (Cerri et al. 2003). In the presence of dihydroxyvitamin D3, osteoclasts, as well as osteoclast precursors, express the receptor activator of NF-κB (RANK). The interaction of RANKL which is produced by osteoblasts with its receptor RANK is crucial for the differentiation of pre-osteoclasts to osteoclasts. If this interaction is blocked by the decoy receptor

OPG, the formation of functionally active osteoclasts is impaired (Simonet et al. 1997; Boyce and Xing 2008; Santini et al. 2011). After binding of RANKL to RANK, the osteoclasts become functionally active and dissolution of HA can occur.

10.4.3 Essential OPG-RANKL Ratio

The balance between bone formation and bone resorption is controlled by the coordinated interaction of the bone-forming osteoblasts and the bone-resorbing osteoclasts. This balance is controlled, in turn, by the cytokine/receptor triad RANKL-RANK-OPG (Boyce and Xing 2008; Santini et al. 2011). Many bone disorders, like osteoporosis and osteopetrosis, are caused by an impairment of this balance, either because of an excessive or reduced bone resorption or an excessive or reduced bone formation. RANKL which is formed by the cells of the osteoblastic lineage is essential for the differentiation of the osteoclasts. This cytokine binds to its cell surface receptor RANK which is expressed by both osteoclast precursors and mature osteoclasts. As a result, an induction of osteoclastogenesis as well as an activation of osteoclasts occurs. The function of RANKL is controlled by OPG (Kearns et al. 2008). OPG is synthesized and secreted by osteoblasts and binds to RANKL with a higher affinity than RANKL binds to its receptor RANK. Consequently, OPG scavenges RANKL by binding to it and thereby prevents RANKL binding to RANK, resulting in an inhibition of osteoclast differentiation and bone resorption (Quinn and Gillespie 2005). Therefore, the tuned expression of RANKL and OPG is of crucial importance for the control of bone mass and strength. Any dysregulation of this system can cause severe bone disorders (Boyce and Xing 2008).

10.4.4 Osteoporosis: Imbalance of the OPG-RANKL Ratio

Osteoporosis is a systemic skeletal disorder which is characterized by a progressive loss of bone tissue and deterioration of bone micro-architecture (Sambrook and Cooper 2006; Lane and Yao 2009). As a consequence of the reduced density of the bone mineral, the risk of bone fractures is increased (Kanis 1994; Felsenberg and Boonen 2005). The cellular basis for osteoporosis is an imbalance of the two cell types involved in bone anabolism and bone catabolism, the osteoblasts and the osteoclasts. The OPG-RANKL ratio has turned out to be a reliable marker for the progression of the disease. This degenerative bone disorder is characterized by a reduced OPG expression and hence an increased bone resorption due to an enhanced osteoclast activity (Raisz 2005). Accordingly, OPG-deficient knockout mice has been found to undergo early-onset osteoporosis (Bucay et al. 1998), while transgenic rats overexpressing the OPG gene show the signs of osteopetrosis (Ominsky et al. 2009).

10.5 Effect of PolyP on Bone-Forming and Bone-Resorbing Cells

10.5.1 Effect on Cell Viability

PolyP is susceptible to hydrolysis by cellular polyphosphatases (Lorenz et al. 1997b). In addition, this polymer forms complexes with divalent cations, such as Ca^{2+} ions. Therefore, for studying the biological effects of polyP, this polymer is preferably used as a complex with Ca^{2+} ions, with a stoichiometric ratio of 2 mol of polyP (with respect to phosphate residues) : 1 mol of Ca^{2+}, in order to avoid any chelating effect that might result in a depletion of Ca^{2+} ions required for HA deposition (Williams and Sallis 1982). In in vitro experiments, this complex of polyP and Ca^{2+} [polyP (Ca^{2+} complex)] as well as polyP (without calcium) turned out to be nontoxic both for osteoblast-like cells (SaOS-2 cells) and osteoclast-like cells (RAW 264.7 cells); the average chain length of the polyP preparation used was 45 phosphate units (Wang et al. 2012a). SaOS-2 cells, used a model system for osteoblasts, are a non-transformed cell line originating from primary osteosarcoma cells (Laitinen et al. 1997) which have a (limited) differentiation capacity (Hausser and Brenner 2005; Kelly et al. 2010) and are capable of mineralization. HA formation of SaOS-2 cells can be initiated by supplementation of the cultures with an activation cocktail, composed of ß-glycerophosphate, ascorbic acid, and dexamethasone; the ß-glycerophosphate which is used as a phosphate source can be replaced by polyP (see below). RAW 264.7 cells are an osteoclast-like murine monocytic cell line (Vincent et al. 2009; Lymperi et al. 2011) which is a suitable model for investigation of osteoclastogenesis in vitro (Kelly et al. 2010). These cells can be induced to differentiate into osteoclasts by cultivation with recombinant RANKL (Tsuda et al. 1997; Teitelbaum 2007; Choi et al. 2010).

10.5.2 Effect on Mineralization of Osteoblasts

PolyP, as Ca^{2+} complex, turned out to be a strong inducer of HA formation in SaOS-2 cells (Wang et al. 2012a). In the experiment shown in Fig. 10.3, the cultures grown in the presence of the activation cocktail (ß-glycerophosphate, ascorbic acid, dexamethasone) were exposed to increasing concentrations of polyP or polyP (Ca^{2+} complex). The extent of mineralization (HA formation) was studied by staining with alizarin red S, an indicator dye for HA formation. A quantitative assessment of HA formation, based on the alizarin red S reaction, using a spectroscopic technique revealed that polyP (Ca^{2+} complex) already at a concentration of 1 μM causes a significant increase in optical density (Wang et al. 2012a); Fig. 10.3. The maximum induction of HA formation is observed at a concentration of 10 μM polyP (Ca^{2+} complex). This effect is only seen if polyP is added to the cells as a calcium

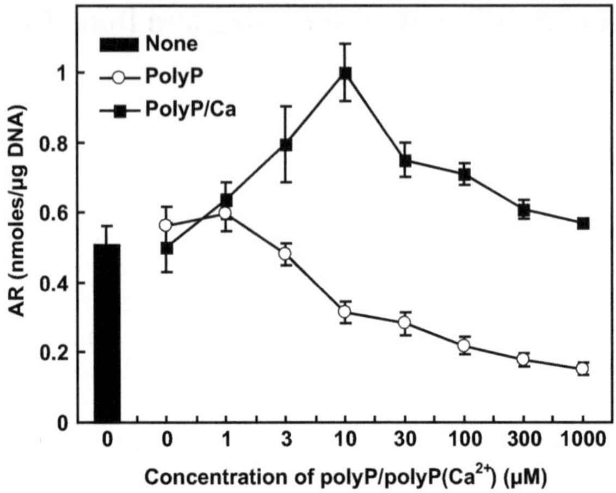

Fig. 10.3 Effect of polyP on HA formation by SaOS-2 cells. The cells were grown in medium/activation cocktail for 7 days in the presence of increasing concentrations of polyP [PolyP] or polyP (Ca^{2+} complex) [PolyP/Ca]. The extent of HA formation was measured spectrophotometrically by staining of cell extracts with alizarin red S [AR] and correlated with the DNA content to normalize for the cell number. Means ± SD ($n = 10$) [Modified after Wang et al. (2012a)]

complex in a 2:1 stoichiometric ratio. Addition of polyP without Ca^{2+} cations results in a decrease in mineralization of the SaOS-2 cells, which becomes significant at concentrations of >10 µM polyP (Wang et al. 2012a). The latter finding has been explained by the fact that the Na^{+} salt of polyP undergoes rapid hydrolytic cleavage into P_i and PP_i by exo- and endophosphatases (Leyhausen et al. 1998). PP_i and polyP are well-known chelators of Ca^{2+} ions (Kulaev et al. 2004) and might deplete this divalent cation required during bone (HA) formation in osteoblasts (Abrams 1998).

The polyP (Ca^{2+} complex) can even be used as a source of P_i in mineralization assays. In SaOS-2 cells, polyP, after hydrolysis by polyP-degrading ALP (Lorenz and Schröder 2001), turned out to be much more effective in HA formation than the organophosphate ß-glycerophosphate (Müller et al. 2011) which is present in the activation cocktail commonly used as a supplement to osteoblast-like cells to induce HA mineralization in vitro (Cheng et al. 1994), even at a low concentrations of 100 µM compared to 1 and 10 mM ß-glycerophosphate (Shioi et al. 1995; Orimo and Shimada 2008). Therefore, polyP (Ca^{2+} salt) can efficiently replace β-glycerophosphate in in vitro HA biomineralization assays (Wang et al. 2012a). The function of polyP as an alternative phosphate source instead of ß-glycerophosphate has also been proposed by Omelon and Grynpas (2008); the latter organophosphate compound is known to be formed intracellularly and released from the cells to the extracellular space (Graff et al. 2003).

SEM analyses revealed a significant increase in the clustering of HA nodules in SaOS-2 cells cultivated in medium supplemented with 10 µM polyP (Ca^{2+} complex) and activation cocktail, compared to cultures grown in the presence of

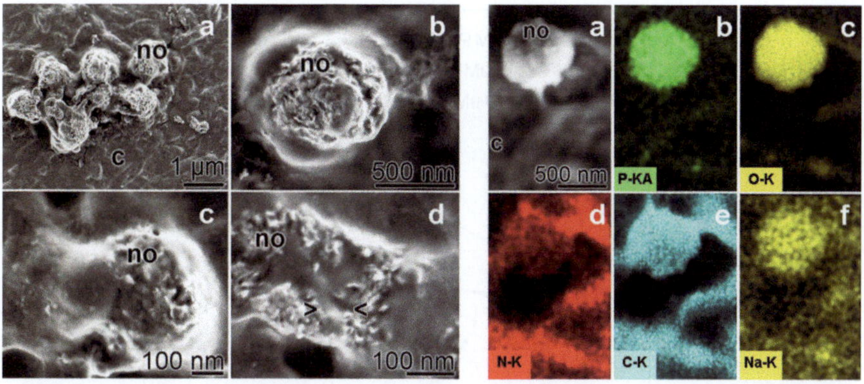

Fig. 10.4 Formation of HA nodules and element mapping. *Left panel*: (**a**), (**b**) Formation of HA nodules on SaOS-2 cells grown in medium supplemented with polyP (Ca^{2+} complex); SEM. Cells were cultivated in medium supplemented with ascorbic acid, dexamethasone, and 100 μM polyP (Ca^{2+} salt) for 4 days. The nodules (no) are surrounded by clusters of cells (c). (**c** and **d**) Higher magnification, showing the fine structure of the HA crystallites on the surface of the cells. In one area the cell lobes that are surrounding and embedding the HA crystallites are highlighted (> <). *Right panel*: Element mapping of one nodule (no) on the surface of a SaOS-2 cell layer; SEM/DEX coupled to a silicon drift detector. Besides the backscattered electron image of the area under study (**a**), the element mappings for P (**b**), O (**c**), N (**d**), C (**e**), and Na (**f**) are shown. Regions with higher accumulation of the respective elements are indicated by a brighter pseudo-color [Modified after Müller et al. (2011)]

activation cocktail without polyP or with 10 μM polyP only (without Ca^{2+}) (Müller et al. 2011). The diameters of the nodules are approximately 1 μm (Fig. 10.4a, b, left panel). The HA crystallites formed in the presence of polyP (Ca^{2+} salt) are intimately associated with the plasma membrane. The analysis of the morphology of the nodules on the surface of the SaOS-2 cells by high-resolution SEM revealed that they consist of crystallites with an average size of 20–40 nm, which are surrounded by cell protrusions, leaving only their tips to be exposed to the extracellular space (Fig. 10.4c, d, left panel).

Two-dimensional EDX mapping of an area within and around a nodule on the SaOS-2 cell surface showed that indeed organic material is interspersed within the crystallites (Müller et al. 2011). In Fig. 10.4a, right panel, the backscattered electron image of the area (comprising one nodule and the surrounding cells) is shown that has been analyzed for the elements P, O, N, C, and Na (Fig. 10.4b–f, right panel). The element distribution pattern over this area revealed that the HA crystallites are embedded in a matrix that contains the elements C and N which are indicative for the presence of organic material (Müller et al. 2011). These results are in line with the results reported by Anderson (1995), indicating that the HA crystallites are initially formed intracellularly, close to the plasma membrane, within vesicles that become exposed and released to the extracellular space, but still remaining attached to the cell surface.

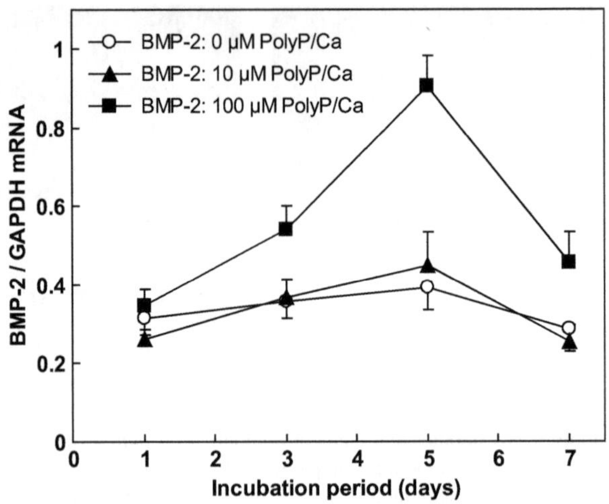

Fig. 10.5 Effect of polyP (Ca^{2+} complex) on expression of *BMP-2* in SaOS-2 cells. The cells were grown either in the absence of polyP (Ca^{2+} complex) [PolyP/Ca] or in the presence of 10 μM polyP (Ca^{2+} complex) or 100 μM polyP (Ca^{2+} complex). The expression levels of *BMP-2* were determined after an incubation period for 1 to 7 days by the qRT-PCR technique. The housekeeping gene *GAPDH* was used as reference for normalization. Means ± SD ($n = 5$) [Modified after Wang et al. (2012a)]

10.5.3 Effect on the Expression of BMP-2

BMP-2 is a well-known inducer of osteoblast differentiation (Nickel et al. 2001). This cytokine is a ligand of a heterotetrameric complex of types I and II transmembrane serine/threonine kinase receptors (BMPR2) (Rosenzweig et al. 1995) and induces osteoblast differentiation via expression of RUNX2-dependent ATF6, a basic leucine zipper transcription factor (Jang et al. 2012). We demonstrated that polyP (Ca^{2+} complex) causes an induction of the expression of the *BMP-2* gene in SaOS-2 cells (Wang et al. 2012a). In quantitative real-time PCR (qPCR) experiments, a strong increase of the steady-state level of *BMP-2* transcripts is seen if the cells are exposed to 10–100 μM polyP (Ca^{2+} complex) (Fig. 10.5). The expression of the housekeeping gene *GAPDH* is used as reference. The highest levels of expression of *BMP-2* are observed 3–5 days after addition of the complex. The expression of BMP-2 then decreases but is still significant after an incubation period of 7 days in the assays containing 100 μM polyP (Ca^{2+} complex) (Wang et al. 2012a).

In further reports, polyP has been described to induce OP, OC, osterix, OPG, COL-I, BSP, and tissue-nonspecific ALP (TNAP) gene expression in the osteoblast-like cell line, MC3T3-E1 (Kawazoe et al. 2004; Hacchou et al. 2007; Usui et al. 2010), and to increase ALP activity in these cells (Hacchou et al. 2007). Moreover, polyP has been found to stimulate the expression of the *BSP* gene in osteoblast-like ROS 17/2.8 cells (Wang et al. 2010). This effect has been proposed to be mediated by a FGF2 response element (FRE) and a homeodomain protein binding site (HOX) in the proximal promoter of the rat BSP gene (Wang et al. 2010). These results confirm that polyP is a potent inductor of osteoblastic differentiation (Usui et al. 2010).

10.5.4 Effect on Osteoclasts/Osteoclastogenesis

Further experiments, on the other hand, showed that polyP (Ca^{2+} complex) impairs osteoclastogenesis. This complex strongly inhibits the progression of RAW 264.7 cells into functional osteoclasts at concentrations >10 μM, as measured by the number of $TRAP^{+}$ cells (Wang et al. 2012a). TRAP is a marker protein for terminally differentiated osteoclasts (Filgueira 2004). The RAW 264.7 cell line, a suitable model to study osteoclastogenesis in vitro, can be induced to differentiate by addition of recombinant RANKL (Choi et al. 2010). As shown in Fig. 10.6, the number of $TRAP^{+}$ cells in cultures, induced by RANKL, becomes strongly reduced in cultures supplemented with polyP (Ca^{2+} complex) at concentrations of 10 μM and higher (Wang et al. 2012a). In the presence of 100 μM polyP (Ca^{2+} complex), the decrease of $TRAP^{+}$ cells even amounts to 50 % after an incubation period of 6 days (Fig. 10.6).

10.5.5 Effect on RANKL-Induced IκBα Phosphorylation

The IκBα kinase (nuclear factor of kappa light polypeptide gene enhancer in B-cell inhibitor, alpha) plays a key role in osteoclast differentiation; it mediates the activation of NF-κB during RANKL-induced differentiation of (pre-)osteoclasts (Lee and Kim 2003). This kinase turned out to be a potential intracellular target for the inhibitory effect of polyP (Ca^{2+} complex) on osteoclast-like RAW 264.7 cells (Wang et al. 2012a). In the experiment shown in Fig. 10.7, RAW 264.7 cultures had been pretreated with polyP (Ca^{2+} complex) for 12 h; in parallel assays, the cells remained untreated. Subsequently the cells were incubated with the peptide ALLN to inhibit ubiquitination and degradation of IκBα (Ghosh and Karin 2002) and finally supplemented with RANKL. Immunoblotting studies revealed an extensive phosphorylation of IκBα in the cytoplasmic fractions of cells that had been incubated in the absence of polyP (Ca^{2+} complex) but the presence of RANKL (Fig. 10.7, upper panel), indicating an activation of the IκBα kinase. The phosphorylation/activation of IκBα kinase is required for NF-κB activation (Lee and Kim 2003; Sung et al. 2009). Addition of polyP (Ca^{2+} complex) to the RAW 264.7 cells strongly inhibits the phosphorylation of IκBα kinase, even at a low concentration of 10–100 μM polyP (Ca^{2+} complex) (Wang et al. 2012a). In control experiments, it was established that the protein level of IκBα remains almost unchanged (Fig. 10.7, lower panel). It has been concluded that the polyP (Ca^{2+} complex) interferes with

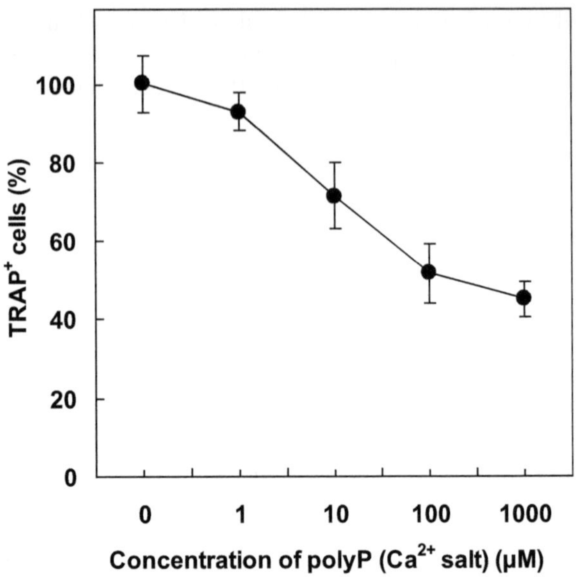

Fig. 10.6 Inhibition of differentiation of RAW 264.7 cells by polyP (Ca^{2+} complex). The cells were incubated for 6 days in the presence of 50 ng/ml of soluble RANKL and increasing concentrations of polyP (Ca^{2+} complex). The percentage of $TRAP^+$ cells is given. Means ± SD ($n = 6$) [Modified after Wang et al. (2012a)]

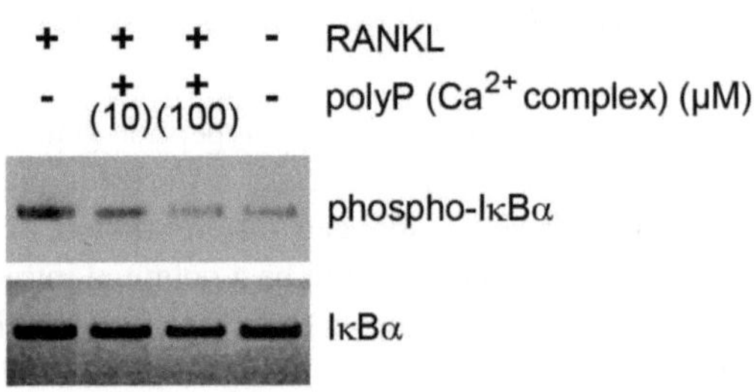

Fig. 10.7 Inhibition of RANKL-induced phosphorylation of IκBα by polyP (Ca^{2+} complex) in RAW 264.7 cells. The cells either remained untreated [– polyP(Ca^{2+} complex)] or were pretreated with 10 or 100 µM polyP (Ca^{2+} complex) [+ polyP(Ca^{2+} complex)] for 12 h. Subsequently, they remained either non-induced (– RANKL) or were exposed to 10 nmol/l of RANKL for 15 min (+ RANKL). Thereafter, cytoplasmic extracts were prepared and subjected to SDS-PAGE and Western blot analysis, using either IκBα or phospho-IκBα antibodies. The immunocomplexes formed were detected using a secondary antibody and visualized by enhanced chemiluminescence reagent [After Wang et al. (2012a)]

the RANKL-mediated NF-κB activation on the level of IκBα kinase (Wang et al. 2012a). As a consequence, the differentiation of pre-osteoclasts to mature osteoclasts is impaired.

10.5.6 Effect on Alkaline Phosphatase

The ALP is a key enzyme involved in the phosphate metabolism in bone tissue (Anderson et al. 2004). This enzyme is particularly found in those areas of bone which show highest mineralization (Lorch 1949). In human, four isoenzymes of ALP are expressed in a tissue-specific pattern (Price 1993). One of these isoforms, the tissue-nonspecific ALP/TNAP, is expressed in bone, kidney, and liver (Henthorn 1996). The three others are present in intestine, placenta, and germ cells. In bone tissue, the ALP (TNAP) has been proposed, on one hand, to generate P_i required for HA precipitation (Millan 2006) and, on the other hand, to hydrolyze PP_i which acts as mineralization inhibitor (Hessle et al. 2002); the monomeric P_i formed in this way can again be used for HA mineralization in the presence of Ca^{2+} ions. The ALP therefore initiates and promotes the formation of HA crystals in the osteoblast matrix vesicles which are then transported to the extracellular matrix (Anderson 1995; Hui and Tenenbaum 1998; Anderson et al. 2004).

PolyP (Ca^{2+} salt) has been demonstrated to be a potent inducer of ALP in SaOS-2 cells (Müller et al. 2011). As shown in Fig. 10.8, exposure of SaOS-2 cells to polyP (Ca^{2+} salt), at a concentration of 100 μM, causes a strong increase in the enzymatic activity of ALP. The highest levels of enzyme activity are observed after an incubation period of 60 h. In extracts from cells grown in the presence of 100 μM polyP (Ca^{2+} salt), the activity of the enzyme increases by about 14-fold, whereas in the absence of the polymer the activity of the enzyme only increases by about 3-fold during that period (Müller et al. 2011). After longer incubation periods, the enzyme activity decreases again. Exposure of the cells to 100 μM P_i instead of polyP (Ca^{2+} salt) only resulted in a small (about threefold) increase in ALP activity between a 12-h and 48-h incubation period. Immunofluorescence analysis revealed a strong increase in the staining of ALP at the cell surface of polyP (Ca^{2+} salt)-treated SaOS-2 cells compared to non-treated cells (Müller et al. 2011) (Fig. 10.8, inset).

Further studies showed that polyP (Ca^{2+} salt) also induces the steady-state expression of the gene encoding the TNAP (Müller et al. 2011). Exposure of SaOS-2 cell cultures to 100 μM polyP (Ca^{2+} salt) resulted in a marked upregulation of *TNAP* expression (Fig. 10.9). The maximal increase of *TNAP* expression is seen after 48 h. Thereafter, a decrease of the steady-state level of *TNAP* expression is found.

It is not yet known if polyP is taken up by the cells. However, it is considered to more likely that the polymer remains outside the cells, where it can bind to cell surface proteins (Müller et al. 2011). At the cell surface, the polyP might be hydrolyzed by the ALP which is bound to the osteoblast cell membrane (Cyboron and Wuthier 1981; Wu et al. 1989). This enzyme is able to hydrolyze polyP (Lorenz and Schröder 2001). The ALP has been localized in membrane fractions of matrix vesicles and plasma membranes at the sites where the HA crystallites grow (Banovac and Koren 2000). It is reasonable to assume that polyP is hydrolyzed to P_i extracellularly at the spot, where P_i is required for HA formation, very likely through the action of the plasma membrane-bound ALP (Müller et al. 2011).

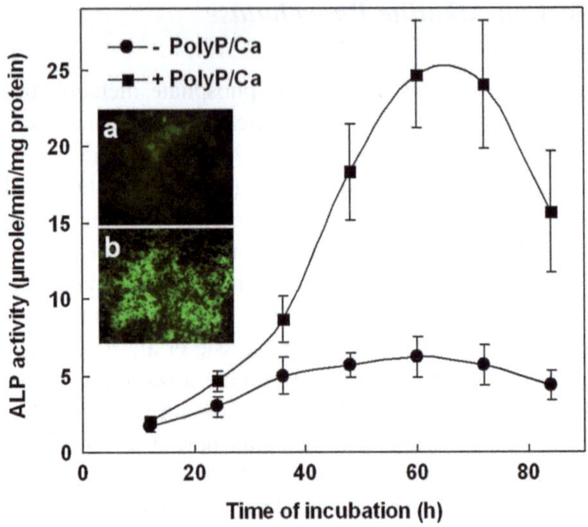

Fig. 10.8 Effect of polyP (Ca^{2+} salt) on ALP activity in SaOS-2 cells. The cells either remained in the absence of polyP (Ca^{2+} salt) [– PolyP/Ca] or were incubated in the presence of 100 µM polyP (Ca^{2+} salt) [+ PolyP/Ca]. After the indicated incubation periods, the cells were broken to solubilize the enzyme. The ALP activity was determined using 4-nitrophenylphosphate as substrate. Mean values ± SD ($n = 6$). Inset: Localization of ALP on SaOS-2 cell layers as analyzed by fluorescence microscopy. SaOS-2 cells were cultured either in medium lacking any additional phosphate component (**a**) or in medium that had been supplemented with 100 µM polyP (Ca^{2+} salt) for 4 days (**b**). Then the cells were exposed to anti-alkaline phosphatase antibodies and then to fluorescent-labeled secondary antibodies [Modified after Müller et al. (2011)]

It should be noted that not only polyP but also monomeric P_i has been described as an inducer of gene expression. Monomeric phosphate causes an upregulation of *OP* gene expression in the osteogenic cell line MC3T3-E1 (Beck et al. 2000). Also in this case, like for polyP, the intracellular pathway through which phosphate species induces gene expression is not known, even though in the case of P_i, several transport systems have been identified, like the osteoblast cell-specific Na-P_i transporter (Veldman et al. 1995; Palmer et al. 1997).

10.5.7 Effect on Intracellular Ca^{2+} Level

The ALP-mediated hydrolysis of polyP (Ca^{2+} salt) results not only in the release of P_i but also in the release of Ca^{2+} ions. In further studies, it has therefore been investigated if the increase in extracellular free calcium concentration ($[Ca^{2+}]_e$) that can be expected as a result of ALP reaction, following exposure of SaOS-2 cells to polyP (Ca^{2+} salt), leads to a change of the intracellular Ca^{2+} level ($[Ca^{2+}]_i$) of the SaOS-2 cells (Müller et al. 2011). The $[Ca^{2+}]_i$ level has been determined by applying the fura-2 AM technique. The results revealed that exposure of SaOS-2 cells to 100 µM polyP (Ca^{2+} salt) causes a significant but transient increase in

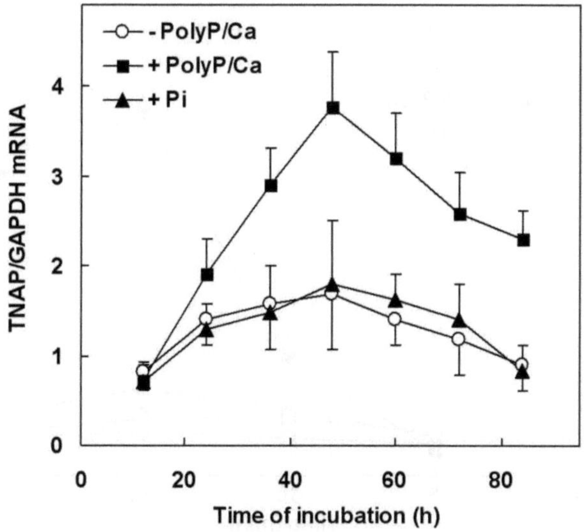

Fig. 10.9 *TNAP* transcript levels in SaOS-2 cells grown either in the absence of any phosphate component [– PolyP/Ca] or in the presence of polyP (Ca^{2+} salt) [+ PolyP/Ca] or in the presence of inorganic phosphate [+ Pi]. The phosphate components were added at a concentration of 100 μM. After the indicated incubation periods, RNA was extracted and subjected to quantitative real-time RT-PCR analysis for *TNAP* and *GAPDH* transcripts. Mean values ± SD ($n = 4$) [Modified after Müller et al. (2011)]

$[Ca^{2+}]_i$ of the cells (Fig. 10.10). As shown in Fig. 10.10a, b, a strong rise of the 340/380 excitation ratio is seen already 10 min after addition of the polymer. The maximum increase is observed at 15 min. Thereafter, the 340/380 ratio dropped to lower values (Fig. 10.10a). In the absence of complexed Ca^{2+}, at the same concentration, polyP causes only a slight increase in $[Ca^{2+}]_i$. P_i and Ca^{2+} have no effect (Fig. 10.10a). These results support the assumption that polyP (Ca^{2+} salt) is locally, on the surface of the cells, hydrolyzed, resulting in an increase of intracellular Ca^{2+} level (Müller et al. 2011).

In this context, it should be noted that in bone, the extracellular calcium level ($[Ca^{2+}]_e$) is comparably high (1.1–1.3 mM; Dvorak and Riccardi 2004). At this concentration, osteoblasts (and SaOS-2 cells) respond with an increase in intracellular calcium level ($[Ca^{2+}]_i$; Chang et al. 1999).

10.6 Effect on Mesenchymal Stem Cells

The polyP-caused activation of the FGF signaling pathway has been found to result in an induction of differentiation of mesenchymal stem cells as demonstrated by Kawazoe et al. (2008). Investigations of the effect of polyP on the osteogenic

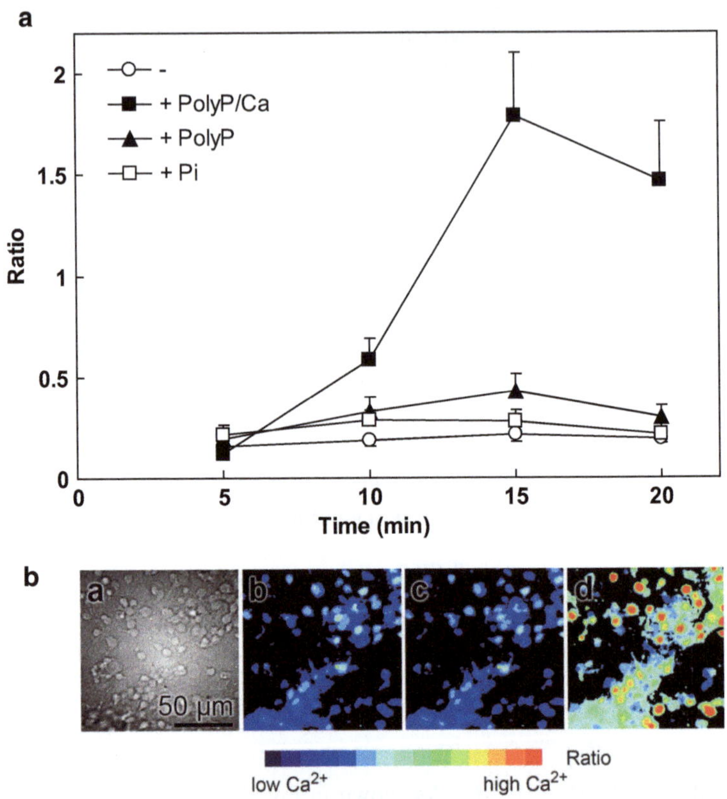

Fig. 10.10 Influence of polyP (Ca^{2+} salt) on the intracellular Ca^{2+} level ($[Ca^{2+}]_i$) in SaOS-2 cells. The cells either remained in medium without addition of any phosphate (–) or were supplemented with 100 µM polyP (Ca^{2+} salt) (+ PolyP/Ca), 100 µM polyP (+ PolyP), or inorganic phosphate (+ Pi). (**a**) Change of the 340/380 nm ratio of fura-2 fluorescence of cells exposed to 100 µM polyP (Ca^{2+} salt), to 100 µM polyP, or to P_i after an incubation period of 5 min, 10 min, 15 min, and 20 min. Mean values ± S.E. ($n = 50$). (**b**) Fura-2 AM fluorescence imaging of the cells during incubation with 100 µM polyP (Ca^{2+} salt). (**b–a**) Nomarski phase contrast image. (**b–b–d**) Fluorescence images recorded 5 min (**b**), 10 min (**c**), and 15 min (**d**) after addition of polyP (Ca^{2+} salt). The color scale ranges from low levels of Ca^{2+} (*blue*) to high Ca^{2+} levels (*red*) [Modified after Müller et al. (2011)]

differentiation of human dental pulp cells (HDPCs) and human mesenchymal stem cells (hMSCs) revealed an enhanced expression of COL-I, OP, OC, OPG, and matrix metalloproteinase-1, suggesting that polyP induces the activation of both proliferation and mineralization of stem cells via the FGF signaling pathway (Kawazoe et al. 2008). Differentiation of hMSCs from patients with osteoarthritis and rheumatoid arthritis into osteoblastic cells by polyP has been reported by Morimoto et al. (2010). Besides an induction of ALP activity, a significant change in expression of COL-I, ALP, OC, and BSP has been found (Morimoto et al. 2010).

10.7 Effect on Chondrocytes

Treatment of primary chondrocyte cultures with polyP has been shown to result in an induction of accumulation of glycosaminoglycan and collagen (St-Pierre et al. 2012). The increased glycosaminoglycan and collagen production was found to be transient, most likely because of the polyP-degrading exopolyphosphatase activities present in chondrocytes (St-Pierre et al. 2012).

10.8 Effect on Fibroblasts

Shiba et al. (2003) reported that polyP enhances the mitogenic activity of the fibroblast growth factors FGF-1 and FGF-2, leading to an increased proliferation of human fibroblasts. They demonstrated that this effect is caused by a polyP-caused stabilization and a polyP-induced increase of the binding affinity of the FGFs to their cell surface receptors (Shiba et al. 2003). Topical application of polyP has been shown to induce remodeling of gingival connective tissue in a rat periodontitis model (Kasuyama et al. 2012).

10.9 Inhibition of Exopolyphosphatase Activity by Bisphosphonates

Bisphosphonates are analogs of PP_i that contain a P–C–P bond instead of the P–O–P bond. These compounds cannot be degraded by the known (poly)phosphatase activities, ALP, pyrophosphatase, and (yeast) exopolyphosphatase (Sperow et al. 1973; Leyhausen et al. 1998). Investigations on (poly)phosphatase activity present in osteoblast-like cells revealed a strong inhibition by bisphosphonates (Leyhausen et al. 1998) (Fig. 10.11). Etidronate is most effective, followed by clodronate and pamidronate. The IC_{50} for etidronate amounts to 2.04 mM. Similar values have been found for the inhibition of the exopolyphosphatase from yeast and rat liver; the IC_{50} values are 1.75 and 1.31 mM, respectively. The IC_{50} values for clodronate and pamidronate on the yeast enzyme are 3.06 and 3.24 mM, respectively. The inhibitory effect of the bisphosphonates on the inorganic pyrophosphatase (yeast) is different from that on exopolyphosphatase activity; clodronate is the most potent inhibitor of pyrophosphatase (Leyhausen et al. 1998).

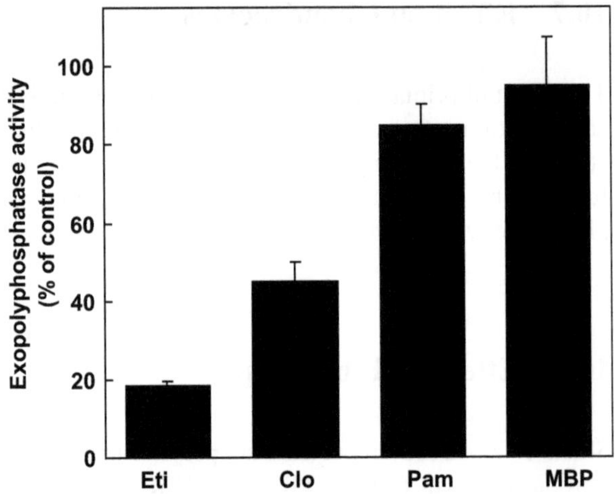

Fig. 10.11 Effect of different bisphosphonates on exopolyphosphatase activity in human osteoblast-like cells; *Eti* etidronate, *Clo* clodronate, *Pam* pamidronate, *MBP* methylene-1,1-bisphosphonate (1 mg/ml each). Substrate: polyP$_{35}$. Mean values ± S.E. ($n = 3$) [Modified after Leyhausen et al. (1998)]

10.10 Coaction of PolyP with Other Biopolymers: Biosilica

Another biopolymer that has been recognized as an inducer of bone (HA) formation and a promising component of morphologically active bone substitution materials (see also Chap. 9 of this volume) is biosilica (Schröder et al. 2005, 2012b; Wang et al. 2012b). This material forms the inorganic skeleton of sponges (Morse 1999). One surprising property of the sponge biosilica is the fact that it can be formed enzymatically, like polyP. The protein that is involved in biosilica formation, termed silicatein (Shimizu et al. 1998; Krasko et al. 2000), has been demonstrated to be an enzyme that mediates the formation of biosilica from monomeric (orthosilicate) precursors (Schröder et al. 2012c; Müller et al. 2013a). Recent results revealed that silicatein does not only catalyze the formation of the siloxane linkages between the silicic acid units but also has structure-guiding activity and hence influences the morphology of the biosilica product (Schröder et al. 2012a). The initially formed biosilica product, formed by silicatein, undergoes a hardening/aging process during which the water that is generated during the polycondensation reaction is removed (syneresis) and new siloxane bonds are formed (Müller et al. 2011). Biosilica is a biocompatible polymer (Müller et al. 2009b), like polyP, and first animal experiments have been performed (Wiens et al. 2010a). Biosilica has been proposed as a promising material that can be applied in bone tissue engineering or in treatment of bone diseases like osteoporosis (Schröder et al. 2011), in combination with polyP.

In in vitro experiments, it has been demonstrated that exposure of bone-forming cells (SaOS-2 cells) to biosilica results in an upregulation of the expression of the *OPG* gene, while the level of *RANKL* remains unaffected (Wiens et al. 2010c). As a consequence, biosilica causes an increase in the OPG-RANKL ratio (Wiens et al. 2010c). This result suggests that this biopolymer is of potential therapeutic importance for the treatment of bone diseases which are characterized by an

OPG-RANKL imbalance, in particular osteoporosis (Jabbar et al. 2011). In addition, biosilica formed by silicatein enhances the expression of *BMP-2* as well as HA formation in SaOS-2 cells (Wiens et al. 2010b) and stimulates the growth of these cells, as determined in incorporation studies of [^{3}H]dT into DNA (Wiens et al. 2010b).

10.11 Differential Effect of PolyP and Pyrophosphate

It should be noted that PP_i acts as an inhibitor of the mineralization process (Fleisch et al. 1966b; Register and Wuthier 1985), while polyP stimulates HA formation. On the other hand, PP_i is an essential source of phosphate required for HA deposition when it is hydrolyzed by tissue-nonspecific ALP (TNAP). PP_i is formed by an ectonucleotide pyrophosphatase/phosphodiesterase-1 (ENPP1) activity. ENPP1 is expressed in matrix vesicles and plasma membranes of osteoblasts, like TNAP (Johnson et al. 2000), and acts as regulator of matrix mineralization by increasing extracellular levels of PP_i. Investigations of the function of ENPP1 in pre-osteoblasts revealed that the expression of this enzyme is crucial for the differentiation into osteoblasts (Nam et al. 2011). It has been demonstrated that treatment of cells with PP_i inhibits the expression of several genes associated with osteoblast differentiation and mineralization, including OC, BSP, and TNAP gene (Nam et al. 2011), in contrast to polyP that causes an increased expression of these genes (see above).

Recently, experimental evidence has been presented that HA mineralization in bone tissue is preceded by Ca carbonate precipitation (Müller et al. 2013b). The enzymatically formed Ca carbonate deposits thereby act as bioseeds. In one intriguing model, it has been proposed (Müller et al. 2013b) that the transformation of Ca carbonate into Ca phosphate is facilitated by dissolution of the Ca carbonate deposits through complex formation of calcium with polyP or PP_i, present in bone tissue. PolyP and PP_i are known to form soluble complexes with Ca^{2+} ions (Fleisch and Neuman 1961; Fleisch et al. 1966a; Fleisch 1999). The chelated Ca^{2+} ions are liberated after hydrolysis of the polyP/PP_i component of the polyP·Ca/PP_i·Ca complexes formed, mediated by bone ALP (Müller et al. 2011), followed by precipitation in the presence of P_i ions under formation of Ca phosphate deposits. This model implies that the initial formation of Ca carbonate in bone tissue might support Ca phosphate deposition by a localized accumulation of Ca, in the form of Ca carbonate deposits and subsequently formed soluble polyP·Ca/PP_i·Ca complexes, which are then transformed into Ca phosphate deposits, even if only limited amounts of phosphate are available.

10.12 PolyP Containing Bone Regeneration Materials

PolyP, because of its morphogenetic activity, has been proposed to be a promising biopolymer for the design of novel biomimetic scaffold materials for bone tissue engineering, like biosilica (Wang et al. 2013). In contrast to its Na^+ and K^+ salts which are soluble at physiological pH, polyP becomes insoluble, at higher concentrations, if complexed with Ca^{2+} and with increasing chain length of the polymer. Therefore, it is possible to harden the polymer, if complexed with Ca^{2+}, a prerequisite for a material to be used for bone tissue engineering applications.

As biosilica is able to increase the OPG-RANKL ratio (enhancement of OPG expression without affecting RANKL expression), a property not found for polyP, the combination of the two biopolymers, polyP and biosilica, in a biomimetic scaffold material for bone tissue engineering might provide additional benefits, in particular for the treatment of osteoporotic patients (Schröder et al. 2011).

Moreover, polyP can be combined with other materials to generate new bone substitution/regeneration materials. For example, polyP can be adsorbed onto porous HA; the generated material was found to enhance the initial bone regeneration (Morita et al. 2010). Nevertheless, the development of morphogenetically active bone substitution/regeneration materials based on polyP is still at a very preliminary stage, but it is assumed that biomimetic materials based on this polymer, either alone or in combination with biosilica, will provide suitable templates for the design of novel bioactive scaffold materials that are able to induce or to promote the growth and differentiation of cells involved in bone formation.

10.13 PolyP Coating of Metal Implants

PolyP treatment of titanium implants has been described to significantly enhance the attachment and proliferation of human bone marrow-derived mesenchymal stem cells (Maekawa et al. 2007), as well as the attachment, proliferation, and expression of *COL-I* gene of mouse osteoblast-like MC3T3-E1 cells (Maekawa et al. 2008). Moreover, polyP treatment promotes bone regeneration around titanium implants on rat tibiae (Maekawa et al. 2009).

10.14 Antifouling Through PolyP Polymers

Recently, we described a new approach to eliminate fouling bacteria (Müller et al. 2012). In this bio-inspired approach, copper is one component in the antifouling system, in combination with polyP. PolyP, also alone, is known to display antibacterial activity (Kulakovskaya et al. 2012). Even more, polyP shows anticorrosion properties (Wang and Yang 2010). In an attempt to elucidate the potential usefulness of polyP as antifouling agent, combination experiments of polyP together with copper have been performed (Müller et al. 2012). It was found that exogenously added polyP, as well as the non-hydrolyzable analog, the bisphosphonate dichloromethylene diphosphonic acid (DMDP), causes a strong synergistic effect on copper-induced toxicity on biofilm-producing *Streptococcus mutans* cells. Further it was found that prior to be functionally active, polyP that is present as an intracellular bacterial metabolite has to be hydrolyzed to orthophosphate. The P_i then binds to and exports the toxic copper out of the cell; this reaction results in a reduction of the copper toxicity for the bacteria. In order to avoid the export of copper, inhibitors of enzymatic hydrolysis of polyP, e.g., DMDP, have been added. Exposure of bacteria to polyP together with DMDP results in a strong enhancement of polyP-caused inhibition of biofilm formation. This new strategy, the application of polyP/polyphosphatase inhibitors along with copper, has been proposed as a new principle for developing bio-inspired antifouling compounds (Müller et al. 2012).

10.15 Antiviral Activity

PolyP has been reported to display antiviral activity in vitro against human immunodeficiency virus type 1 (HIV-1) infection (Lorenz et al. 1997a). Infection of cells with HIV-1 can be inhibited by polyP of various chain lengths (4, 15, 34, and 91 phosphate residues) at concentrations of 300 μM and higher (based on phosphate residues). PolyP molecules with longer chain lengths (15 and 34 phosphate residues) also inhibit HIV-1-induced syncytium formation, while tripolyphosphate and tetrapolyphosphate are not effective. Based on the results of the syncytium assay and of cell-virus binding experiments, it has been concluded that the anti-HIV effect of polyP is most likely caused by an inhibition of virus adsorption (Lorenz et al. 1997a).

10.16 Effect on Blood Coagulation and Fibrinolysis

Platelets have important functions in hemostasis, thrombosis, and inflammation. The dense granules of human platelets contain substantial amounts of polyP, with chain lengths of 70–75 (Ruiz et al. 2004) or 60–100 phosphate units (Müller et al. 2009b), which is released upon platelet activation (Ruiz et al. 2004; Smith et al. 2006). The platelet dense granules show similar characteristics like acidocalcisomes or volutin granules. These electron dense and acidic organelles of bacteria and unicellular eukaryotes have a high content of polyP. Similar polyP- and calcium "storage" organelles have been found in eggs of insects (Ramos et al. 2011). Smith et al. (2006) reported that polyP secreted by platelets acts as a hemostatic regulator. PolyP turned out as a procoagulant agent that accelerates blood clotting by promoting the activation of factor V and activation of the contact pathway (Smith et al. 2006; Smith and Morrissey 2008). On the other hand, polyP delays clot lysis by enhancing the thrombin-activatable fibrinolysis inhibitor (Smith et al. 2006). In addition, polyP secreted from activated human platelets binds to and activates factor XII (Müller et al. 2009a). Activation of this plasma protease in turn triggers the release of bradykinin, an inflammatory mediator, formed by processing of kininogen mediated by plasma kallikrein (Müller et al. 2009a). PolyP released from platelets hence acts as an endogenous factor XII activator that initiates fibrin formation via the intrinsic pathway (Müller and Renné 2011). The various effects of polyP on coagulation and fibrinolysis (Caen and Wu 2010) depend on the size of the polymer (Smith et al. 2010). PolyP molecules with chain lengths $\geq$500 phosphate units such as those formed in microorganisms exhibit optimal activation of the contact pathway, while those with shorter chain lengths of around 100 phosphate units, as present in platelets, are less active but still sufficient to activate factor V and to abolish the anticoagulant function of the tissue factor pathway inhibitor (Smith et al. 2010). Investigation of the interaction of polyP with fibrinogen and fibrin revealed that polyP downregulates fibrinolysis by reducing the binding of plasminogen and tissue plasminogen activator (tPA) to fibrin surfaces (Mutch et al. 2010). PolyP also acts as a cofactor for the thrombin-mediated activation of factor XI. Choi et al. (2011) reported that polyP accelerates the activation of factor XI by α-thrombin, β-thrombin, and factor XIa. These results show that polyP is a potent modulator of the blood clotting that might play an important role in platelet-driven proinflammatory and procoagulant disorders (Morrissey 2012).

10.17 Conclusion

In recent years, polyP has turned out to be a promising biomedical polymer that raised increasing interest in bone tissue engineering (Wang et al. 2012b). PolyP can be formed enzymatically both in microorganisms and in animals (e.g., sponges). This biopolymer is biocompatible like biosilica (Wiens et al. 2010a) and has—at least in vitro—the potential to contribute to or to potentiate the beneficial effects of biosilica on bone formation. PolyP, given as a Ca^{2+} complex, displays a dual effect on bone-forming osteoblasts and bone-resorbing osteoclasts (1) it promotes—through upregulation of *BMP-2* expression—the maturation of osteoblasts, resulting in an enhanced mineralization (HA deposition), and (2) it inhibits the differentiation of osteoclast precursors to functionally active osteoclasts through interfering with the NF-κB signaling pathway. In addition, polyP (Ca^{2+} salt) stimulates bone ALP. It is not yet known whether polyP is taken up by the cells. On the other hand, it is well established that polyP exists intracellularly (Lorenz et al. 1994b; Imsiecke et al. 1996; Kornberg 1999). The biological effects of polyP on osteoblasts and osteoclasts, either alone or in combination with biosilica, demonstrate that this biopolymer may have the potential also to show beneficial activity in vivo, in particular towards therapy/prophylaxis of osteoporosis. First animal experiments are ongoing to demonstrate if the morphogenic effects of polyP, either alone or in combination with biosilica, can be applied in humans.

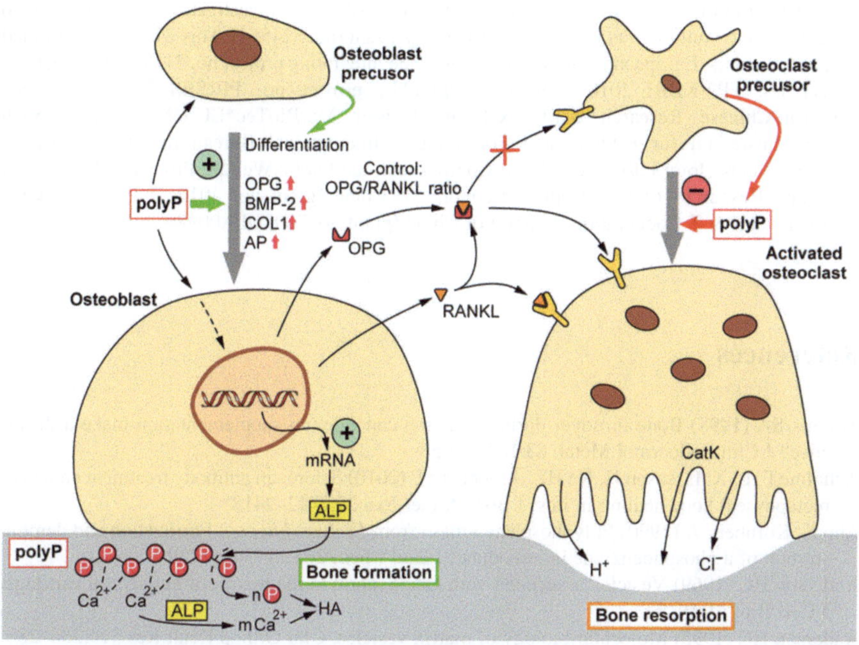

Fig. 10.12 Schematic summary of the effect of polyP (Ca^{2+} salt) on the bone-forming osteoblasts and bone-dissolving osteoclasts. The interaction between these cells and the targets of polyP during osteoblast and osteoclast differentiation are marked. PolyP supports the progression of precursor osteoblasts to mature osteoblasts by induction of the BMP-2 gene. In addition, it induces the expression/activation of the bone-specific ALP, resulting in an increased release of ALP. This enzyme is able to hydrolyzing not only β-glycerophosphate (the commonly used substrate in bone mineralization studies) but also polyP-Ca^{2+}-complexes. The inorganic phosphate (P_i) produced during this reaction might act as a phosphate source for HA crystallite formation, along with the liberated Ca^{2+} ions which might be involved in the rise in intracellular free calcium level, induced by polyP (Ca^{2+} salt). In addition, the cytokine/receptor triad RANKL-RANK-OPG is outlined that can be affected by the second biopolymer, biosilica [not shown here; for details, see Sect. 10.10 and Schröder et al. (2011)]. The bone resorption by osteoclasts is mediated both by chemical (e.g., HA dissolution through decrease in pH) and enzymatic (e.g., cathepsin K) processes

A schematic presentation summarizing the modes of action of polyP (Ca^{2+} salt) is shown in Fig. 10.12. PolyP shifts the tuned balance between osteoblasts (bone anabolism) and osteoclasts (bone catabolism) towards new bone formation. Therefore, this polymer might be applicable under disease conditions which are characterized by an increased HA degradation due to an imbalance between bone formation and bone degradation. There are good reasons to assume that this polymer, based on its unique properties, either alone or in combination with biosilica, opens new ways not only for prophylaxis and/or treatment of osteoporosis but also for the design of novel morphogenetically active scaffold materials which can be applied for rapid prototyping (3D printing) procedures (Wang et al. 2013).

Acknowledgments W.E.G.M. is a holder of an ERC Advanced Investigator Grant (no. 268476 BIOSILICA) as well as of an ERC Proof-of-Concept grant (no. 324564). This work was supported by grants from the European Commission (large-scale integrating project no. 311848, BlueGenics; project no. FP7-KBBE-2010-4-266033, SPECIAL; project no. PIRSES-GA-2009-246987, European-Chinese Research Staff Exchange Cluster MarBioTec*EU-CN*), the German Bundesministerium für Bildung und Forschung – International Bureau (no. CHN 09/1AP, German-Chinese Joint Lab on Bio-Nano-Composites), the Public Welfare Project of Ministry of Land and Resources of the People's Republic of China (grant no. 201011005-06), and the International S & T Cooperation Program of China (grant no. 2008DFA00980).

References

Abrams SA (1998) Bone turnover during lactation-can calcium supplementation make a difference? J Clin Endocrinol Metab 83:1056–1058

Agholme F, Li X, Isaksson H, Ke HZ, Aspenberg P (2010) Sclerostin antibody treatment enhances metaphyseal bone healing in rats. J Bone Miner Res 25:2412–2418

Ahn K, Kornberg A (1990) Polyphosphate kinase from *Escherichia coli*. Purification and demonstration of a phosphoenzyme intermediate. J Biol Chem 265:11734–11739

Anderson HC (1969) Vesicles associated with calcification in the matrix of epiphyseal cartilage. J Cell Biol 41:59–72

Anderson HC (1995) Molecular biology of matrix vesicles. Clin Orthop Relat Res 314:266–280

Anderson HC, Sipe JB, Hessle L, Dhanyamraju R, Atti E, Camacho NP, Millán JL (2004) Impaired calcification around matrix vesicles of growth plate and bone in alkaline phosphatase-deficient mice. Am J Pathol 164:841–847

Banovac K, Koren E (2000) Triiodothyronine stimulates the release of membrane-bound alkaline phosphatase in osteoblastic cells. Calcif Tissue Int 67:460–465

Beck GR, Zerler B, Moran E (2000) Phosphate is a specific signal for induction of osteopontin gene expression. Proc Natl Acad Sci USA 97:8352–8357

Bellido T, Plotkin LI (2011) Novel actions of bisphosphonates in bone: preservation of osteoblast and osteocyte viability. Bone 49:50–55

Bolander ME, Young MF, Fisher LW, Yamada Y, Termine JD (1988) Osteonectin cDNA sequence reveals potential binding regions for calcium and hydroxyapatite and shows homologies with both a basement membrane protein (SPARC) and a serine proteinase inhibitor (Ovomucoid). Proc Natl Acad Sci USA 85:2919–2923

Boskey A, Gadaleta S, Gundberg C, Doty S, Ducy P, Karsenty G (1998) Fourier transform infrared microspectroscopic analysis of bones of osteocalcin-deficient mice provides insight into the function of osteocalcin. Bone 23:187–196

Boyce BF, Xing L (2008) Functions of RANKL/RANK/OPG in bone modeling and remodeling. Arch Biochem Biophys 473:139–146

Boyle WJ, Simonet WS, Lacey DL (2003) Osteoclast differentiation and activation. Nature 423: 337–342

Bruedigam C, Eijken M, Koedam M, van de Peppel J, Drabek K, Chiba H, van Leeuwen JP (2010) A new concept underlying stem cell lineage skewing that explains the detrimental effects of thiazolidinediones on bone. Stem Cells 28:916–927

Bucay N, Sarosi I, Dunstan CR, Morony S, Tarpley J, Capparelli C, Scully S, Tan HL, Xu W, Lacey DL, Boyle WJ, Simonet WS (1998) Osteoprotegerin-deficient mice develop early onset osteoporosis and arterial calcification. Genes Dev 12:1260–1268

Caen J, Wu Q (2010) Hageman factor, platelets and polyphosphates: early history and recent connection. J Thromb Haemost 8:1670–1674

Cerri PS, Boabaid F, Katchburian E (2003) Combined TUNEL and TRAP methods suggest that apoptotic bone cells are inside vacuoles of alveolar bone osteoclasts in young rats. J Periodontol Res 38:223–226

Chang W, Tu C, Chen TH, Komuves L, Oda Y, Pratt SA, Miller S, Shoback D (1999) Expression and signal transduction of calcium-sensing receptors in cartilage and bone. Endocrinology 140:5883–5893

Chen D, Zhao M, Mundy GR (2004) Bone morphogenetic proteins. Growth Factors 22:233–241

Cheng SL, Yang JW, Rifas L, Zhang SF, Avioli LV (1994) Differentiation of human bone marrow osteogenic stromal cells *in vitro*: induction of the osteoblast phenotype by dexamethasone. Endocrinology 134:277–286

Choi HJ, Park YR, Nepal M, Choi BY, Cho NP, Choi SH, Heo SR, Kim HS, Yang MS, Soh Y (2010) Inhibition of osteoclastogenic differentiation by Ikarisoside A in RAW 264.7 cells via JNK and NF-κB signaling pathways. Eur J Pharmacol 636:28–35

Choi SH, Smith SA, Morrissey JH (2011) Polyphosphate is a cofactor for the activation of factor XI by thrombin. Blood 118:6963–6970

Cyboron GW, Wuthier RE (1981) Purification and initial characterization of intrinsic membrane-bound alkaline phosphatase from chicken epiphyseal cartilage. J Biol Chem 156:7262–7268

Dvorak MM, Riccardi D (2004) Ca^{2+} as an extracellular signal in bone. Cell Calcium 35:249–255

Felsenberg D, Boonen S (2005) The bone quality framework: determinants of bone strength and their interrelationships, and implications for osteoporosis management. Clin Ther 27:1–11

Filgueira L (2004) Fluorescence-based staining for tartrate-resistant acidic phosphatase (TRAP) in osteoclasts combined with other fluorescent dyes and protocols. J Histochem Cytochem 52: 411–414

Fleisch H (1999) From polyphosphates to bisphosphonates and their role in bone and calcium metabolism. In: Schröder HC, Müller WEG (eds) Inorganic polyphosphates biochemistry, biology, biotechnology. Springer, Heidelberg, pp 197–215

Fleisch H, Neuman WF (1961) Mechanisms of calcification: role of collagen, polyphosphates, and phosphatase. Am J Physiol 200:1296–1300

Fleisch H, Russel R, Straumann F (1966a) Effect of pyrophosphate on hydroxyapatite and its implications in calcium homeostasis. Nature 212:901–903

Fleisch H, Straumann F, Schenk R, Bisaz S, Allgöwer M (1966b) Effect of condensed phosphates on calcification of chick embryo femurs in tissue culture. Am J Physiol 211:821–825

Garimella R, Tague SE, Zhang J, Belibi F, Nahar N, Sun Ben H, Insogna K, Wang J, Anderson HC (2008) Expression and synthesis of bone morphogenetic proteins by osteoclasts: a possible path to anabolic bone remodeling. J Histochem Cytochem 56:569–577

Ghosh S, Karin M (2002) Missing pieces in the NF-κB puzzle. Cell 109(Suppl):S81–S96

Graff RD, Picher M, Lee GM (2003) Extracellular nucleotides, cartilage stress, and calcium crystal formation. Curr Opin Rheumatol 15:315–320

Griffith EJ (1995) Phosphate fibers, Topics in applied chemistry. Springer, Berlin

Hacchou Y, Uematsu T, Ueda O, Usui Y, Uematsu S, Takahashi M, Uchihashi T, Kawazoe Y, Shiba T, Kurihara S, Yamaoka M, Furusawa K (2007) Inorganic polyphosphate: a possible stimulant of bone formation. J Dent Res 86:893–897

Han KY, Hong BS, Yoon YJ, Yoon CM, Kim YK, Kwon YG, Gho YS (2007) Polyphosphate blocks tumour metastasis via anti-angiogenic activity. Biochem J 406:49–55

Hausser HJ, Brenner RE (2005) Phenotypic instability of SaOS-2 cells in long-term culture. Biochem Biophys Res Commun 333:216–222

Hayase Y, Muguruma Y, Lee MY (1997) Osteoclast development from hematopoietic stem cells: apparent divergence of the osteoclast lineage prior to macrophage commitment. Exp Hematol 25:19–25

Henthorn PS (1996) Alkaline phosphatase. In: Bilezikian JP, Raisz LG, Rodan GA (eds) Principles of bone biology, 1st edn. Academic, San Diego, CA, pp 197–206

Hernandez-Ruiz L, González-García I, Castro C, Brieva JA, Ruiz FA (2006) Inorganic polyphosphate and specific induction of apoptosis in human plasma cells. Haematologica 91: 1180–1186

Hessle L, Johnson KA, Anderson HC, Narisawa S, Sali A, Goding JW, Terkeltaub R, Millan JL (2002) Tissue-nonspecific alkaline phosphatase and plasma cell membrane glycoprotein-1 are central antagonistic regulators of bone mineralization. Proc Natl Acad Sci USA 99:9445–9449

Hu R, Liu W, Li H, Yang L, Chen C, Xia ZY, Guo LJ, Xie H, Zhou HD, Wu XP, Luo XH (2011) A RUNX2/MIR-3960/MIR-2861 regulatory feedback loop during mouse osteoblast differentiation. J Biol Chem 286:12328–12339

Hui M, Tenenbaum HC (1998) New face of an old enzyme: alkaline phosphatases may contribute to human tissue aging by inducing tissue hardening and calcification. Anat Rec 253:91–94

Imsiecke G, Münkner J, Lorenz B, Bachinski N, Müller WEG, Schröder HC (1996) Inorganic polyphosphates in the developing freshwater sponge *Ephydatia muelleri*: effect of stress by polluted waters. Environ Toxicol Chem 15:1329–1334

Jabbar S, Drury J, Fordham JN, Datta HK, Francis RM, Tuck SP (2011) Osteoprotegerin, RANKL and bone turnover in postmenopausal osteoporosis. J Clin Pathol 64:354–357

Jang WG, Kim EJ, Kim DK, Ryoo HM, Lee KB, Kim SH, Choi HS, Koh JT (2012) BMP2 protein regulates osteocalcin expression via Runx2-mediated Atf6 gene transcription. J Biol Chem 287:905–915

Jimenez-Nunez MD, Moreno-Sanchez D, Hernandez-Ruiz L, Benitez-Rondan A, Ramos-Amaya-A, Rodriguez-Bayona B, Medina F, Brieva JA, Ruiz FA (2012) Myeloma cells contain high inorganic polyphosphate levels that are associated with nucleolar transcription. Haematologica 97:1264–1271

Johnson KA, Hessle L, Vaingankar S, Wennberg C, Mauro S, Narisawa S, Goding JW, Sano K, Millan JL, Terkeltaub R (2000) Osteoblast tissue-nonspecific alkaline phosphatase antagonizes and regulates PC-1. Am J Physiol Regul Integr Comp Physiol 279:R1365–R1377

Kanis JA (1994) WHO Study Group. Assessment of fracture risk and its application to screening for postmenopausal osteoporosis: synopsis of a WHO report. Osteoporos Int 4:368–381

Kasuyama K, Tomofuji T, Ekuni D, Azuma T, Irie K, Endo Y, Morita M (2012) Effects of topical application of inorganic polyphosphate on tissue remodeling in rat inflamed gingiva. J Periodontal Res 47:159–164

Katchman BJ, Smith HE (1958) Diffusion of synthetic and natural polyphosphates. Arch Biochem Biophys 76:396–402

Kawazoe Y, Shiba T, Nakamura R, Mizuno A, Tsutsumi K, Uematsu T, Yamaoka M, Shindoh M, Kohgo T (2004) Induction of calcification in MC3T3-E1 cells by inorganic polyphosphate. J Dent Res 83:613–618

Kawazoe Y, Katoh S, Onodera Y, Kohgo T, Shindoh M, Shiba T (2008) Activation of the FGF signaling pathway and subsequent induction of mesenchymal stem cell differentiation by inorganic polyphosphate. Int J Biol Sci 4:37–47

Kearns AE, Khosla S, Kostenuik PJ (2008) Receptor activator of nuclear factor kappaB ligand and osteoprotegerin regulation of bone remodeling in health and disease. Endocr Rev 29:155–192

Kelly SE, Di Benedetto A, Greco A, Howard CM, Sollars VE, Primerano DA, Valluri JV, Claudio PP (2010) Rapid selection and proliferation of CD133(+) cells from cancer cell lines: chemotherapeutic implications. PLoS One 5:e10035

Kornberg A (1999) Inorganic polyphosphate: a molecule of many functions. In: Schröder HC, Müller WEG (eds) Inorganic polyphosphates: biochemistry, biology, biotechnology, vol 23, Progress in molecular subcellular biology. Springer, Berlin, pp 1–26

Krasko A, Batel R, Schröder HC, Müller IM, Müller WEG (2000) Expression of silicatein and collagen genes in the marine sponge *Suberites domuncula* is controlled by silicate and myotrophin. Eur J Biochem 267:4878–4887

Kulaev IS (1979) The biochemistry of inorganic polyphosphates. Wiley, New York, NY

Kulaev IS, Vagabov VM, Kulakovskaya TV (2004) The biochemistry of inorganic polyphosphates. Wiley, Chichester, pp 1–277

Kulakovskaya TV, Vagabov VM, Kulaev IS (2012) Inorganic polyphosphate in industry, agriculture and medicine: modern state and outlook. Process Biochem 47:1–10

Kumble KD, Kornberg A (1995) Inorganic polyphosphate in mammalian cells and tissues. J Biol Chem 270:5818–5822

Laitinen M, Jortikka L, Halttunen T, Böhling T, Marttinen A, Lindholm TS (1997) Soluble factors from human Saos-2 osteosarcoma cells induce ectopic bone formation and osteoblastic differentiation of cultured mesenchymal cells. J Musculoskelet Res 1:21–32

Landis WJ, Hodgens KJ, Song MJ, Arena J, Kiyonaga S, Marko M, Owen C, McEwen BF (1996) Mineralization of collagen may occur on fibril surfaces: evidence from conventional and high-voltage electron microscopy and three-dimensional imaging. J Struct Biol 117:24–35

Lane NE, Yao W (2009) Developments in the scientific understanding of osteoporosis. Arthritis Res Ther 11:228

Lee ZH, Kim HH (2003) Signal transduction by receptor activator of nuclear factor kappa B in osteoclasts. Biochem Biophys Res Commun 305:211–214

Lee J, Park JB, Herr Y, Chung JH, Kwon YH (2008) The effect of polyphosphate on exophytic bone formation. J Korean Acad Periodontol 38:59–66

Lee BH, Kim MC, Choi SH, Lee YK (2009) Amorphous calcium polyphosphate bone regenerative materials based on calcium phosphate glass. Key Eng Mater 396–398:209–212

Lee SJ, Kang SW, Do HJ, Han I, Shin DA, Kim JH, Lee SH (2010) Enhancement of bone regeneration by gene delivery of BMP2/Runx2 bicistronic vector into adipose-derived stromal cells. Biomaterials 31:5652–5659

Leyhausen G, Lorenz B, Zhu H, Geurtsen W, Bohnensack R, Müller WEG, Schröder HC (1998) Inorganic polyphosphate in human osteoblast-like cells. J Bone Miner Res 13:803–812

Lorch IJ (1949) Alkaline phosphatase and the mechanism of ossification. J Bone Joint Surg 31: 94–99

Lorenz B, Schröder HC (1999) Methods for investigation of inorganic polyphosphates and polyphosphate-metabolizing enzymes. In: Schröder HC, Müller WEG (eds) Inorganic polyphosphates—biochemistry, biology, biotechnology, Prog Mol Subcell Biol 23: 217–239

Lorenz B, Schröder HC (2001) Mammalian intestinal alkaline phosphatase acts as highly active exopolyphosphatase. Biochim Biophys Acta 1547:254–261

Lorenz B, Marmé S, Müller WEG, Unger K, Schröder HC (1994a) Preparation and use of polyphosphate-modified zirconia for purification of nucleic acids and proteins. Anal Biochem 216:118–126

Lorenz B, Müller WEG, Kulaev IS, Schröder HC (1994b) Purification and characterization of an exopolyphosphatase activity from *Saccharomyces cerevisiae*. J Biol Chem 269:22198–22204

Lorenz B, Batel R, Bachinski N, Müller WEG, Schröder HC (1995) Purification and characterization of two exopolyphosphatases from the marine sponge *Tethya lyncurium*. Biochim Biophys Acta 1245:17–28

Lorenz B, Leuck J, Köhl D, Müller WEG, Schröder HC (1997a) Anti-HIV-1 activity of inorganic polyphosphates. J Acquir Immune Defic Syndr Hum Retrovirol 14:110–118

Lorenz B, Münkner J, Oliveira MP, Kuusksalu A, Leitão JM, Müller WEG, Schröder HC (1997b) Changes in metabolism of inorganic polyphosphate in rat tissues and human cells during development and apoptosis. Biochim Biophys Acta 1335:51–60

Lorenz B, Münkner J, Oliveira MP, Leitão JM, Müller WEG, Schröder HC (1997c) A novel method for determination of inorganic polyphosphates using the fluorescent dye fura-2. Anal Biochem 246:176–184

Lymperi S, Ersek A, Ferraro F, Dazzi F, Horwood NJ (2011) Inhibition of osteoclast function reduces hematopoietic stem cell numbers *in vivo*. Blood 117:1540–1549

Maekawa K, Yoshida Y, Mine A, Fujisawa T, Van Meerbeek B, Suzuki K, Kuboki T (2007) Chemical interaction of polyphosphoric acid with titanium and its effect on human bone marrow derived mesenchymal stem cell behavior. J Biomed Mater Res A 82:195–200

Maekawa K, Yoshida Y, Mine A, van Meerbeek B, Suzuki K, Kuboki T (2008) Effect of polyphosphoric acid pre-treatment of titanium on attachment, proliferation, and differentiation of osteoblast-like cells (MC3T3-E1). Clin Oral Implants Res 19:320–325

Maekawa K, Shimono K, Oshima M, Yoshida Y, Van Meerbeek B, Suzuki K, Kuboki T (2009) Polyphosphoric acid treatment promotes bone regeneration around titanium implants. J Oral Rehabil 36:362–367

Millan JL (2006) Alkaline phosphatases. Structure, substrate specificity and functional relatedness to other members of a large superfamily of enzymes. Purinergic Signal 2:335–341

Morgan EF, Barnes GL, Einhorn TA (2008) The bone organ system: form and function. In: Marcus R, Feldman D, Nelson D, Rosen CJ (eds) Osteoporosis, 3rd edn. Elsevier Academic, Boston, MA, pp 3–25

Morimoto D, Tomita T, Kuroda S, Higuchi C, Kato S, Shiba T, Nakagami H, Morishita R, Yoshikawa H (2010) Inorganic polyphosphate differentiates human mesenchymal stem cells into osteoblastic cells. J Bone Miner Metab 28:418–423

Morita K, Doi K, Kubo T, Takeshita R, Kato S, Shiba T, Akagawa Y (2010) Enhanced initial bone regeneration with inorganic polyphosphate-adsorbed hydroxyapatite. Acta Biomater 6: 2808–2815

Morrissey JH (2012) Polyphosphate: a link between platelets, coagulation and inflammation. Int J Hematol 95:346–352

Morse DE (1999) Silicon biotechnology: harnessing biological silica production to construct new materials. Trends Biotechnol 17:230–232

Müller F, Renné T (2011) Platelet polyphosphates: the nexus of primary and secondary hemostasis. Scand J Clin Lab Invest 71:82–86

Müller F, Mutch NJ, Schenk WA, Smith SA, Esterl L, Spronk HM, Schmidbauer S, Gahl WA, Morrissey JH, Renné T (2009a) Platelet polyphosphates are proinflammatory and procoagulant mediators in vivo. Cell 139:1143–1156

Müller WEG, Wang XH, Cui FZ, Jochum KP, Tremel W, Bill J, Schröder HC, Natalio F, Schloßmacher U, Wiens M (2009b) Sponge spicules as blueprints for the biofabrication of inorganic-organic composites and biomaterials. Appl Microbiol Biotechnol 83:397–413

Müller WEG, Wang XH, Diehl-Seifert B, Kropf K, Schloßmacher U, Lieberwirth I, Glasser G, Wiens M, Schröder HC (2011) Inorganic polymeric phosphate/polyphosphate as an inducer of alkaline phosphatase and a modulator of intracellular Ca^{2+} level in osteoblasts (SaOS-2 cells) *in vitro*. Acta Biomater 7:2661–2671

Müller WEG, Wang XH, Guo YW, Schröder HC (2012) Potentiation of the cytotoxic activity of copper by polyphosphate on biofilm-producing bacteria: a bioinspired approach. Mar Drugs 10:2369–2387

Müller WEG, Schröder HC, Burghard Z, Pisignano D, Wang XH (2013a) Silicateins: a paradigm shift in bioinorganic chemistry. Enzymatic synthesis of inorganic polymeric silica. Chem Eur J 19:5790–5804

Müller WEG, Schröder HC, Schlossmacher U, Grebenjuk VA, Ushijima H, Wang XH (2013b) Induction of carbonic anhydrase in SaOS-2 cells, exposed to bicarbonate and consequences for calcium phosphate crystal formation. Biomaterials 34:8671–8680

Mutch NJ, Engel R, Uitte de Willige S, Philippou H, Ariëns RA (2010) Polyphosphate modifies the fibrin network and down-regulates fibrinolysis by attenuating binding of tPA and plasminogen to fibrin. Blood 115:3980–3988

Nam HK, Liu J, Li Y, Kragor A, Hatch NE (2011) Ectonucleotide pyrophosphatase/ phosphodiesterase-1 (ENPP1) protein regulates osteoblast differentiation. J Biol Chem 286: 39059–39071

Nickel J, Dreyer MK, Kirsch T, Sebald W (2001) The crystal structure of the BMP-2–BMPR-IA complex and the generation of BMP-2 antagonists. J Bone Joint Surg Am 83-A(Suppl 1): S7–S14

Oldberg A, Franzen A, Heinegard D (1986) Cloning and sequence analysis of rat bone sialoprotein (osteopontin) cDNA reveals an Arg-Gly-Asp cell-binding sequence. Proc Natl Acad Sci USA 83:8819–8823

Oldberg A, Franzen A, Heinegard D (1988) The primary structure of a cell binding bone sialoprotein. J Biol Chem 263:19430–19432

Omelon SJ, Grynpas MD (2008) Relationships between polyphosphate chemistry, biochemistry and apatite biomineralization. Chem Rev 108:4694–4715

Omelon S, Georgiou J, Henneman ZJ, Wise LM, Sukhu B, Hunt T, Wynnyckyj C, Holmyard D, Ryszard B, Grynpas MD (2009) Control of vertebrate skeletal mineralization by polyphosphates. PLoS One 4:e5634

Ominsky MS, Stolina M, Li X, Corbin TJ, Asuncion FJ, Barrero M, Niu QT, Dwyer D, Adamu S, Warmington KS, Grisanti M, Tan HL, Ke HZ, Simonet WS, Kostenuik PJ (2009) One year of transgenic overexpression of osteoprotegerin in rats suppressed bone resorption and increased vertebral bone volume, density, and strength. J Bone Miner Res 24:1234–1246

Omoto M, Imai T, Seki K, Nomura R, Otahara Y (1997) The effect on the bones of condensed phosphate when used as food additives: its importance in relation to preventive medicine. Environ Health Prev Med 2:105–116

Orimo H, Shimada T (2008) The role of tissue-nonspecific alkaline phosphatase in the phosphate-induced activation of alkaline phosphatase and mineralization in SaOS-2 human osteoblast-like cells. Mol Cell Biochem 282:101–108

Palmer G, Bonjour JP, Caverzasio J (1997) Expression of a newly identified phosphate transporter/retrovirus receptor in human SaOS-2 osteoblast-Like cells and its regulation by insulin-like growth factor I. Endocrinology 138:5202–5209

Pavlov E, Aschar-Sobbi R, Campanella M, Turner RJ, Gómez-García MR, Abramov AY (2010) Inorganic polyphosphate and energy metabolism in mammalian cells. J Biol Chem 285: 9420–9428

Price CP (1993) Multiple forms of human serum alkaline phosphatase: detection and quantitation. Ann Clin Biochem 30:355–372

Quinn JM, Gillespie MT (2005) Modulation of osteoclast formation. Biochem Biophys Res Commun 328:739–745

Raisz LG (2005) Pathogenesis of osteoporosis: concepts, conflicts, and prospects. J Clin Invest 115:3318–3325

Ramos I, Gomes F, Koeller CM, Saito K, Heise N, Masuda H, Docampo R, de Souza W, Machado EA, Miranda K (2011) Acidocalcisomes as calcium- and polyphosphate-storage compartments during embryogenesis of the insect *Rhodnius prolixus* Stahl. PLoS One 6:e27276

Rao NN, Gómez-Garcıa MR, Kornberg A (2009) Inorganic polyphosphate: essential for growth and survival. Annu Rev Biochem 78:605–647

Register TC, Wuthier RE (1985) Effect of pyrophosphate and two diphosphonates on ^{45}Ca and $^{32}P_i$ uptake and mineralization by matrix vesicle-enriched fractions and by hydroxyapatite. Bone 6:307–312

Rezende LA, Ciancaglini P, Pizauro JM, Leone FA (1998) Inorganic pyrophosphate-phosphohydrolytic activity associated with rat osseous plate alkaline phosphatase. Cell Mol Biol (Noisy-le-grand) 44:293–302

Robey PG (2002) Bone matrix proteoglycans and glycoproteins. In: Bilezikan JP, Raisz LG, Rodan GA (eds) Principles of bone biology. Academic, San Diego, CA, p 225

Robins SP, Bilezikian JP, Seibel MJ (2006) Dynamics of bone and cartilage metabolism: principals and clinical applications. Academic, San Diego, CA

Rohde M, Mayer H (2007) Exocytotic process as a novel model for mineralization by osteoblasts in vitro and in vivo determined by electron microscopic analysis. Calcif Tissue Int 80:323–336

Rosenzweig BL, Imamura T, Okadome T, Cox GN, Yamashita H, ten Dijke P, Heldin CH, Miyazono K (1995) Cloning and characterization of a human type II receptor for bone morphogenetic proteins. Proc Natl Acad Sci USA 92:7632–7636

Ruiz FA, Lea CR, Oldfield E, Docampo R (2004) Human platelet dense granules contain polyphosphate and are similar to acidocalcisomes of bacteria and unicellular eukaryotes. J Biol Chem 279:44250–44257

Sambrook P, Cooper C (2006) Osteoporosis. Lancet 367:2010–2018

Santini D, Schiavon G, Vincenzi B, Gaeta L, Pantano F, Russo A, Ortega C, Porta C, Galluzzo S, Armento G, La Verde N, Caroti C, Treilleux I, Ruggiero A, Perrone G, Addeo R, Clezardin P, Muda AO, Tonini G (2011) Receptor activator of NF-kB (RANK) expression in primary tumors associates with bone metastasis occurrence in breast cancer patients. PLoS One 6: e19234

Schröder HC, Müller WEG (eds) (1999) Inorganic polyphosphates: biochemistry. biology. biotechnology, Prog Mol Subcell Biol, vol 23. Springer, Berlin

Schröder HC, Lorenz B, Kurz L, Müller WEG (1999) Inorganic polyP in eukaryotes: enzymes, metabolism and function. In: Schröder HC, Müller WEG (eds) Inorganic polyphosphates—biochemistry, biology, biotechnology, Prog Mol Subcell Biol 23: 45–81

Schröder HC, Kurz L, Müller WEG, Lorenz B (2000) Polyphosphate in bone. Biochemistry (Moscow) 65:296–303

Schröder HC, Borejko A, Krasko A, Reiber A, Schwertner H, Müller WEG (2005) Mineralization of SaOS-2 cells on enzymatically (Silicatein) modified bioactive osteoblast-stimulating surfaces. J Biomed Mater Res B Appl Biomater 75B:387–392

Schröder HC, Wiens M, Wang XH, Schloßmacher U, Müller WEG (2011) In: Müller WEG (ed) Molecular biomineralization, Prog Mol Subcell Biol 52: 283–312

Schröder HC, Wang XH, Manfrin A, Yu SH, Grebenjuk VA, Korzhev M, Wiens M, Schloßmacher U, Müller WEG (2012a) Silicatein: acquisition of structure-guiding and structure-forming properties during maturation from the pro-silicatein to the silicatein form. J Biol Chem 287:22196–22205

Schröder HC, Wang XH, Wiens M, Diehl-Seifert B, Kropf K, Schloßmacher U, Müller WEG (2012b) Silicate modulates the cross-talk between osteoblasts (SaOS-2) and osteoclasts (RAW 264.7 cells): inhibition of osteoclast growth and differentiation. J Cell Biochem 113: 3197–3206

Schröder HC, Wiens M, Schloßmacher U, Brandt D, Müller WEG (2012c) Silicatein-mediated polycondensation of orthosilicic acid: modeling of a catalytic mechanism involving ring formation. Silicon 4:33–38

Seidlmayer LK, Gomez-Garcia MR, Blatter LA, Pavlov E, Dedkova EN (2012) Inorganic polyphosphate is a potent activator of the mitochondrial permeability transition pore in cardiac myocytes. J Gen Physiol 139:321–331

Shahidi F, Rubin LJ, Diosady LL, Kassam N, Fong JC, Li S, Wood DF (1986) Effect of sequestering agents on lipid oxidation in cooked meats. Food Chem 21:145–152

Shiba T, Nishimura D, Kawazoe Y, Onodera Y, Tsutsumi K, Nakamura R, Ohshiro M (2003) Modulation of mitogenic activity of fibroblast growth factors by inorganic polyphosphate. J Biol Chem 278:26788–26792

Shimizu K, Cha J, Stucky GD, Morse DE (1998) Silicatein alpha: cathepsin L-like protein in sponge biosilica. Proc Natl Acad Sci USA 95:6234–6238

Shioi A, Nishizawa Y, Jono S, Koyama H, Hosoi M, Morii H (1995) ß-Glycerophosphate accelerates calcification in cultured bovine vascular smooth muscle cells. Arterioscler Thromb Vasc Biol 15:2003–2009

Simonet WS, Lacey DL, Dunstan CR, Kelley M, Chang MS, Lüthy R, Nguyen HQ, Wooden S, Bennett L, Boone T, Shimamoto G, DeRose M, Elliott R, Colombero A, Tan HL, Trail G, Sullivan J, Davy E, Bucay N, Renshaw-Gegg L, Hughes TM, Hill D, Pattison W, Campbell P, Sander S, Van G, Tarpley J, Derby P, Lee R, Boyle WJ (1997) Osteoprotegerin: a novel secreted protein involved in the regulation of bone density. Cell 89:309–319

Sinha KM, Yasuda H, Coombes MM, Dent SYR, de Crombrugghe B (2010) Regulation of the osteoblast-specific transcription factor Osterix by NO66, a Jumonji family histone demethylase. EMBO J 29:68–79

Smith SA, Morrissey JH (2008) Polyphosphate as a general procoagulant agent. J Thromb Haemost 6:1750–1756

Smith SA, Mutch NJ, Baskar D, Rohloff P, Docampo R, Morrissey JH (2006) Polyphosphate modulates blood coagulation and fibrinolysis. Proc Natl Acad Sci USA 103:903–908

Smith SA, Choi SH, Davis-Harrison R, Huyck J, Boettcher J, Rienstra CM, Morrissey JH (2010) Polyphosphate exerts differential effects on blood clotting, depending on polymer size. Blood 116:4353–4359

Sodek J, Ganss B, McKee MD (2000) Osteopontin. Crit Rev Oral Biol Med 11:279–303

Sperow JW, Moe OA, Ridlington JW, Butler LG (1973) Yeast inorganic pyrophosphatase. VI. Studies on specificity and mechanism. J Biol Chem 248:2062–2065

St-Pierre JP, Wang Q, Li SQ, Pilliar RM, Kandel RA (2012) Inorganic polyphosphate stimulates cartilage tissue formation. Tissue Eng Part A 18:1282–1292

Sun L, Blair HC, Peng Y, Zaidi N, Adebanjo OA, Wu XB, Wu XY, Iqbal J, Epstein S, Abe E, Moong BS, Zaidi M (2005) Calcineurin regulates bone formation by the osteoblast. Proc Natl Acad Sci USA 102:17130–17135

Sung B, Murakami A, Oyajobi BO, Aggarwal BB (2009) Zerumbone abolishes RANKL-induced NF-κB activation, inhibits osteoclastogenesis, and suppresses human breast cancer-induced bone loss in athymic nude mice. Cancer Res 69:1477–1484

Tammenkoski M, Koivula K, Cusanelli E, Zollo M, Steegborn C, Baykov AA, Lahti R (2008) Human metastasis regulator protein H-prune is a short-chain exopolyphosphatase. Biochemistry 47:9707–9713

Teitelbaum SL (2007) Osteoclasts: what do they do and how do they do it? Am J Pathol 170:427–435

Tsuda E, Goto M, Mochizuki S, Yano K, Kobayashi F, Morinaga T, Higashio K (1997) Isolation of a novel cytokine from human fibroblasts that specifically inhibits osteoclastogenesis. Biochem Biophys Res Commun 234:137–142

Usui Y, Uematsu T, Uchihashi T, Takahashi M, Takahashi M, Ishizuka M, Doto R, Tanaka H, Komazaki Y, Osawa M, Yamada K, Yamaoka M, Furusawa K (2010) Inorganic polyphosphate induces osteoblastic differentiation. J Dent Res 89:504–509

Vaisman DN, McCarthy AD, Cortizo AM (2005) Bone-specific alkaline phosphatase activity is inhibited by bisphosphonates; role of divalent cations. Biol Trace Elem Res 104:131–140

Van Wazer JR (ed) (1958) Phosphorus and its compounds: chemistry, vol 1. Interscience Publishers Inc., New York, NY

Veldman CM, Markovich D, Schmid C, Murer H (1995) Expression of sodium dependent phosphate (Na_dPi) transport in *Xenopus laevis* oocytes induced by mRNA from 1α,25-dihydroxyvitamin D_3-treated rat osteoblast-like cells. Pflügers Arch 430:64–67

Vincent C, Kogawa M, Findlay DM, Atkins GJ (2009) The generation of osteoclasts from RAW 264.7 precursors in defined, serum-free conditions. J Bone Miner Metab 27:114–119

Wang G, Yang J (2010) Influences of binder on fire protection and anticorrosion properties of intumescent fire resistive coating for steel structure. Surf Coat Technol 204:1186–1192

Wang Z, Li X, Li Z, Yang L, Sasaki Y, Wang S, Zhou L, Araki S, Mezawa M, Takai H, Ogata Y (2010) Effects of inorganic polyphosphate on bone sialoprotein gene expression. Gene 452: 79–86

Wang XH, Schröder HC, Diehl-Seifert B, Kropf K, Schloßmacher U, Wiens M, Müller WEG (2012a) Dual effect of inorganic polymeric phosphate/polyphosphate on osteoblasts and osteoclasts *in vitro*. J Tissue Eng Regen Med 7:767–776

Wang XH, Schröder HC, Wiens M, Ushijima H, Müller WEG (2012b) Bio-silica and bio-polyphosphate: applications in biomedicine (bone formation). Curr Opin Biotechnol 23: 570–578

Wang XH, Schröder HC, Feng QL, Draenert F, Müller WEG (2013) The deep-sea natural products, biogenic polyphosphate (bio-polyP) and biogenic silica (bio-silica) as biomimetic scaffolds for bone tissue engineering: fabrication of a morphogenetically-active polymer. Mar Drugs 11:718–746

Wiens M, Wang X, Natalio F, Schröder HC, Schloßmacher U, Wang S, Korzhev M, Geurtsen W, Müller WEG (2010a) Bioinspired fabrication of bio-silica-based bone substitution materials. Adv Eng Mater 12:B438–B450

Wiens M, Wang XH, Schloßmacher U, Lieberwirth I, Glasser G, Ushijima H, Schröder HC, Müller WEG (2010b) Osteogenic potential of bio-silica on human osteoblast-like (SaOS-2) cells. Calcif Tissue Int 87:513–524

Wiens M, Wang XH, Schröder HC, Kolb U, Schloßmacher U, Ushijima H, Müller WEG (2010c) The role of biosilica in the osteoprotegerin/RANKL ratio in human osteoblastlike cells. Biomaterials 31:7716–7725

Williams G, Sallis JD (1982) Structural factors influencing the ability of compounds to inhibit hydroxyapatite formation. Calcif Tissue Int 34:169–177

Wood HG, Clark JE (1988) Biological aspects of inorganic polyphosphates. Annu Rev Biochem 57:235–260

Wu LNY, Valhmu WB, Lloyd GC, Genge BR, Wuthier RE (1989) Isolation of two glycosylated forms of membrane-bound alkaline phosphatase from avian growth plate cartilage matrix vesicle-enriched microsomes. Bone Miner 7:113–125

Wuthier RE, Makjeska RJ, Collins GM (1977) Biosynthesis of matrix vesicles in epiphyseal cartilage. I. In vivo incorporation of 32P orthophosphate into phospholipids of chondrocyte, membrane, and matrix vesicle fractions. Calcif Tissue Res 23:135–139

Yuan Q, Kubo T, Doi K, Morita K, Takeshita R, Katoh S, Shiba T, Gong P, Akagawa Y (2009) Effect of combined application of bFGF and inorganic polyphosphate on bioactivities of osteoblasts and initial bone regeneration. Acta Biomater 5:1716–1724

Zakharian E, Thyagarajan B, French RJ, Pavlov E, Rohacs T (2009) Inorganic polyphosphate modulates TRPM8 channels. PLoS One 4:e5404

Zhou Z, Han JY, Xi CX, Xie JX, Feng X, Wang CY, Mei L, Xiong WC (2008) HMGB1 regulates RANKL-induced osteoclastogenesis in a manner dependent on RAGE. J Bone Miner Res 23:1084–1096

Index [1]

[1] Title: Progress in Molecular and Subcellular Biology
Author: Werner E.G. Muller
Year : 2013

W.E.G. Müller et al. (eds.), *Biomedical Inorganic Polymers*, Progress in Molecular and Subcellular Biology 54, DOI 10.1007/978-3-642-41004-8,

MIX
Papier aus verantwortungsvollen Quellen
Paper from responsible sources
FSC® C105338

If you have any concerns about our products,
you can contact us on
ProductSafety@springernature.com

In case Publisher is established outside the EU,
the EU authorized representative is:
Springer Nature Customer Service Center GmbH
Europaplatz 3, 69115 Heidelberg, Germany

Printed by Libri Plureos GmbH
in Hamburg, Germany